# Lecture Notes on Data Engineering and Communications Technologies

Volume 19

**Series editor**

Fatos Xhafa, Technical University of Catalonia, Barcelona, Spain
e-mail: fatos@cs.upc.edu

The aim of the book series is to present cutting edge engineering approaches to data technologies and communications. It publishes latest advances on the engineering task of building and deploying distributed, scalable and reliable data infrastructures and communication systems.

The series has a prominent applied focus on data technologies and communications with aim to promote the bridging from fundamental research on data science and networking to data engineering and communications that lead to industry products, business knowledge and standardisation.

More information about this series at http://www.springer.com/series/15362

Hari Vasudevan · Amit A. Deshmukh
K. P. Ray
Editors

# Proceedings of International Conference on Wireless Communication

ICWiCom 2017

*Editors*
Hari Vasudevan
D J Sanghvi College of Engineering
Mumbai, Maharashtra
India

Amit A. Deshmukh
Department of Electronics and
 Telecommunication Engineering
SVKM's, D J Sanghvi College of
 Engineering
Mumbai, Maharashtra
India

K. P. Ray
Department of Electronics Engineering
Defence Institute of Advance Technology
 (DIAT)
Pune, Maharashtra
India

ISSN 2367-4512 ISSN 2367-4520 (electronic)
Lecture Notes on Data Engineering and Communications Technologies
ISBN 978-981-10-8338-9 ISBN 978-981-10-8339-6 (eBook)
https://doi.org/10.1007/978-981-10-8339-6

Library of Congress Control Number: 2018932524

© Springer Nature Singapore Pte Ltd. 2018
This work is subject to copyright. All rights are reserved by the Publisher, whether the whole or part of the material is concerned, specifically the rights of translation, reprinting, reuse of illustrations, recitation, broadcasting, reproduction on microfilms or in any other physical way, and transmission or information storage and retrieval, electronic adaptation, computer software, or by similar or dissimilar methodology now known or hereafter developed.
The use of general descriptive names, registered names, trademarks, service marks, etc. in this publication does not imply, even in the absence of a specific statement, that such names are exempt from the relevant protective laws and regulations and therefore free for general use.
The publisher, the authors and the editors are safe to assume that the advice and information in this book are believed to be true and accurate at the date of publication. Neither the publisher nor the authors or the editors give a warranty, express or implied, with respect to the material contained herein or for any errors or omissions that may have been made. The publisher remains neutral with regard to jurisdictional claims in published maps and institutional affiliations.

Printed on acid-free paper

This Springer imprint is published by the registered company Springer Nature Singapore Pte Ltd. part of Springer Nature
The registered company address is: 152 Beach Road, #21-01/04 Gateway East, Singapore 189721, Singapore

# Preface

In the modern era of communication, a lot of research is being carried out in diversified areas of communication technologies like signal and image processing, wireless networks, computer communication networks, biomedical applications using telecommunications (telemedicine systems), radio frequency design, and microwave and antennas. In the recent past, many conferences across the globe were organized on a wide range of topics. In continuation with institute's earlier National Conference (NCCT 2011) and International Conference (ICCT 2013 and 2015) on Communication Technologies, with a view to focus on specific topics, this International Conference on Wireless Communication 2017 (ICWiCOM 2017) is being organized. This conference targets the research in the areas of wireless communication domains like microwave and antennas, and networking. It is expected to provide a common platform for researchers working in domains of wireless communication so as to share their ideas.

| | |
|---|---|
| Mumbai, India | Hari Vasudevan |
| Mumbai, India | Amit A. Deshmukh |
| Pune, India | K. P. Ray |

# ICWiCOM—2017 Committee

Shri Amrish R. Patel, Chief Patron, President, SVKM
Shri Bhupesh R. Patel, Joint President, SVKM
Shri Bharat M. Sanghvi, Vice President and Trustee, SVKM & I/C, DJSCE
Shri Sunandan R. Divatia, Secretary, SVKM
Shri Utpal H. Bhayani, Treasurer, SVKM
Shri Shalin S. Divatia, Joint Secretary, SVKM
Shri Jayant P. Gandhi, Joint Secretary, SVKM
Shri Harshad H. Shah, Joint Treasurer, SVKM
Shri Harit H. Chitalia, Joint Treasurer, SVKM

## International Advisory Committee

Dr. W. Ross Stone, Hon. Life Member, IEEE APS AdCom
Dr. Banmali Rawat, University of Nevada, USA
Dr. S. K. Sharma, San Diego University, USA
Dr. Sheel Aditya, NTU, Singapore
Dr. Manoj Patankar, SLU, USA

## National Advisory Committee

Dr. Asoke Basak, CEO, SVKM
Prof. U. B. Desai, Director, IIT Hyderabad
Prof. R. K. Shevgaonkar, IIT Bombay
Prof. Girish Kumar, IIT Bombay
Prof. V. M. Gadre, IIT Bombay
Prof. D. Manjunath, IIT Bombay
Prof. S. P. Duttagupta, IIT Bombay

Dr. K. P. Ray, DIAT, Pune
Prof. K. Vasudevan, CUSAT, Kochi, India
Dr. U. D. Kolekar, Principal, APSIT
Dr. Suresh Ukarande, Principal, KJSIEIT
Dr. U. Bhosle, Principal, RGIT

## Committee

Dr. Hari Vasudevan, General Chair, Principal, DJSCE
Dr. A. C. Daptardar, General Co-Chair, V. Principal (Admin.), DJSCE
Dr. M. J. Godse, General Co-chair, Vice-Principal (Acad.), DJSCE
Dr. Amit A. Deshmukh, Conference Chair, Professor and Head, EXTC Department, DJSCE
Prof. T. D. Biradar, Finance Chair
Prof. S. B. Deshmukh, Finance Chair
Prof. R. S. Taware, Finance Chair
Dr. M. H. Patwardhan, Tech. Program Chair
Prof. A. A. Odhekar, Tech. Program Chair
Prof. A. G. Ambekar, Tech. Program Chair
Prof. V. V. Kelkar, Organizing Chair
Prof. P. A. Kadam, Organizing Chair
Prof. A. A. Chaudhary, Organizing Chair
Prof. S. S. Bhattacharjee, Publication Chair
Prof. V. A. P. Chavali, Publication Chair
Prof. Revathi A. S., Publication Chair
Prof. A. A. Kadam, Publicity Chair
Prof. R. Pal, Publicity Chair
Dr. S. H. Karamchandani, Publicity Chair
Prof. M. S. Pimpale, Sponsorship Chair
Prof. Y. S. Bandi, Sponsorship Chair

# Contents

**Part I  Networking**

**Congestion Control in Wireless Sensor Network Using Innovative Modification to the Increase Decrease Algorithm** .................. 3
Sanu Thomas and Thomaskutty Mathew

**Analysis on Injection Vulnerabilities of Web Application** ............ 13
Nilesh Yadav and Narendra Shekokar

**MAC-Based Group Management Protocol for IoT [MAC GMP-IoT]** ................................................ 23
Yeole Anjali and D. R. Kalbande

**A Secured Two-Factor Authentication Protocol for One-Time Money Account** ................................... 29
Devidas Sarang and Narendra Shekokar

**Comparative Analysis of Methods for Monitoring Activities of Daily Living for the Elderly People** .......................... 39
D. R. Kalbande, Anushka Kanawade and Smruti Varvadekar

**Smart Farming System: Crop Yield Prediction Using Regression Techniques** ...................................................... 49
Ayush Shah, Akash Dubey, Vishesh Hemnani, Divye Gala and D. R. Kalbande

**Part II  Antenna Design**

**Elliptical Planar Dipole Antenna with Harmonic Rejection** .......... 59
P. Jishnu, Arnab Pattanayak, Siddhartha P. Duttagupta and Amit A. Deshmukh

**Resonance Frequency Estimation for Equilateral Triangular Microstrip Antennas Using Artificial Neural Network Model**......... 67
Amit A. Deshmukh, Megh Shukla, Stuti Patel, Saurabh Labde and A. P. C. Venkata

**Analysis of Circular Microstrip Antenna with Single Shorting Post for 50 Ω Microstrip-Line Feed**........................... 75
S. M. Rathod, R. N. Awale and K. P. Ray

**Gap-Coupled Designs of Compact F-Shape Microstrip Antennas for Wider Bandwidth**.................................. 85
Amit A. Deshmukh and Shefali Pawar

**Naturally Tapered Series Fed Arrays for First Sidelobe Level Reduction**........................................ 95
Bharati Singh, Nisha Sarwade and K. P. Ray

**Variations of Slot Cut Multiband Isosceles Microstrip Antennas for Dual-Polarized Response**......................... 103
Amit A. Deshmukh, Anish Mishra, Foram Shah, Pooja Patil, Hetvi Shah and Aarti G. Ambekar

**Design of a Novel CPW Band Stop Filter Using Asymmetric Meander-Line Defected Ground Structure**..................... 111
Makarand G. Kulkarni, A. N. Cheeran, K. P. Ray and S. S. Kakatkar

**Multi-Resonant Wide Band Rectangular Microstrip Antenna with U-Shape and Rectangular Slots**........................ 119
Amit A. Deshmukh, Poonam A. Kadam and Akshay Doshi

**Spanner Shape Monopole MIMO Antenna with High Gain for UWB Applications**............................... 129
Vandana Satam, Shikha Nema and Sanjay Singh Thakur

**Analysis of Multi-resonant Rectangular Microstrip Antenna Embedded with Multiple Slots**............................ 139
Amit A. Deshmukh, A. P. C. Venkata and Aarti G. Ambekar

**Broadband Rectangular Microstrip Antennas Embedded with Pairs of Rectangular Slots**.............................. 151
Amit A. Deshmukh and Divya Singh

**CPW-Fed Printed Monopole with Plus Shaped Fractal Slots for Wider Bandwidth**................................ 161
Ameya A. Kadam and Amit A. Deshmukh

**Multi-resonator Variations of 120° Sectoral Microstrip Antennas for Wider Bandwidth**.............................. 169
Amit A. Deshmukh, Pritish Kamble, Akshay Doshi and A. P. C. Venkata

**Partial Corner Edge-Shorted Rectangular Microstrip Antenna Embedded with U-slot for Dual-Band Response** .................. 177
Amit A. Deshmukh and Mohil Gala

**Novel π-Shape Microstrip Antenna Design for Multi-band Response** ........................................................... 185
Amit A. Deshmukh, Archana Nishad, Gauri Gosavi, Priyanka Narayanan, Siddharth Nayak and Aarti G. Ambekar

**The Design of Wideband E-shape Microstrip Antennas on Varying Substrate Thickness** ................................. 195
Amit A. Deshmukh and Divya Singh

**Modified Circular Shape Microstrip Antenna for Circularly Polarized Response** ......................................... 207
Amit A. Deshmukh, Anuja Odhekar, Akshay Doshi and Pritish Kamble

**Wideband Designs of 60° Sectoral Microstrip Antenna Using Parasitic Angular Sectoral Patches** ..................... 217
Amit A. Deshmukh and S. B. Deshmukh

**Part III  Embedded Systems/Communication**

**Performance Evaluation of Transform Domain Methods for Satellite Image Resolution Enhancement** ................ 227
Mansing Rathod and Jayashree Khanapuri

**Test Case Analysis of Android Application Analyzer** ............... 237
Bushra Almin Shaikh

**Effect of Windowing in the Performance of OFDM Systems** ......... 247
Ranjushree Pal

**Telemedicine: Making Health Care Accessible** .................... 255
Akshita V. Nichani, Shruti T. Pistolwala, Amit A. Deshmukh and Manali J. Godse

**Virtual Piano** .................................................. 265
Aditi Patel, Abhishek Satpute, Mital Pattani and V. Venkataramanan

**Automatic Garbage Collector Bot Using Arduino and GPS** .......... 273
Niharika Mehta, Shikhar Verma and Shivani Bhattacharjee

**Cellulose Acetate Substrates for Design and Calibration of Strain Gauges in Angle Measurement** ..................... 281
Megh Doshi, Maitri Fafadia, Charmi Gandhi and Sunil Karamchandani

**Segmentation Techniques for Differential Variations in Fingerprint Images** ........................................... 291
Aniruddha Garge, Sunil Karamchandani and Sweta Suhasaria

**Smart Traffic Density Management System Using Image Processing** ... 301
Jeet D. Sanghavi, Alay M. Shah, Saurabh S. Rane and V. Venkataramanan

**Energy-Efficient Solar-Powered Weather Station and Soil Analyzer** ... 313
Aniket Kalkar, Abhiroop Mattiyil, Krupa Modi, Sagar Moharir and Archana Chaudhari

**An Algorithm to Extract Handwriting Feature for Personality Analysis** ... 323
Anamika Sen, Harsh Shah, Jessie Lemos and Shivani Bhattacharjee

**Author Index** ... 331

# About the Editors

**Dr. Hari Vasudevan** obtained his M.E. in Production Engineering from VJTI, Mumbai, and Ph.D. from IIT Bombay. He has also done a 3-month full-time certificate program (ERP-BaaN) from S. P. Jain Institute of Management and Research, Mumbai, under the University Synergy Programme of Baan Institute, the Netherlands. His areas of interest include manufacturing engineering, manufacturing systems and strategy, market orientation of manufacturing firms, and world-class manufacturing. He is an approved Ph.D. guide of Mumbai University and NMIMS (Deemed to be University). He is the President of Indian Society of Manufacturing Engineering (ISME); Life Member of ISTE, New Delhi; Fellow of the Institution of Engineers (India); Fellow of ISME; and Senior Member of IEDRC. He has 25 years of teaching experience and 2 years of industry experience. Presently, he is working as the Principal of D. J. Sanghvi College of Engineering. He has published over 70 papers in international conferences and journals as well as in national conferences and journals.

# About the Editors

**Dr. Amit A. Deshmukh** obtained his B.E. (Electronics) from VIT, Pune University, in 1997. He obtained his M.Tech. in 2000 and Ph.D. in 2004 from Department of Electrical Engineering, IIT Bombay. His thesis work during masters and doctoral was on compact broadband and dual-band microstrip antennas. He worked as Research Assistant in Department of Electrical Engineering, IIT Bombay. Further, he worked as Assistant Professor at Sardar Patel Institute of Technology, Mukesh Patel School of Technology and Management (NMIMS—DU), and D J Sanghvi College of Engineering. He also worked as Member of Technical Staff in R&D division of Air Tight Network Pvt. Ltd., Pune. Currently, he is working as Professor and Head of the Department of Electronics and Telecommunication Engineering at D J Sanghvi College of Engineering, Vile Parle (W), Mumbai, India. He has together (teaching, industry, and research) more than 17 years of experience. He has published more than 200 research papers in various international and national journals and conferences. He is on reviewer list of many international journals like IEEE Magazine on Antennas and Propagation, IEEE Transaction of Antennas and Propagation, IEEE Antennas and Wireless Propagation Letters, IET Microwave Antennas and Propagation, PIERS journals, International Journal of Electronics, IETE Journal of Research, Elsevier Electronics and Communication Journal. He has delivered many lecture talks in national/international conferences and workshops in the areas of patch antennas.

**Dr. K. P. Ray** obtained his M.Tech. in Microwave Electronics from University of Delhi and Ph.D. from Department of Electrical Engineering, IIT Bombay. He joined SAMEER (TIFR) in 1985 and have been working in the areas of RF and microwave systems/components and developed expertise in the design of antenna elements/arrays and high power RF/microwave sources for RADAR and industrial applications. He has successfully executed over 40 projects sponsored by government agencies/industries in the capacity of a designer, a chief investigator, and a project manager. He was Guest/Invited//Adjunct Faculty in Electrical Department at IIT Bombay, University of Mumbai, and CEERI (CSIR), Pilani, for postgraduate courses. Since November 2016, he is working as Professor and Head of the Department of Electronics Engineering at DIAT, Pune. He has co-authored a book with Prof. G. Kumar for Artech House, USA, and published over 300 research papers in international/national journals and conference proceedings. He holds three patents and filed three patents. He has been in an advisory capacity for many international/national conferences, engineering colleges, chaired many sessions, and delivered several invited talks. He has been on the Editorial Board of International Journal on RF and Microwave Computer-Aided Engineering, John Wiley, and a Reviewer of IEEE AWPL, IET (formerly IEE, UK), International Journal of Antennas and Propagation (USA), PIERS (USA), International Journal of Electronics (USA), Journal of Electromagnetic Waves and Applications (USA), International Journal of Microwave and Optical Technology (IJMOT), USA, International Journal of Microwave Science and Technology, Hindawi, AEU Elsevier, SADHNA Springer, IETE (India), and many international/national conferences and chaired many sessions. He is a Senior Member of IEEE (USA), a Fellow of IETE, and a Life Member of Instrument Society of India and Engineers of EMI/EMC Society of India.

# Part I
# Networking

# Congestion Control in Wireless Sensor Network Using Innovative Modification to the Increase Decrease Algorithm

Sanu Thomas and Thomaskutty Mathew

**Abstract** A new method of congestion control in wireless sensor network is described. We make a simple but innovative modification to the increase–decrease algorithm to improve its performance. The modification proposed uses additive as well as multiplicative operation for both increase and decrease phases. Basically, we consider a single-link multisource system. The sending rate of each source is optimally controlled to prevent the congestion of the common link. The rate control at sources is adjusted based on the feedback signal from the destination. The proposed method provides efficiency, fairness, and fast convergence.

**Keywords** Congestion control · Additive multiplicative increase
Additive multiplicative decrease · Fast convergence · Shortest path trajectory

## 1 Introduction

In a modern clustered Wireless Communication Network (WSN) [1], the major data flow is from the sensors toward the Base Station (BS) or sink through the Cluster Heads (CHs). In general, the geographically disbursed sensor nodes acquire data from their surroundings and send them to the BS. This is basically a many-to-one type communication. Therefore, the occurrence of congestion is higher here compared to one-to-one or one-to-many type of communication systems. When the WSNs are deployed for event-driven applications, the sudden spurt in the communication traffic, triggered by the detection of events, may cause severe congestion and result in data loss [2]. Therefore, proper congestion control measures are

---

S. Thomas (✉)
School of Technology and Applied Sciences, Mahatma Gandhi University, Regional Center, Pullarikunnu, Kottayam, Kerala, India
e-mail: thomas.sanu@gmail.com

T. Mathew
School of Technology and Applied Sciences, Mahatma Gandhi University, Regional Center, Edappally, Kochi, Kerala, India

© Springer Nature Singapore Pte Ltd. 2018
H. Vasudevan et al. (eds.), *Proceedings of International Conference on Wireless Communication*, Lecture Notes on Data Engineering and Communications Technologies 19, https://doi.org/10.1007/978-981-10-8339-6_1

very much needed in such scenarios. Another major cause for congestion is the lower available bandwidths prevalent in wireless networks. In this paper, we consider the case of event-driven WSN where the traffic density is high. For WSNs, we have several congestion control protocols [3, 4] like PSFQ, CODA, ESRT, and so on. We use a modified version of additive increase and multiplicative decrease method which is popular in TCP congestion control.

Additive Increase Multiplicative Decrease (AIMD) is the standard algorithm used in TCP for congestion control. AIMD can be easily adapted for wireless communication also [5–8]. Our paper is mainly inspired by the work of Chiu and Jain [9]. In AIMD, the increase is purely additive but the decrease is purely multiplicative. In the existing AIMD, the convergence is slower and oscillatory. In our proposed algorithm, we use additive and multiplicative increase as well as additive and multiplicative decrease. We control the relevant flow rate in a multisource closed-loop system to avoid the congestion at the receiver. This is mainly due to the limited bandwidth of the shared channel that carries the data from the multiple sources to a single receiver. In our control system, we achieve faster convergence using the *shortest path approach* without oscillations to reach the convergence point. Instead of binary feedback as in [9], we use full feedback.

## 2 Network Model and Assumptions

Consider a cluster-based Wireless Sensor Network (WSN) having $N$ nodes. We assume a single Cluster Head (CH) which collects data from the sensor nodes and then forwards it to the Base Station (BS). The basic model is shown in Fig. 1. The CH collects data from $N$ sources which are the respective sensor nodes. The data sources are named as $S_1, S_2, \ldots, S_N$. The corresponding flow rates from these sources are represented by $x_1(t), x_2(t), \ldots, x_N(t)$ in packets per second. Here, $x_i(t)$ is the transmission rate in the present time slot $t$, from source $S_i$ for $i = 1$ to $N$. The present rate vector $x(t)$ is written as

$$x(t) = [x_1(t), \ldots, x_N(t)] \quad (1)$$

The capacity of the forward link from the CH to the BS is taken as $C$ packets per second. $C$ mainly depends on the bandwidth and quality of the forward link. It also depends on the receiver buffer size and the processing delay at the BS. For the specified session, we assume that $C$ is constant. Congestion in the link is avoided when the total incoming rate is less than or equal to $C$. Therefore, at a time slot $t$, this condition is represented as

$$\sum_{i=1}^{N} x_i(t) \leq C \quad (2)$$

**Fig. 1** Cluster head with N sources

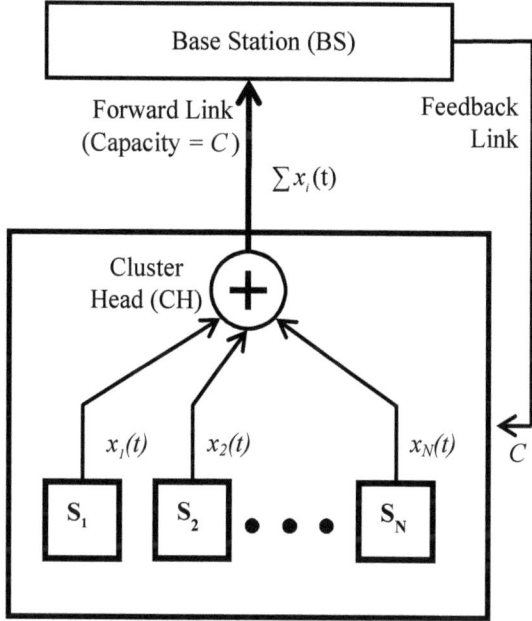

The CH receives the value of C at regular intervals from the BS through the feedback link and the CH in turn controls $x_i(t)$'s such that constraint (2) is satisfied.

## 2.1 Basic Working of the Rate Control

We propose a linear closed-loop negative feedback control system to control the rates. The control system block diagram for $x_1(t)$ is shown in Fig. 2. In Fig. 2, the reference input is $b_1$ which is the final desired output, multiplicative factor is $a_1$, error term is $e_1(t)$, and the present controlled output is $x_1(t)$.

For the sake of clarity, we introduce a two-source rate control system which can be extended for multisource. Let the two sources be $S_1$ and $S_2$ with the respective flow rates $x_1(t)$ and $x_2(t)$. Now with $N = 2$, constraint (2) becomes

**Fig. 2** Control system block diagram for the single source rate control

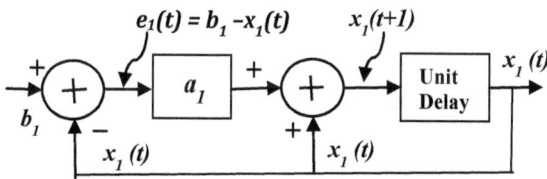

Governing Eq. : $x_1(t+1) = x_1(t) + a_1(b_1 - x_1(t))$

$$x_1(t) + x_2(t) \leq C \tag{3}$$

The sum $(x_1(t) + x_2(t))$ is called the utilization [10]. Therefore, the maximum possible utilization is $C$. At steady state both $x_1(t)$ and $x_2(t)$ remain stable. They do not vary with respect to time. Therefore, the desirable steady-state value corresponding to the cumulative maximum $C$ is

$$b_1 + b_2 = C \tag{4}$$

Here, $b_1$ and $b_2$ are the desired steady-state (equilibrium) values of $x_1(t)$ and $x_2(t)$, respectively.

**Fairness between $x_1(t)$ and $x_2(t)$.** Assuming equal priority, between $S_1$ and $S_2$, the condition for fairness at equilibrium requires, $b_1 = b_2$ and then, $b_1 = b_2 = C/2$.

## 2.2 Control Equations

The discretized control equations for $x_1(t)$ and $x_2(t)$ can be expressed as

$$x_1(t+1) - x_1(t) = a_1(b_1 - x_1(t)) \tag{5}$$

$$x_2(t+1) - x_2(t) = a_2(b_2 - x_2(t)) \tag{6}$$

The flow rates $x_1(t)$ and $x_2(t)$ are governed by (5) and (6), respectively.

**Additive–Multiplicative Increase and Additive–Multiplicative Decrease.** The RHS of (5) has $a_1 b_1$, an additive term and $-a_1 x_1(t)$, a multiplicative term. Therefore, the increase/decrease in $x_1(t)$ is both additive and multiplicative. (The same property holds true for $x_2(t)$.) Therefore, our method is called the Additive–Multiplicative Increase and Additive–Multiplicative Decrease (**AMIAMD**) algorithm.

**Equilibrium at $b_1$.** From Eq. (5), we see that whenever $x_1(t) > b_1$, the RHS of (5) is negative. Therefore, $x_1(t+1) - x_1(t)$ is also negative. This means $x_1(t)$ decreases as $t$ increases. That is, $x_1(t)$ moves toward $b_1$. On the other hand, if $x_1(t) < b_1$, the RHS of (5) is positive. Hence, $x_1(t)$ increases toward $b_1$. Thus, whenever $x_1(t)$ is away from its equilibrium value $b_1$, it moves toward $b_1$. Finally, when $x_1(t) = b_1$, the RHS of (5) is zero and $x_1(t)$ neither increases nor decreases. Thus, $b_1$ is the equilibrium or the steady-state value of $x_1(t)$.

## 2.3 Pictorial Representation

From (5) and (6), $x_1(t)$ and $x_2(t)$ are calculated iteratively for $t = 0, 1, 2,\ldots$, and so on. To see the control action pictorially, we represent the variations in $x_1(t)$ and

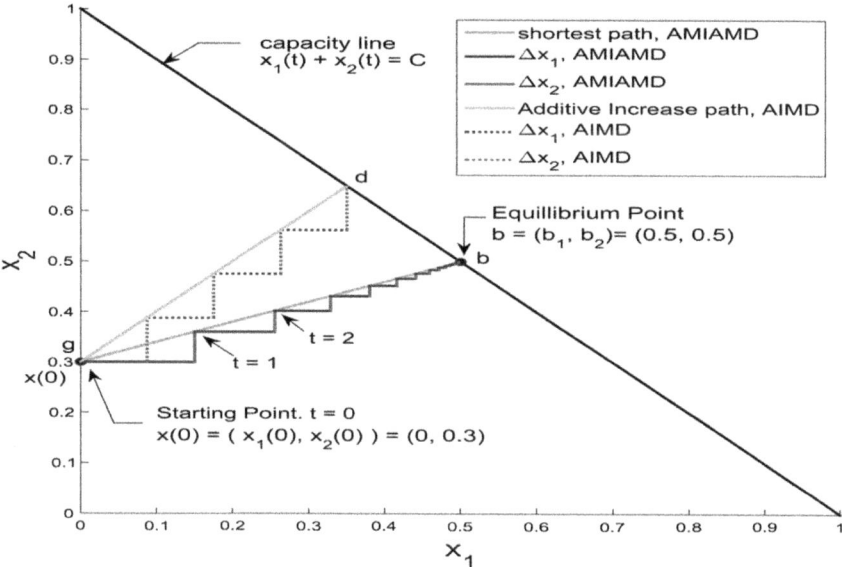

**Fig. 3** State space diagram for $x_1$ and $x_2$

$x_2(t)$ as a trajectory of vector $x(t)$, where $x(t) = (x_1(t), x_2(t))$, in a two-dimensional space (state space) as shown in Fig. 3.

In Fig. 3, the capacity parameter $C$ is taken as 1. The capacity line (also called efficiency line [9]) is drawn from (0, 1) to (1, 0). On this line, $x_1(t) + x_2(t) = C$. The region above this line represents the overloaded (congested) link and the region below this line represents the underloaded (uncongested) link. The trajectory of $x(t)$ is shown in the region where $x_1(t) + x_2(t) \leq C$. This is the region below the capacity line. In Fig. 3, $x(0) = (x_1(0), x_2(0)) = (0, 0.3)$, is the starting point (initial values of $x(t)$) for $t = 0$. The equilibrium point (final value) is $b = (b_1, b_2) = (0.5, 0.5)$. The trajectory gb, for our AMIAMD, is shown in green. The other trajectory gd, in cyan, is for AIMD which is shown for the purpose of comparison. The LHS of (5) gives $\Delta x_1(t)$ and the LHS of (6) gives $\Delta x_2(t)$ which are also shown in Fig. 3.

## 3 AMIAMD Multisource Rate Control Algorithm

Here, we have $N$ sources. The rate vector $x(t)$ is given by (1) subjected to the constraint (2). Then, the discretized multisource rate control is given by (5) which is generalized and rewritten as

$$x_i(t+1) = x_i(t) + a_i(b_i - x_i(t)) \tag{7}$$

for $i = 1$ to $N$ and for $t = 0, 1, 2,\ldots$ etc. For equal priority sources, $b_i$ is calculated for all $i$'s as, $b_i = (C/N)$. In (7), $a_i$'s are equal and taken as, $a_1 = a_2 = \cdots = a_N = a$ with $0 < a < 1$. The value of $a$ is chosen by the designer. Then, (7) is further rewritten as

$$x_i(t+1) = x_i(t) + a(b_i - x_i(t)) = ab_i + x_i(t)(1-a) \tag{8}$$

The rate control is governed by (8) for $i = 1$ to $N$. The error $e_i(t)$ is defined as

$$e_i(t) = b_i - x_i(t) \tag{9}$$

When $e_i(t)$ reaches a value less than certain threshold, $e_{min}$, we assume that the system has reached the steady state. $e_{min}$ is selected by the designer. In general, $e_{min}$ is in the range 0.1–0.5% of $b_i$. Hence, at steady state, $e_i(t) < e_{min}$ for $i = 1$ to $N$.

**AMIAMD Algorithm.** Inputs : $C, N, a, e_{min}$ and $x_i(0)$'s.
Output : Optimal rate $x_i(t)$ for $i=1$ to $N$ and $t = 1, 2, \ldots$, etc.

1. Start with time $t = 0$.
2. Set $i = 1$. //start with source $S_1$.
3. Get the present values of $C$ and $N$.
4. Get $b_i$ as, $b_i = (C/N)$.  // taking equal fairness.
5. Calculate error $e_i(t)$ using (9).
    If $e_i(t) \geq e_{min}$  // $x_i(t)$ has not reached steady state yet.
      Get $x_i(t+1)$ using the governing Equation (8).
    Else  // $x_i(t)$ has reached steady state.
      Get $x_i(t+1)$ as,
      $x_i(t+1) = x_i(t)$. //In steady state $x_i(t+1)$ is same as $x_i(t)$.
    Endif
    If $i = N$  // all the sources covered and updated.
      Go to step 6.  //next time slot.
    Else  // $i < N$, all the sources not covered.
      $i = i+1$.  //go for next source.
      Go to step 3
    Endif
6. $t = t+1$.  // next time slot.
7. Go to step 2.  // go to the next iteration for the sources.

AMIAMD is a continuous process. It works continuously as long as the communication process between the CH and the BS is active. If there are no changes in $C$ or $N$, it continues in the steady state. The controller receives the feedback value $C$

from the BS, periodically (or asynchronously in an emergency). The control system at CH knows $N$. If $C$ or $N$ changes, then at step 4, algorithm AMIAMD calculates the new $b_i$ and the control process continues accordingly.

## 3.1 Comparison Between AMIAMD and AIMD

We compare our AMIAMD with the standard increase–decrease rate control algorithm AIMD [9]. The criteria used for the comparison are transient response, fairness, network utilization, convergence, and distributedness.

**Transient Response.** The variation of $x_1(t)$ for AMIAMD and AIMD with respect to time is shown in Fig. 4. For AMIAMD plot, $a_1 = 0.24$, $x_1(0) = 0.3, x_2(0) = 0$ and $b_1 = 0.5$. For AIMD plot, the additive parameter = 0.06 and the multiplicative factor = 0.08. Initial and final values of AIMD are same as in AMIAMD.

**Fairness.** When sources have equal priority, fairness means the equality among $x_i(t)$'s for all $i$'s and $t$'s. In AMIAMD, fairness is achieved during the steady state but not during the transient state. In AIMD, fairness is maintained with respect to increments during the increase phase. The differential behavior between AMIAMD and AIMD is shown in Fig. 5.

In AIMD described by Chiu and Jain [9], for the two-source scenario, the transition is along the line $\{x(0) - d\}$ which is purely additive and then along $\{d - b\}$ where the trajectory is oscillatory. The path $\{x(0) - d\}$ is a 45°-degree line from $x(0)$ to $d$ that represents equal increments in $x_1(t)$ and $x_2(t)$. This happens in AIMD because, when underloaded (congestion-free zone), the increase is purely additive. Therefore, the transient path in AIMD is $\{x(0) - d\}$. Along this path, the horizontal and vertical increments are equal $(\Delta x_2(t) = \Delta x_1(t))$. Thus, in AIMD, fairness is maintained for increment phase only but not for actual values $x_1(t)$ and

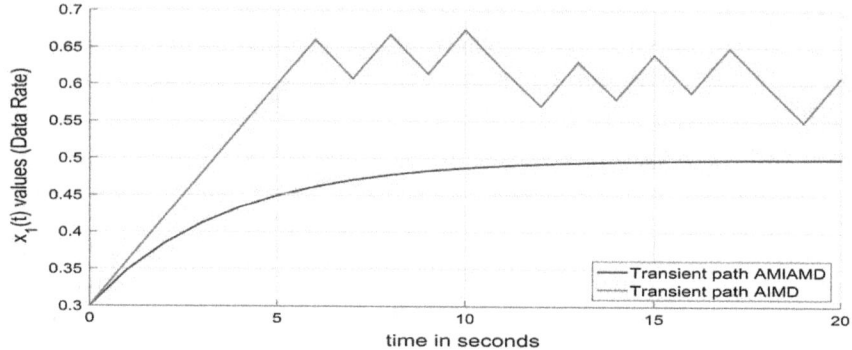

**Fig. 4** Transient response of $x_1(t)$ in AMIAMD and AIMD

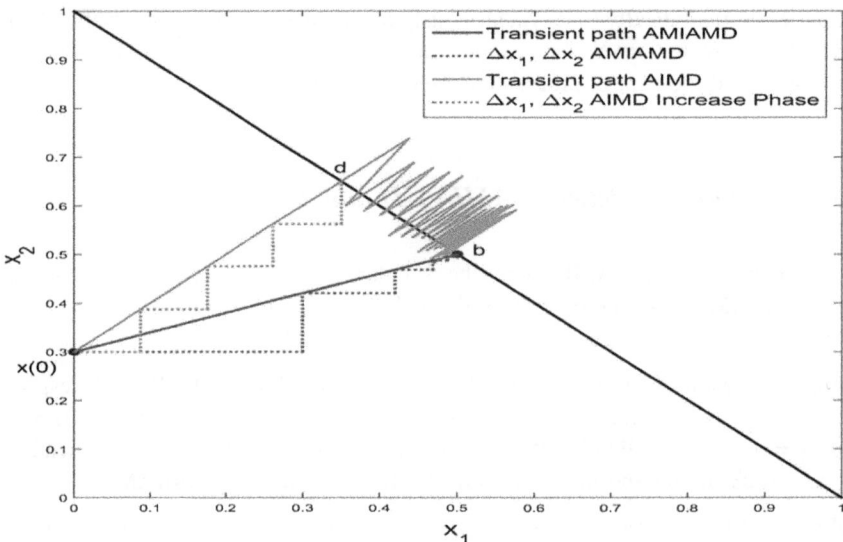

**Fig. 5** Transient responses of AMIAMD and AIMD in state space

$x_2(t)$. After the increase phase, the path crosses point $d$ and then reaches $b$ along $\{d - b\}$ with oscillations as shown in Fig. 5. In AMIAMD, during the transient phase, it is not possible to maintain actual or incremental fairness, because AMIAMD follows the shortest path $\{x(0) - b\}$ to reach $b$ as fast as possible.

**Network utilization.** It is the sum of all $x_i(t)$'s at the steady state [10]. In AMIAMD, when $x_i(t)$'s reach their goal, $e_i(t)$'s $< e_{\min}$ and the channel utilization is maximum and is very near to $C$. In AIMD, also the utilization reaches the maximum value $C$.

**Convergence.** AMIAMD achieves fast convergence, because it follows the shortest path $\{x(0) - b\}$. In AIMD, the path traversed is $\{x(0) - d\}$ and then $\{d - b\}$. During the $\{d - b\}$ phase traversal, AIMD produces oscillations as shown in Fig. 5. This two-segment traversal increases the convergence time of AIMD, whereas the single straight shortest path $\{x(0) - b\}$ traversed by AMIAMD scheme naturally converges faster. This is the strongest advantage of AMIAMD.

**Distributedness.** AMIAMD is a centralized algorithm, unlike AIMD. In AMIAMD, the control system is implemented in the CH to whom the sensor nodes report. The CH receives full feedback instead of binary feedback (as in AIMD) [9]. Therefore, all the information, regarding the number of active sources and initial/ starting rates is known to the CH. Therefore, the centralized control operation is not difficult or expensive. If required, AMIAMD can be altered into a distributed algorithm.

## 4 Conclusions

An innovative modification is made to the popular AIMD algorithm. The modification incorporates both additive and multiplicative increase as well as additive and multiplicative decrease. Basically, it is a centralized algorithm which uses negative feedback to achieve optimal data rate allocations while avoiding congestion. It can be modified into a distributed one. The strength of the algorithm is fast convergence and non-oscillatory transient response. It provides *affirmative corrective action* among different sources while incrementing (or decrementing) the rates originating from unequally placed starting points. The proposed algorithm can be modified to implement weighted and max-min fairness rate control.

## References

1. Akyildiz, I.F., Su, W., Sankarasubramaniam, Y., Cayirci, E.: Wireless sensor networks: a survey. Comput. Netw. **38**(4), 393–422 (2002)
2. Tilak, S., Abu-Ghazaleh, N.B., Heinzelman, W.: Infrastructure tradeoffs for sensor networks. In: Proceeding of WSNA 2002, September 2002, Atlanta, GA, USA
3. Sergio, C., Antoniou, P., Vassiliou, V.: A comprehensive survey of congestion control protocols in wireless sensor networks. In: IEEE Communications Surveys & Tutorials, vol. 16, no. 4, pp. 1839–1859 (Fourth Quarter 2014)
4. Kafi, M.A., Djenouri, D., Ben-Othman, J., Badache, N.: Congestion control protocols in wireless sensor networks: a survey. In: IEEE Communications Surveys & Tutorials, vol. 16, no. 3, pp. 1369–1390 (Third Quarter 2014)
5. Li, K., Nikolaidis, I., Harms, J.: The analysis of the additive-increase multiplicative-decrease MAC protocol. In: 2013 10th Annual Conference on Wireless On-demand Network Systems and Services (WONS), Banff, AB, 2013, pp. 122–124
6. Lai, C., Leung, K.C., Li, V.O.K.: Design and analysis of TCP AIMD in wireless networks. In: 2013 IEEE Wireless Communications and Networking Conference (WCNC), Shanghai, 2013, pp. 1422–1427
7. Stuedli, S., Corless, M., Middleton, R., Shorten, R.: On the AIMD algorithm under saturation constraints. In: IEEE Transactions on Automatic Control, no. 99, pp. 1–8, Jan 2017
8. Shah, S.A., Nazir, B., Khan, I.A.: Congestion control algorithms in wireless sensor networks: trends and opportunities. J King Saud Univ. Comput. Inf. Sci. **29**(3), 236–245 (2016, July 2017)
9. Chiu, D.M., Jain, R.: Analysis of the increase and decrease algorithms for congestion avoidance in computer networks. J. Comput. Netw. **17**(1), 1–14 (1989)
10. Sridharan, A., Krishnamachari, B.: Maximizing network utilization with max-min fairness in wireless sensor networks. In: 2007 5th International Symposium on Modeling and Optimization in Mobile, Ad Hoc and Wireless Networks and Workshops, Limassol, 2007, pp. 1–9

# Analysis on Injection Vulnerabilities of Web Application

Nilesh Yadav and Narendra Shekokar

**Abstract** The number of Internet users has incredible grown. Web applications are normally utilized in various sectors like Ecommerce, Banking, and Military. It is collection of thousands of lines of program, which habitually contain some bugs. Part of them have impact on security and can lead to complete control of the application by an attacker. While in client–server communication, the attacker inputs the vulnerable content into the application, these unnoticed vulnerabilities cause financial losses to organizations. Thus, mitigating such an attack is vital to evade mischievous penalties. An enormous research work on application security has been continuously going on but every defense has its own advantages and disadvantages. The aim of this paper is to study and consolidate the understanding of injection vulnerabilities and its mitigation technique. Different approaches proposed by researchers are analyzed here and discussed about the observed pitfalls present in the existing solutions.

**Keywords** Web application · Injection vulnerability · Attack · Security OWASP

## 1 Introduction

The web applications are made up of millions of systems and services. However, the big challenge of using this is how to increase the confidence of people to use these environments? The web content has become critical to firms, consumers, society, and countries. Web applications handle an extensive form of information including social security data, medical reports, intellectual property, national

---

N. Yadav (✉) · N. Shekokar (✉)
Department of Computer Engineering, D. J. Sanghvi College of Engineering,
Vile Parle, Mumbai, India
e-mail: nileshyadav2004@gmail.com

N. Shekokar
e-mail: narendra.shekokar@djsce.ac.in

© Springer Nature Singapore Pte Ltd. 2018
H. Vasudevan et al. (eds.), *Proceedings of International Conference on Wireless Communication*, Lecture Notes on Data Engineering and Communications Technologies 19, https://doi.org/10.1007/978-981-10-8339-6_2

defense data, and economic data. Vulnerabilities [1] such as SQL Injection, XSS (Cross Site Scripting), File Inclusion (FI), Directory Traversal (DT), Source code disclosure (SCD), Command Injection (OSCI), Code/LDAP Injection, etc. have been used to exploit and damage these applications. The unnoticed web vulnerabilities cause significant financial losses to organizations. Professionals state [2] that the groups of most important companies are still blind for the cybersecurity risks. The vulnerabilities came across here often rely on compound client input situations that are really hard to define with an intrusion detection system. Security professionals recognize that technology has changed dramatically. These fast technological changes have created a number of challenges [3] like multiple devices creating numerous entry points, lack of continuous visibility, lack of resources, difficulty in accessing organization level risk, no clear strategy, and no clear strategy for response. As per OWASP latest report [4], improper input validation/ injection vulnerabilities are very critical vulnerability.

The paper representation is as follows: the Sect. 2 talks about injection vulnerabilities. The primary goal is to show technical analysis of the existing solutions, presented in Sect. 3 with findings/research gaps, and finally Sect. 4 concludes the paper.

## 2 Injection Vulnerabilities

The SQLI attack objects the interactive application that utilizes DB services. Allows an assailant to create, read, update, modify, or delete data stored in the DB. Thus, at the DB layer SQLI exploits security vulnerabilities. XSS defects arise when it gives approval to attacker to execute the scripts in client side, which can capture its sessions, spoil websites, or transmit the victim to malevolent sites. Non-persistent and persistent are the two main types of XSS attack.

Other vulnerabilities are like RFI vulnerability, which allows the attackers to insert a file containing language code in the weak exposed program. Local file inclusions are different from RFI because it includes a file from the local file scheme of the application. A (PT/DT) traversal attack in which an invader uses subjective local files, probably exterior to the web directory. SCD exposés configuration files and program code. OSCI means OS command injection where the application to run a command forwarded by the intruder. (RCE) Code injection vulnerability permits an attacker to insert a code which one is executed at server. LDAP injection as similar to SQLI occurs when an adversary is able to manipulate the value of user-input parameters used as part of an LDAP query.

Numerous techniques have been recommended, mostly at the time of SDLC [5]. Some solutions used static and some are used both dynamic and static techniques.

## 3 Analysis of Injection Vulnerabilities

Below tables are showing the techniques by discussing respective paper with help of common comparison factors. The tabular analysis will give clear idea of existing systems and serve motivation to design new approach. Here, below the respective table, we have identified some research gaps related to each solution.

**SQLI**: Table 1 shows the analysis of SQLIA.

The machine learning methods combined with another technique to detect SQLI attacks. The solution [6] contains multistage log analysis with supervised ML approach. However, the results show that it is not a real-time log analysis system and manual log classification efforts are more with having more false alarms. The author's [8] approach uses the query parse tree with SVM classification detection. However, here the unknown attacks cannot be detected; the data collection process is lengthy and is impacting performance. The technique presents [10] a numerical encoding scheme to tame SQLIA. ML and ANN algorithms implemented in it but the solution is at the complex proto stage with big challenge of proper mapping of random decimal values to attack features. The paper [26] demonstrates the HIPS system. But this solution requires significant CPU resources and decreases the classifier performances in multi-gigabits rate networks; also, the firewall configuration/rule setup is a big issue here.

Unsupervised ML-based novel method [9] for SQLI detection uses the twin HMM, but the process is really complex and time-consuming hence impacting the performance, also end results giving more false alarms. The proposed [16] naïve Bayes ML method combined with role-based access control mechanism to mitigate SQLI. However, the results show that it cannot detect the new attack signatures. The all mentioned ML-based results may contain more false alarms.

In non-ML-based solutions, the proposed architecture [18] uses the resource tree technique for SQLIA mitigation but the generation of correct regular expressions is a challenging risky task and vitally it is not detecting new attack signatures. A formal approach [19] detects the tautology/encoding attacks using regX and finite automata. However, it depends on coding language and it is not giving assurance of new attack signature detection. The authors [20] proposed an intention-oriented approach but the problem is with automatic generation of SQLIDL programs from the logged SQL requests and no assurance about new attack signature detection here. The papers [7, 13, 17] present a novel approach based on information theory but the results shows that it is lengthy process and failed sometimes to detect new attack signatures.

A diversification is implemented in paper [21]. However, it is a challenging task due to variety of language and large volume of DB libraries. Also, queries created dynamically by an application may be hard to diversify automatically. The approach "AMNESIA" is suggested [22] but the construction of SQL query model using NFA is a challenging complex task and the solution is unable to detect new attack signatures. The authors [12] proposed lexical feature comparison and

**Table 1** Analysis based on existing technique for SQLIA

| Tech. name | Params ||||||
|---|---|---|---|---|---|
| | Extra burden on server | Code-based change | Complexity of solution | Accuracy factor | Protection/ Response time |
| Multistage log analysis [6] | High | No | High | Effective | Offline process |
| Expectation criterion [7] | Medium | Yes | Less | Not much effective | Less |
| Fragmented query parse tree detection [8] | High | No | High | Much effective for existing patterns | Medium |
| SQLID using HMM [9] | High | No | High | Much effective | More |
| Numerical encoding [10] | High | Yes | High | Much effective | Medium |
| SQLID a safer app. [11] | High | No | High | Not much effective | Offline process |
| SQLID static and dynamic [12] | High (offline) | No (code access) | High | Effective | Vulnerability detection process |
| Chromatography [13] | High | No | Less | Not much effective | Offline process |
| Novel approach [14] | Medium | Yes | Medium | Not much effective | Less |
| Header sanitization tech. [15] | Medium | Yes (one liner) | High | Effective | Medium |
| SQLID using ML [16] | Medium | No | Medium | Not much effective | Medium |
| SQLIP using entropy [17] | High | Yes | Less | Much effective | More |
| Auto white-list learning tech. [18] | High | Yes | Medium | Effectiveness decrease as list increases | Medium |
| Formal approach [19] | Medium | Yes | More | Effective | Medium |
| Intention-oriented [20] | High | No | More | Not much effective | Medium |
| Diversifying SQL to prevent injection [21] | High (DB server) | Yes | More | Much effective | More |
| AMNESIA [22] | High | No | More | Effective | More |
| SQLDOM [23] | Medium | Yes | More | Not much effective | Less |
| Decision tree tech. [24] | High | No | Medium | Effective | More |

(continued)

**Table 1** (continued)

| Tech. name | Params | | | | |
|---|---|---|---|---|---|
| | Extra burden on server | Code-based change | Complexity of solution | Accuracy factor | Protection/ Response time |
| Joza [25] | High | No | High | Much effective | More |
| Improving firewalls [26] | More | No | High | Much effective | Medium |
| MySql injector [27] | Medium | No | More | Effective for blind injection | More |
| DIWeDA [28] | More | Yes | More | Effective | More |
| SQLRand [29] | Medium | Yes | More | Effective | More |

behavior model but the model not able to detect the unknown attack signatures; here we need to improve bug test algorithm as well as the behavior model set. The proposed model [15] which works by sanitizing received variables inside HTTP header request methods but the current escaping algorithms are not 100% sufficient and secure.

The authors' implemented strong input distillation using pattern matching [11, 14]. However, it is not detecting unknown attack signatures, and validating them inside the validation code is practically not feasible. Also, this resolution needs improvement in network analyzer. In SQLDOM, author [23] considers the existing flaws. However, in this solution, usage of parser at database level is a challenging task. The novel technique [24] uses the decision tree classification for mitigation. However, the solution is not detecting new attack signatures. Authors presented the taint analysis technique Joza [25] for prevention. However, the output shows more false alarms and needs to improve on application-specific encodings. MYSQL injector [27] is detecting mostly blind timing attack but it failed to detect the new attack signatures. An IDS is developed using DIWEDA [28] but this system failed to detect new attack signatures. The SQLrand approach [29] is proposed for detection but this unified solution needs to integrate object mapper which is complex task and impacting performance.

**XSS**: The most addressed XSS type in the literature is the reflected XSS. An analysis is mentioned in Table 2.

CSP is the first-known policy to mitigate [30] XSS. However, enabling CSP for a web application would affect the application's behavior and disrupt its functionality. The proposed method [31] uses the attack vector repertory concept but this requires more trained optimization model and unable to prevent the new type of signature. The approach [32] implements a technique called XSS-Me but this tool is using manual testing to identify the XSS, which is inflexible and also unable to detect new signatures. The author [33] proposes the detection approach based on boundary injection. However, it is lengthy, time-consuming, and language dependent. Authors [34] propose a linear-based automaton approach called XSS Chaser. However, the solution has issues like more complexity, slow processing, extra

**Table 2** Analysis based on existing technique for XSS

| Tech. | Params | | | | |
|---|---|---|---|---|---|
| | Extra burden on server | Code-based change | Complexity of solution | Accuracy factor | Protection/ Response time |
| CSP [30] | No | No | More | Effective | Medium |
| Attack repertory [31] | Medium | No | More | Not much effective | Medium |
| XSS-me [32] | Low | No | More | Effective | Less |
| XSS-S2Xs2 [33] | High | Yes | More | Not much effective | More |
| Automaton tech. [34] | High | No | More | Effective | More |
| Positive security [35] | High | No | More | Not much effective | More |

software/hardware, and configurations required for instrumentisation of code and giving more false alarms. The paper [35] proposed a solution based on sanitization and cleaning the inputs. However, the output shows more false alarms and impacted the performance.

**Other Vulnerabilities**: The other injection attacks are FI, DT, RCE, LDAPI, CommandI, etc. The analysis of these attacks is mentioned in Table 3.

The authors [36] combined two approaches for injection vulnerability detection. However, the solution depends on the taint analysis having its own basic issues with more complexity. The framework [37] comprises hybrid program analysis and classifications based on ML schemes but the issues are already present in static analysis technique and its complexity. In this paper [40], authors put forward a new dynamic taint flow based solution in which they use the "source-derivation-sink" with the AOP technology but this solution is unable to detect the new type of signature.

The author examines logging text using Shannon entropy [38] for detection of SQLI, XSS, DT, and RFI. However, the results fail sometimes to detect new attack signatures and giving more false alarms. Anomaly-based IDS is deployed [39] to detect some attacks. However, the results failed to detect new attack signatures with giving more false alarms. The paper [41] focused on a dynamic and sound framework for preventing LFI. However, the process shows that it is complex, lengthy, and language-specific solution. In paper [42], researchers' expressed constraints in OCL form to check LDAPI but still the solution is unable to detect the new type of attack signature. The proposed approach [43] called VSCM is giving more false alarms, and its language dependability is really high. The authors proposed [44] a way to detect injection vulnerabilities but still this solution is unable to detect the new type of attack signature and also it is giving more false alarms.

**Table 3** Analysis based on existing technique for other types of injection attacks

| Tech. | Params | | | | |
|---|---|---|---|---|---|
| | Extra burden on server | Code-based change | Complexity of solution | Accuracy factor | Protection/ Response time |
| Static analysis_mining [36] | High | No | More | Much effective | Less |
| Predicting hybrid prg [37] | High | No | More | Much effective | Medium |
| Entropy [38] | High | No | Medium | Not much effective | Offline |
| Information theory [39] | High | No | More | Effective | More |
| Dynamic taint tracking [40] | High | No | More | Much effective | More |
| A framework for LFI [41] | High | Yes | More | Not much effective | Medium |
| OCLFI-LDAP [42] | High | No | More | Not much effective | Offline |
| VSCM [43] | High | Yes | More | Not much effective | Medium |
| Detection in Java code [44] | High | Yes | High | Not much effective | Medium |

All the above approaches have been highlighted in literature reveal that engineers attitude for validation is controlling each input in separation. Such a tactic leaves major openings for attacks. No auto-technique is present in the coding language, runtime system, or compiler that alerts the developer about escaping the validation. Hence, static or dynamic analysis is required to confirm that entire untrusted i/ps. Lots of recommended techniques for i/p confirmation are based on signature, computing well-known wicked strings essential for injection attacks but restriction for the length of i/p strings. Anomaly-based techniques are generating an application's behavior model during legitimate use; this model is then utilized at runtime to prevent injection attacks. However, it may fail during run time because it needs to access the source code, it is heavily dependent on the normal use model and it is only applicable to a single web application.

Nowadays, the SQLI and XSS attacks are 3 times more hazardous because the methods based on Information Theory, ANN, ML, etc. are smart and proficient but these approaches cannot 100% deal with new attack signature. The proxy solutions are complex in their rule setup, configurations, and maintenance. Some solutions are based on attack signatures DB and regular expressions, they looked simple but automatic maintenance and updating of these repositories with fine-tuned regEx coding is a little bit challenging tasks. In security, employing either dynamic taint analysis or forward symbolic execution, or a mix of the two can do automatic input filter generation and attack analysis but there are several challenges for using taint analysis technique properly such as under/over tainting, tainted addresses problem, false alarm, etc. However, it is discovered that the above existing detection and prevention solutions are insufficient to protect against advanced attackers.

## 4 Conclusion

Improper input validation/injection vulnerabilities have been reported as the top internet security threats in recent years. We have attempted to analyze and classify different mitigation techniques of SQLIA, XSS, FI, DT, RCE, LDAPI, and SCD attacks. Our observations reveal that every defense has its own limitations, and these are not sufficient to provide well proof solutions. Several types of themes and techniques are present to mitigate the web attacks, having their own advantages and disadvantages, and hence identified some research gaps related to each approach. This analysis will serve as a motivation to design new approach. It will help to develop a solution considering the existing issues and recommendations so that this detection system would handle new/unknown type of attack signatures and it will become a performance-oriented system with fewer false alarms.

## References

1. Deepa, G., Thilagam, P.S.: Securing web applications from injection and logic vulnerabilities: approaches & challenges. ACM Inf. Soft. Technol. (India) **74**, 160–180 (2016)
2. European Personal Security Blog. [online] Available: http://securityaffairs.co/
3. CYREN, August 2015: How to Overcome Web Security Challenges. [online] Available: http://pages.cyren.com/WP_Overcome_WebSec_Challenges.html
4. OWASP Group: Top 10 Most Critical Web Application Security Vulnerabilities. [online] Available: https//www.Owasp.org/index.php
5. WAP Tool Website: [Online] Available: http://awap.sourceforge.net/
6. Moh, M., Pininti, S., Doddapaneni, S., Moh, T.S.: Detecting web attacks using multi-stage log analysis. In: 2016 IEEE 6th International Conference on Advanced Computing (IACC), Bhimavaram, pp. 733–738 (2016)
7. Xiao, L., Ishikawa, T., Sakurai, K.: SQL Injection Attack Detection Method Using Expectation Criterion. CANDAR, IEEE, Hiroshima, Japan, pp. 649–654 (2016)
8. Deva Priyaa, B., Devi, M.I.: Fragmented query parse tree based SQL injection detection system for web applications. ICCTIDE, IEEE, Kovilpatti, pp. 1–5 (2016)
9. Kar, D., Agarwal, K., Sahoo, A., Panigrahi, S.: Detection of SQL injection attacks using Hidden Mar kov Model. IEEE-ICETECH, Coimbatore, pp. 1–6 (2016)
10. Uwagbole, S.O., Buchanan, W.J., Fan, L.: Numerical encoding to Tame SQL injection attacks. NO-MS 2016 IEEE/IFIP, pp. 1253–1256 (2016)
11. Pramod, A., Ghosh, A., Mohan, A., Shrivastava, M., Shettar, R.: SQLI detection system for a safer web application. Int Conf IEEE (IACC) (2015)
12. Wang, Y., Wang, D., Liu, Y.: Detecting SQL vulnerability attack based on the dynamic and static analysis. COMPSAC, IEEE, Taichung, pp. 604–607 (2015)

13. Watcharapupong, A., Threepak, T.: Web attack detection using chromatography-like entropy analysis. In: Recent Advances in Information and Communication Technology, 361. Springer, Switzerland (2015)
14. Sonewar, P., Mhetre, N.: A novel approach for detection of SQL injection and cross site scripting attacks. In: ICPC. IEEE, Pune, India (2015)
15. Sadeghian, A., Zamani, M., Abd. Manaf, A.: SQL injection vulnerability general patch using header sanitization. In: IEEE I4CT, Langkawi, Malaysia (2014)
16. Joshi, A., Geetha, V.: SQL injection detection using machine learning. In: ICCICCT, pp. 1111–1115. IEEE, Kanyakumari (2014)
17. Shahriar, H., Zulkernine, M.: Information-theoretic detection of SQL injection attacks. In: IEEE 14th ISHASE, Omaha, USA (2012)
18. Murtaza, S., Abid, A.S.: Automated white-list learning technique for detection of malicious attack on web application. In: IBCAST, pp. 416–420. IEEE, Islamabad (2016)
19. Qbea'h, M., Alshraideh, M., Sabri, K.E.: Detecting and preventing SQL injection attacks: a formal approach. In: IEEE/CCC, Amman, pp. 123–129 (2016)
20. Chenyu, M., Fan, G.: Defending SQL injection attacks based-on intention-oriented detection. In: 11th International Conference ICCSE, Nagoya, pp. 939–944 (2016)
21. Rauti, S., Teuhola, J., Leppänen, V.: Diversifying SQL to prevent injection attacks. In: IEEE Trustco-m/BigDataSE/ISPA (2015)
22. Halfond, W.G.J., Orso, A.: AMNESIA: analysis and monitoring for neutralizing SQL injection attacks. In: ACM, ASE'05, Long Beach, California, USA (2005)
23. McClure, R.A., Krüger, I.H.: SQL DOM: compile time checking of dynamic SQL statements. In: ACM ICSE'05, St. Louis, Missouri, USA (2005)
24. Hanmanthu, B., Ram, B.R., Niranjan, P.: SQL injection attack prevention based on decision tree classification. In: IEEE ISCO, Coimbatore, India, pp. 1–5 (2015)
25. Afooshteh, A.N., Tuong, A.N., Hiser, J.D., Davidson, J.W.: Joza: hybrid taint inference for defeating web application SQL injection attacks. IEEE/IFIP, Brazil (2015)
26. Makiou, A., Begriche, Y., Serhrouchni, A.: Improving web application firewalls to detect advanced SQL injection attacks. In: IEEE/IAS, Okinawa, Japan (2014)
27. Liban, A., Hilles, S.M.S.: Enhancing Mysql injector vulnerability checker tool (Mysql Injector) using inference binary search algorithm for blind timing-based attack. In: IEEE 5th, Shah Alam/ICSGRC, Shah Alam, Malaysia, pp. 47–52 (2014)
28. Roichman, A., Gudes, E.: DIWeDa—detecting intrusions in web databases. In: Atluri, V. (ed.) IFIP International Federation for Information Processing 2008. DAS 2008, LNCS 5094, pp. 313–329 (2008)
29. Boyd, S.W., Keromytis, A.D.: SQLrand: preventing SQL injection attacks. In: Proceedings of the 2nd ACNS Conference, pp. 292–302 (2004)
30. Yusof, I., Pathan, A.-S.K.: Mitigating cross-site scripting attacks with a content security policy. Computer **49**, 56–63 (IEEE) (2016)
31. Guo, X., Jin, S., Zhang, Y.: XSS vulnerability detection using optimized attack vector repertory. In: 2015 International Conference on Xi'an of Cyber-Enabled Distributed Computing and Knowledge Discovery (CyberC), pp. 29–36 (2015)
32. Mewara, B., Bairwa, S., Gajrani, J., Jain, V.: Enhanced browser defense for reflected cross-site scripting. In: 2014 3rd International Conference on Reliability, Infocom Technologies and Optimization (ICRITO), Noida, pp. 1–6 (2014)
33. Shahriar, H., Zulkernine, M.: S2XS2: a server side approach to automatically detect XSS attacks. In: IEEE/DASC, Sydney, NSW, pp. 7–14 (2011)
34. Suju, D.A., Gandhi, G.M.: An automaton based approach for forestalling cross site scripting attacks in web application. In: ICoAC, Chennai, India, pp. 1–6 (2015)
35. Maurya, S.: Positive security model based server-side solution for prevention of cross-site scripting attacks. In: IEEE/INDICON, New Delhi, pp. 1–5 (2015)
36. Medeiros, I., Neves, N., Correia, M.: Detecting and removing web application vulnerabilities with static analysis and data mining. In: IEEE Transactions on Reliability, vol. 65, no. 1, pp. 54–69. IEEE Reliability Society (2016)

37. Shar, L.K., Briand, L.C., Tan, H.B.K.: Web application vulnerability prediction using hybrid program analysis and machine learning. In: IEEE Transactions on Dependable and Secure Computing, vol. 12, no. 6, pp. 688–707. IEEE (2015)
38. Threepak, T., Watcharapupong, A.: Web attack detection using entropy-based analysis. In: COIN2014, pp. 244–247. IEEE, Phuket (2014)
39. Bronte, R., Shahriar, H., Haddad, H.: Information theoretic anomaly detection framework for web application. In: COMPSAC, pp. 394–399. IEEE, Atlanta (2016)
40. Zhao, J., Qi, J., Zhou, L., Cui, B.: Dynamic taint tracking of web application based on static code analysis. In: IMIS, pp. 96–101. IEEE, Fukuoka (2016)
41. Tajbakhsh, M.S. Bagherzadeh, J.: A sound framework for dynamic prevention of local file inclusion. In: IKT, pp. 1–6. IEEE, Urmia (2015)
42. Shahriar, H., Haddad, H., Bulusu, P.: OCL fault injection-based detection of LDAP query injection vulnerabilities. In: COMPSAC, pp. 455–460. IEEE, Atlanta (2016)
43. Hussein, O., Hamza, N., Hefny, H.: A proposed approach to detect and thwart previously unknown code injection attacks. In: 7th International Conference on Intelligent Computing and Information Systems (ICICIS), pp. 336–342. IEEE, Cairo (2015)
44. Pasaribu, S., Asnar, Y., Liem, M.M.I.: Input injection detection in Java code. In: 2014 International Conference on Data and Software Engineering (ICODSE), pp. 1–6. IEEE, Bandung (2014)

# MAC-Based Group Management Protocol for IoT [MAC GMP-IoT]

**Yeole Anjali and D. R. Kalbande**

**Abstract** Generally, 6LoWPAN network is used for communication in IoT. All the 6LOWPAN nodes in the radio range of 6LoWPAN boarder router will start sending or reading data. There is no authentication present; hacker can take advantage of this situation and can be able to place his own sensor in the network which will start generating some random data at high speed and of huge size. This will result in Distributed Denial of Service [IoT-DDOS]. To avoid this problem, group management protocol has been proposed in this paper, "MAC-based Group Management Protocol-IoT" [MAC GMP-IoT].

**Keywords** IoT · 6LoWPAN · Group management protocol

## 1 Introduction

The IoT is the Internet of Things, where sensors are connected to the Internet, which collect data for analysis to make our world more interconnected and intelligent. A common person carries on average one or two mobile devices nowadays. Hence, by taking advantage of the increasing presence of mobile devices, the cost of equipment can be reduced significantly in many industries like health care.

In traditional TCP/IP networks, main goal of security is to protect the confidentiality, integrity, and availability (CIA) of network data. As the characteristic of node and application environment, wireless sensor network security not only needs traditional security protection but also the special requirements of trust, security, and privacy (TSP).

---

Y. Anjali (✉)
Computer Engineering, VES Institute of Technology, Mumbai, India
e-mail: anjali.yeole@ves.ac.in

D. R. Kalbande
Computer Engineering, Sardar Patel Institute of Technology, Mumbai, India
e-mail: drkalbande@spit.ac.in

## 2 Literature Survey

"IPv6 over Low-Power Wireless Personal Area Networks"—6LoWPAN—is a networking technology or adaptation layer that allows IPv6 packets to be carried efficiently within small link-layer frames, such as those defined by IEEE 802.15.4. 6LoWPAN adopts the physical (PHY) and Media Access Control (MAC) layer protocols defined in IEEE 802.15.4 standard as its lower layer protocols. The IPv6 protocol is used as the network layer protocol in 6LoWPAN. Since the IPv6 network layer's maximum transmission unit (MTU) is not compatible with the MAC layer of IEEE 802.15.4, an adaptation layer is introduced between the network and MAC layers. It performs fragmentation, reassembling, IPv6 header compression, and addressing mechanism to enable compatibility [1].

### 2.1 6LoWPAN Addressing and Auto-configuration

6LoWPAN supports Stateless Address Auto-configuration (SAA) mechanism which reduces the configuration overhead on the nodes. The auto-configuration process includes generation of link-local address and global address. Further, it also performs Duplicate Address Detection (DAD) procedure to verify the unique nature of the addresses on a link. The stateless method is used, when the consideration for the exact addresses that the nodes use is taken in a random manner. While Dynamic Host Configuration Protocol for IPv6 (DHCPv6) is used, there is a strong consideration for exact address assignments. Both methods can be used simultaneously [2]. To test the unique nature of the address on a given link, nodes run Duplicate Address Detection (DAD) algorithm on addresses before assigning them to an interface. This DAD algorithm is performed on all addresses even if they are obtained through DHCPv6 or stateless auto-configuration. This auto-configuration process is applied only to nodes and not to routers. The advantage of stateless address auto-configuration includes the reduction of manual configuration of individual nodes [1].

### 2.2 Addressing in 6LoWPAN

The link-layer address is not a routable address and it cannot predict the nodes present in the network. In 6LoWPAN, data frames carry both the source and the destination addresses. This destination address is used by a receiver to decide whether the frame is actually intended for this receiver or not [1]. The source address plays a vital role in mesh forwarding. 6LoWPAN nodes are permanently identified by Extended Unique Identifier (EUI)-64 identifier. Also, it defines a 16-bit short address format which is dynamically assigned during the bootstrapping of the network. 6LoWPAN requires that both the source and destination addresses be included in the frame header. Thus, IPv6 addresses are compressed in 6LoWPAN [2].

## 2.3 Neighbor Discovery (ND)

6LoWPAN Neighbor Discovery (ND) describes network auto-configuration and operations of hosts, routers, and edge routers in LoWPANs. ND is used to bootstrap the whole LoWPAN. Bootstrapping is the process of assigning IPv6 addresses to nodes that are within radio range to enable basic communication. In order to reduce the cost involved in multicast flooding, a registry of the nodes in each LoWPAN is maintained in the white board database of the Edge Router (ER) [2]. Using Neighbor Discovery (ND) protocol, each node present in the LoWPAN discovers its neighbors as shown in Fig. (1). It also determines its link-layer addresses, to find routers. ND helps to maintain reachability information about the paths to neighbors that the node is actively communicating with others. ND can be combined with other protocols such as DHCPv6 to obtain additional node configuration information. The ND protocol classifies the nodes into two category namely hosts and routers [1]. Host nodes are basically the source node, while router nodes are the intermediate node that forwards data to the destination. Edge router is considered to be the destination node that interfaces with Internet to formulate 6LoWPAN. The edge router compared to other routers performs complex functions such as maintaining the whiteboard database about the nodes and routers inside the LoWPAN. Thus, centralized administration is maintained by the Edge Router (ER) [1].

## 3 Proposed Algorithm

Propose group management protocol is going to follow guidelines of group membership service. The group membership service is responsible for adding members to the group, removing members from the group, and maintaining the correct membership list at all correct processes.

### 3.1 Weaknesses Needs to Be Removed in Existing Working

- Any device can become part of 6lowpan network and can start sending random data which can result in DDOS [3].
- If any external device can be part of this network is threat to privacy.
- IP address can be spoofed [4].

## 3.2 Reasons for Using MAC Address

- Machines IP can change depending on which network it is using.
- Any external entity (like Dr's mobile in IoT based healthcare system [5]) IP address can change but MAC address will be same.

## 3.3 Modules

1. Dictionary creation module: This module is responsible for creation of MAC address list of valid recourses and shared secrete.
2. Validation module: This will take incoming and outgoing request on each 6LoWPAN network that will check whether the address is present in dictionary. If it is there, then it will have challenge–response protocol to check whether MAC is authenticated.
3. Group maintenance module: If MAC address is valid and authenticated that packet can be forwarded in or out of network.

## 3.4 Algorithm

The sensors will use the concept of nonce. To authenticate each other, they will have to encrypt a nonce given by the other using DES and send it back; proper encryption will authenticate the motes and then data share can happen. The process will need to save randomly generated nonce only for some time, so there is no need for extra memory to be mounted.

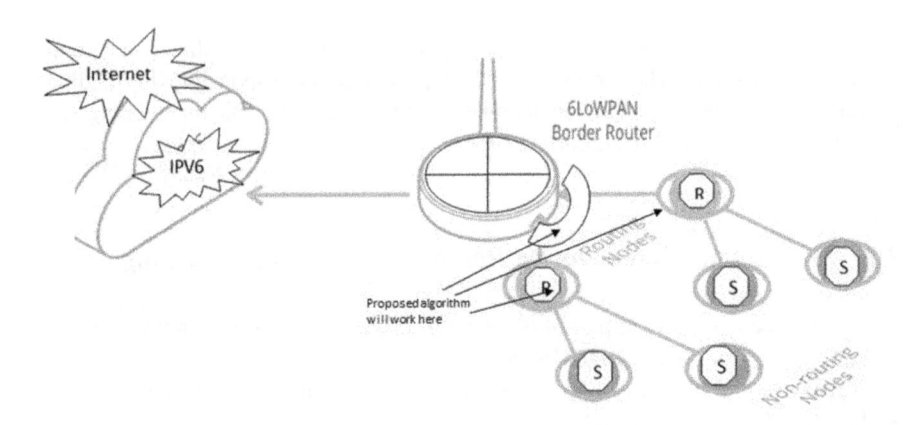

**Fig. 1** 6LoWPAN network with proposed algorithm

# MAC-Based Group Management Protocol for IoT [MAC GMP-IoT]

1. If any sensor which is 6LoWPAN node wants to become part of network need to register its MAC address and Shared secrete
2. A router sends request for data
3. The responding router returns a random nonce to requesting node.
4. The requesting node (first router) then encrypts and sends it back,
    If the encryption is correct,
        The responding router will treat this response as a nonce from the first router, encrypt it and send it back.
            If this encryption is also correct,
                Checked by the first router, then both routers have authenticated each other and can now send data to each other. Bidirectional authentication will be achieved
            else
                Bidirectional authentication is not achieved
        else Node is not authenticated communication can't start
5. Till an encrypted response arrives, the routers keep the nonce saved in a linked list. This memory is freed when the response check is done.

In the proposed algorithm, nonce are not saved in the memory permanently which overcomes the problem of limited memory size in IoT. As DES is hardcoded in IoT device, it will be fast and will require less processing power.

## 3.5 Outcomes

- Proposed algorithm will take care of limited memory size and limited processing power in IoT device.
- Group will be maintained according to MAC list.
- MAC spoofing is not possible because of challenge–response protocol.

# References

1. Renuka Venkata Ramani, C.: Two Way Firewall for Internet of Things Degree Project in Information and Communication Technology, Second Cycle, 30 Credits Stockholm, Sweden (2016)
2. 6LoWPAN: The Wireless Embedded Internet Zach Shelby Sensinode, Finland Carsten Bormann University Bremen TZI, Germany
3. DDOS Attack that Made Enterprises Rethink IoT Security. http://www.crn.com/slide-shows/internet-of-things/300084663/8-ddos-attacks-that-made-enterprises-rethink-iot-security.htm
4. Pongle, P., Chavan, G.: Survey Attacks on 6LOWPAN and RPL in IoT. Computer Engineering Department, Sinhgad College of Engineering, Pune, India
5. Gia, T.N., Thanigaivelan, N.K., Rahmani, A.-M., Westerlund, T., Liljeberg, P., Tenhunen, H.: Customizing 6LoWPAN Networks Towards Internet of-Things Based Ubiquitous Healthcare Systems, Department of Information Technology University of Turku, Finland

# A Secured Two-Factor Authentication Protocol for One-Time Money Account

**Devidas Sarang and Narendra Shekokar**

**Abstract** Credentials information stealing and online banking fraud are common problem in today's world. Two-factor authentications are used to overcome online banking frauds. But it can be easily broken by fraudster using different phishing techniques and synchronization vulnerabilities. These vulnerabilities weaken the security guarantees of smartphone based on two-factor authentication. Once authentication is broken fraudster has a direct online access of bank account with all access privileges. In this paper, we have attempted to minimize banking fraud by proposing OTM protocol for virtualization of bank account. Virtualization gives indirect and partial online access to bank account at the time of online financial transaction. OTM protocol derives virtual sub-accounts (VSA) from user bank account at the ATM machine using respective credit/debit card. Each virtual sub-account has assigned limit of maximum amount and used only one time for online banking.

**Keywords** OTM (one-time money) · VSA (virtual sub-account)
ATM · Virtualization · Synchronization vulnerability

## 1 Introduction

Development in e-commerce leads to rapid growth in e-business, and lot of organizations have established online platform for online shopping, ticket reservations, etc. mostly e-business uses the online banking to complete e-commerce. Credentials stealing and channel breaking attacks [1] are two big threads in e-commerce/online banking. The solution for this problem is QR-based visual authentication protocol [2].

---

D. Sarang (✉)
Department of Computer Engineering, YTCEM, Chandai, Raigad, India
e-mail: sarang.fri@gmail.com

N. Shekokar (✉)
Department of Computer Engineering, DJSCE, Vile Parle, Mumbai, India
e-mail: narendra.shekokar@djsce.ac.in

© Springer Nature Singapore Pte Ltd. 2018
H. Vasudevan et al. (eds.), *Proceedings of International Conference on Wireless Communication*, Lecture Notes on Data Engineering and Communications Technologies 19, https://doi.org/10.1007/978-981-10-8339-6_4

But visual authentication is not useful if user complete online banking only using smart mobile because Trojan viruses easily steal user credential by video recording.

Key loggers can get full control of the PC and steal the credential by video buffering and logging each event. It is similar to the hidden shoulder surfing attack and directly observes the behavior of the user i/p on client PC. Many graphical password schemes are introduced in [3–5] to avoid shoulder surfing attack but very few schemes are effectively usable. Key logger/shoulder surfing problem is effectively solved by visual authentication but limitations of these techniques are that two devices are required for QR code scan and description QR code [2].

## 2 Existing System

Today's online banking transition is based on net-banking or credit/debit cards [6]. Most of the client cannot change net-banking password regularly. So, mostly client credentials are constant from one transition to another. These credentials are easily stolen by the Trojan virus/key loggers.

Financial organization commonly uses the second factor of client authentication is time-stamp-based one-time password (TOTP) [2]. Synchronization vulnerability is a new class of vulnerabilities that weaken the security-based two-factor authentication (2FA) [7] of mobile phone. TOTP is mostly used in plain-text format. So, any malfunctioning app [8, 9] on smart mobile easily read the TOTP [10] and forward to fraudster for online banking fraud [11].

The rest of the paper organization is as follows: Sect. 3 deals with proposed system for online banking. Section 4 is defining OTM (One-Time Money) protocol by driving VSA from client bank account. Section 5 shows detailed steps for online banking using OTM/VSA protocol. In Sect. 6, we proposed new flexible authentication procedure using smart mobile and keyboard layout. Section 7 shows color code generation steps. Section 8 describes mitigation of attacks at the time of online banking. We conclude this paper under Sect. 9.

## 3 Proposed Model for Online Banking

In this paper, we proposed OTM protocol for virtualization of bank account. It gives indirect access to bank account with virtual controller OTM server. OTM protocol has two components. First component is the generation of VSA (Virtual Sub-account), and another component is online banking using OTM-QR code. VSA generation depends on four entities ATM machine, credit/debit cards, cloud-based OTM server, and printing paper at ATM machine. Second component is online banking using OTM-QR code and it has three entities, client smart mobile, printout of OTM-QR code with index mapping keyboard, and OTM-PIN. OTM-PIN is the

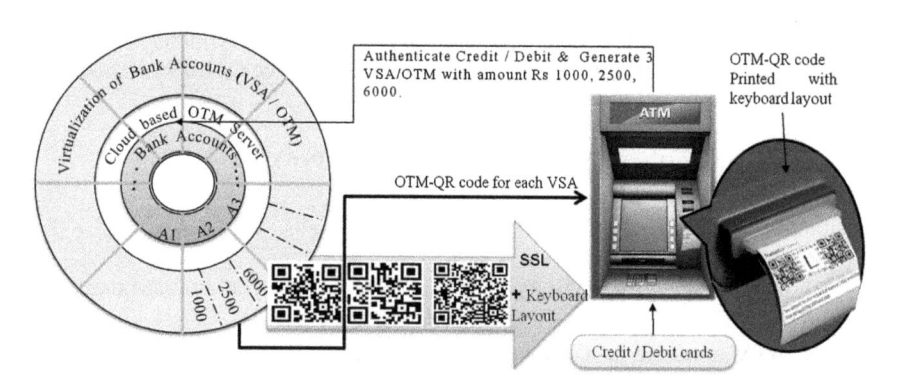

**Fig. 1** Proposed system model for OTM protocol

five digit number and users remember it for indexing OTM password. OTM password is changed from transaction to transaction.

OTM/VSA (virtual sub-account) generation initiated at the A ATM machine using client credit/debit card as shown in Fig. 1. If client authenticated, then ATM machine forward OTM generation request with OTM amount to bank server. Bank server validates client request; if request is valid, then forward request to cloud-based OTM server. Once request is received at OTM server, it derives the VSA account from user base account and generates encrypted OTM-QR code for each VSA account.

Finally, OTM server establishes SSL and sends encrypted OTM-QR codes with index mapping keyboard. Ones received at ATM machine, then it is directly printed on paper.

## 4 Deriving Virtual Sub-account from User Bank Account

Proposed system has two steps: First is deriving OTM/VSA (Virtual Sub-accounts) from bank base account. And, second is obtaining printed Encr (OTM-QR) code with keyboard layout from ATM machine. These two steps are completed by the following procedures:

- Initially, user has to visit bank ATM machine with valid credit/debit card.
- User initiates with credit/debit card at machine and select OTM protocol.
- Now user enters details of OTM/VSA amount and number of accounts.
- ATM machine forwards OTM protocol request to bank server.

Message[ATMno, ATMpin, OTMcount, OTMamounts]

- Bank server validates ATM-machine request then F/W to cloud-based OTM server.

- Cloud-based OTM server derives OTM/VSA (Virtual Sub-accounts) from bank base account and regenerates unique OTM number for each OTM/VSA. Encrypt OTM/VSA number with user Pub key.

$$E(OTM\text{-}QR) = Encr(OTM\text{-}NO, Pvt\text{-}Key).$$

- Cloud-based OTM server replies with all E(OTM-QR) codes and a fresh keyboard layout.
- Bank server F/W all E(OTM-QR) codes and a keyboard layout to ATM machine via SSL.
- Now, ATM machine is ready to print all E(OTM-QR) codes and a keyboard layout.

$$Print[E(OTM\text{-}QR) \text{ codes and a keyboard layout.}]$$

- User received printout of E(OTM-QR) codes and a keyboard layout.

## 5 Online Banking Using OTM Protocol

Following steps describe online banking using OTM/VSA (Virtual Sub-account).

- Client initiates online banking with unique client id. and select OTM amount.
- Now users scan encrypt OTM number [E(OTM-NO)] from QR code and decrypt using public key at mobile bank app.

$$X = E(OTM\text{-}NO, Pvt\ key)$$
$$OTM\text{-}NO = D[X]$$

- Client types password using flexible user login password and sends to OTM server.
- OTM server validates user login credentials reply with encrypted NULL-OTM number E(NULL-OTM) color code.

$$Y = (NULL\text{-}OTM\text{-}NO, Pvt\ key)$$
$$E(NULL\text{-}OTM) = Encr[Y]$$

- Users can encrypt color code as shown in Fig. 2, E(NULL-OTM) and decrypt using public key.

$$Z = E(NULL\text{-}OTM)$$
$$NULL\text{-}OTM = D[Z]$$

**Fig. 2** Scanned or screenshot of color code in smartphone

- After decryption of color code bank app validates Pvt key and NULL-OTM number present in NULL-OTM color code.
- If Pvt key is same, then bank app generates encrypted color code "C" with OTM-NO and NULL-OTM number

$$C = \text{Encr}[\text{OTM-NO}, \text{NULL-OTM NO}, \text{Pvt-key}]$$
$$E(\text{Color code}) = C$$

- Finally, user scans color code "C" E(OTM-NO) and sends to OTM server. Server decrypts OTM-NO and validates NULL-OTM-NO and user Pvt key.
- After validation, OTM server sends 20-digit OTM password to the user mobile by instant SMS service. Once SMS is received, then bank displays it with key mapper.
- User enters OTM password using index mapping keyboard and then clicks on submit button.
- Bank OTM server validates received OTM password and completes the online transaction.

## 6 Flexible Authentication

Flexible authentication is based on index mapping technique. Index mapping gives dynamic solution for user credential like user login password, OTM-PIN, and OTM password. Index mapping ensures that user credential information automatically changed from transition to transaction. So, it is called flexible authentication and following method A and B shows it.

## 6.1 Flexible User Login Password

Client login password is mapped to another character using level 2 index mapping. Client types the mapped password character using key mapper, blank key layout, and printed keyboard. Initially, user has to select correct amount of OTM (2500) account, and then user types password (e.g., bobis568) using key mapper and keyboard layout. Here, user password (e.g., bobis568) is mapped to another eight characters [J6FA4BM9].

## 6.2 Flexible OTM Password

After confirmation of user authentication, OTM server sends OTM password via instant SMS, and at the same time it sends color code of encrypted NULL-OTM-NO to client Pc/mobile. Now client mobile scans/screenshots the NULL-OTM color code from the Pc/mobile. Then, decrypt color code for NULL-OTM-NO and validate it from color code. After validation, bank app displays OTM password with key mapper index (in mobile).

OTM password is decoded using two-level index mapping. Decoding process started with 5 digit OTM-PIN [client has to remember it (e.g., 25768)]. If printed keyboard has center character L, then use left-to-right indexing for starting (top row in the mobile) 10 characters else if R characters at center of printed keyboard, then user starts indexing from right to left for last 10 character (bottom row of mobile). Each digit of OTM-PIN used to address/decode 20 characters encoded OTM password (e.g., bob54). Here, two-level index mapping is shown below diagram:

Level 1: decode five digits from 20-digit OTM password using 5-digit OTM-PIN
Level 2: map the selected five digits from OTM password using key mapper (Fig 3).
Finally, client click/press mapped character (e.g., J6FBG) using keyboard layout.

## 7 Color Code Generation Steps

Color code generation is based on two steps as shown in Fig. 4. In step one, encrypted data bits are encoded by run length encoding technique. In step two, apply color code for run length. If run length is 3, 4, 5, or 6, then color codes are red, green, blue, or yellow, respectively.

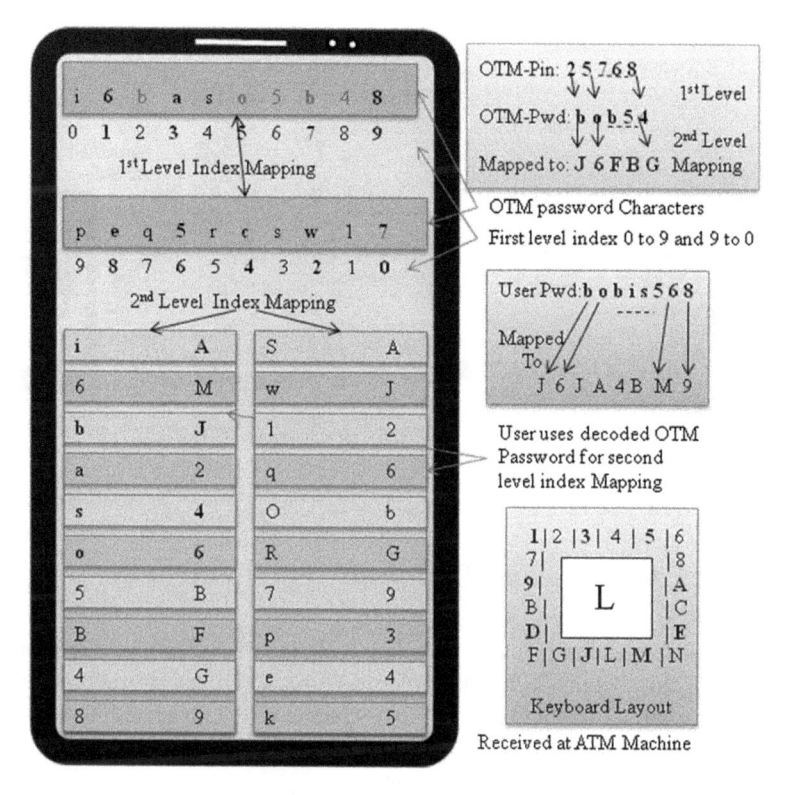

**Fig. 3** Index mapping based authentication

**Fig. 4** Special patterns in color code

Data start synchronization pattern
Data End synchronization pattern
Data padding bytes

## 7.1 Run Length Encoding Technique

Run length encoding technique is lossless character compressing technique [12]. It is widely used for text file compression. Here, we are applying run length encoding technique for bit-level compression using color codes.

e.g.1: Char. aa encoded as 2a    e.g.2: Char. bbb encoded as 3b
e.g.3: Bits 0000 encoded as 40   e.g.4: Bits 11111 encoded as 51

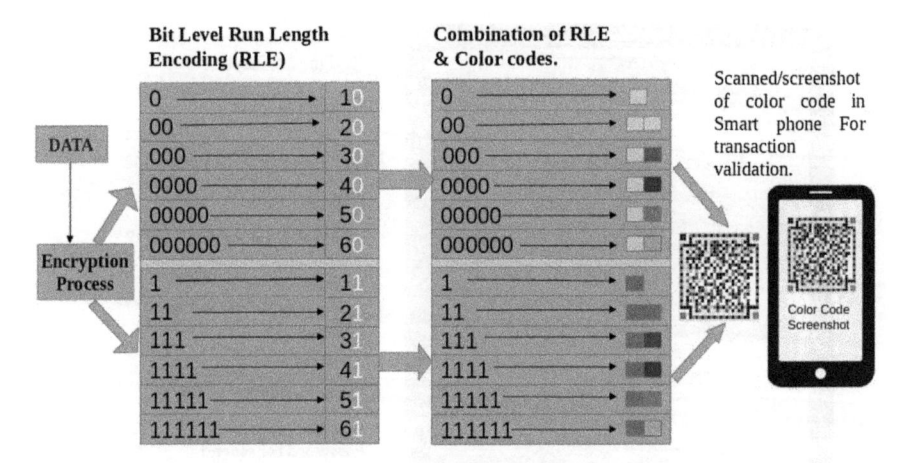

**Fig. 5** Bit-level run length encoding and color code encoding

## 7.2 Color Code Encoding Technique

We have proposed new color code encoding technique by combining run length encoding and colors. In this technique, colors are used to represent group of bits as one color character as shown in Fig. 5. Color characters are generated by applying run length encoding and color codes on binary data. The following pattern (Fig. 4) represents start, end, and padding of the data.

## 8 Mitigation of Attacks

OTM protocol and flexible authentication resist to severe threads like key logger, Trojan, and session hijacking at the time of online banking. OTM protocol ensures that authenticated user has one-time and partial-with-limit online access of bank account.

## 8.1 Prevention for Key Loggers and Trojan

In the previous section, we proposed flexible authentication technique. It shows that client login password and OTM password (session password) are changed automatically from transaction to transaction using index mapping and printed keyboard layout. Flexible authentication uses blank key layout to click client credentials (e.g., login and OTM password); hence, key logger and Trojan are unable to fetch credential information from key logs and screenshots, respectively. Client machine sends only position of the click and server automatically detects clicked character on the basis of OTM number and respective keyboard layout.

## 8.2 Prevention for Session Hijacking

Client scan-printed OTM-QR code using respective bank app is presented in client mobile and waiting for OTM color code for session verification. Client sends login request using flexible authentication technique. Bank servers verify client login credential and response to client request with encrypted OTM color code and blank key layout. OTM-NO and client Pvt key are embedded in OTM color code.

Client scan and decrypt OTM color code using respective bank app present in client mobile. Bank app verifies that OTM number and Pvt key present in OTM color code and OTM-QR code are similar or not. If OTM-NO and Pvt key are same, then session will be continued else session is terminated. Hence, OTM protocol successfully verifies the session at the initial stage and avoids anything wrong.

## 9 Conclusion

Secure color code based client authentication and OTM protocol are mitigating severe threads like key logger, Trojan, and phishing technique at the time of online banking. OTM protocol ensures partial (with Limit) and one-time access of bank account for online banking. OTM protocol guarantees that client authentication credentials are changed automatically from transaction to transaction. Hence, OTM protocol provides secure shield for online banking.

## References

1. Hayashi, E., Dhamija, R., Christin, N., Perrigo, A.: Use your illusion: secure authentication usable anywhere. In: Proceedings of ACM SOUPS (2008)
2. Divya, R., Muthukumarasamy, S.: An impervious QR-based visual authentication protocols to prevent black-bag cryptanalysis. In IEEE Sponsored 9th International Conference on Intelligent Systems and Control (ISCO) (2015)
3. Gao, H., Guo, X., Chen, X., Wang, L., Liu, X.: Yagp: yet another graphical password strategy. In: Proceedings of ACM ACSAC, pp. 121–129 (2008)
4. Goldwasser, S., Micali, S., Rivest, R.L.: A digital signature scheme secure against adaptive chosen-message attacks. SIAM J (1988)
5. Katz, J., Lindell, Y.: Introduction to Modern Cryptography. CRC Press (2008)
6. Bureau of Justice Statistics. Identity Theft Supplement (ITS) to the National Crime Victimization Survey
7. Konoth, R.K., van der Veen, V., Bos, H.: How anywhere computing just killed your phone-based two-factor authentication. In: Financial Crypto (FC) in Bandroid (2016)

8. White, S.N.: Secure mobile-based financial transactions, Feb 2013, US Patent 8,374,916
9. Maggi, F., Volpatto, A., Gasparini, S., Boracchi, G., Zanero, S.: Don't touch a word! a practical input eavesdropping attack against mobile touchscreen devices. Politecnico di Milano, Tech. Rep. TR-2010-59 (2010)
10. M. Labs.: Android Malware spreads through QR code. Kaspersky Secure List Blog (2011)
11. Hsu, J.: How google glass can improve atm banking security. Online at google-glass-can-improve-atm-banking-security, Mar 2014, IEEE Spectrum
12. No Inventor.: Data compression using run length encoding and statistical encoding. US patent US4626829 A publication date DEC 1986

# Comparative Analysis of Methods for Monitoring Activities of Daily Living for the Elderly People

D. R. Kalbande, Anushka Kanawade and Smruti Varvadekar

**Abstract** With the advancement in the field of medicine and technology, the age of living has increased, thus also increasing the population of the elderly people. Currently, there are various healthcare centres for elderly people but this scenario can be improved by developing a system that will monitor their activities remotely, helping them to perform the activities independently. It will also raise an alarm when an anomalous behaviour is detected and will send a report to their relatives. Several methodologies have been proposed and implemented for carrying out the remote monitoring of the ADLS with every method having its own advantages and disadvantages. This review paper focuses on the critical and comparative analysis of various methods employed and also proposes an effective approach for monitoring the ADLS of the elderly people with reference to the methods studied.

**Keywords** ADL · Kinect sensor · SVM classifier · Monitoring

## 1 Introduction

As per the 2011 consensus, the Indian population comprises 103.9 millions of elderly population which forms a significant 8.6% of the total population of 1210.9 millions of people residing in India [1]. The traditional joint family trend has been replaced by the nuclear family and thus there is no one to take care of the elderly people. Due to the developments in technology, it has become possible to

---

D. R. Kalbande (✉)
Sardar Patel Institute of Technology, University of Mumbai, Mumbai, India
e-mail: drkalbande@spit.ac.in

A. Kanawade · S. Varvadekar
Computer Engineering, Sardar Patel Institute of Technology, Mumbai, India
e-mail: anushkakanawade@gmail.com

S. Varvadekar
e-mail: smruts1596@gmail.com

provide effective solutions for monitoring the daily activities of the elderly people when alone at home.

Availability of the various sensors allows capturing everyday data of the person monitored. Also, these systems can be trained to identify the daily activities of the elderly as normal or abnormal (in case of a fall). Hence, several systems are developed which keep track of the activities of the elderly people based on different features selected and send a report or a notification in case of an abnormal activity. This paper presents a comprehensive survey of the various methods used for monitoring the activities of daily living of the elderly people and proposes a novel method based on the methods analysed.

## 2 Review of Various ADL Monitoring Techniques

Activities of daily living (ADL) are of six basic types which include eating, bathing, dressing, toileting, walking and continence [2]. For elderly people, monitoring these ADL is highly important. The following sensors are used in some of the methods described in Fig. 1.

Following is the review of various methodologies employed for monitoring activities of daily living of the elderly people.

### 2.1 Wearable Textronic System for Protecting Elderly People [3]

The textronic system developed in 2016 monitors specific human physiological parameters such as pulse, frequency of breathing, underclothing temperature, and position inside and outside the house. Table 1 categorizes software components used for measuring the parameters.

A special textile clothing interface along with the software for data acquisition was designed. The analysis on the data was used to generate alarm signals which

Fig. 1 Sensors

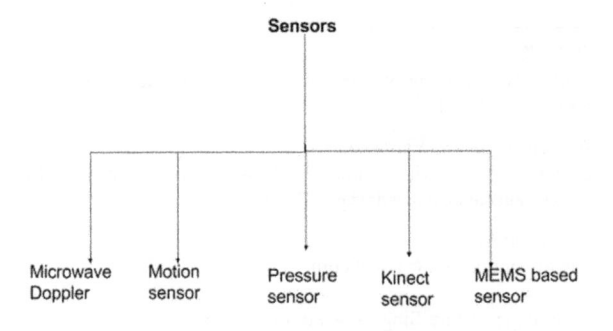

**Table 1** Software components for measuring parameters

| S. No. | Software components | Parameter measured |
|---|---|---|
| 1 | Pulse measurement panel | Pulse—dry textile electrodes are placed on textronic t-shirt |
| 2 | Breath and temperature measurement panel | Frequency of breathing—electrically conductive microscopic elastic fabric |
|   |   | Underclothing temperature |
| 3 | Indoor positioning system | Inside house positioning—RFID technology |
| 4 | Global positioning system | Outside house positioning—GPRS and GSM |

served as a tool for monitoring the health status of the elderly people. Portability is one of the significant features of this system. Thus, this system provides with the combination of sensors which can be inserted in the clothing and a software to keep track of all the parameters monitored to determine the status of the person being observed.

## 2.2 Unsupervised Monitoring of ADLs of an Elderly Person Using Kinect Sensors and a Power Metre [4]

The research provides a framework for monitoring of the physical and the instrumental ADLs of the elderly people using unsupervised fuzzy logic with the help of Kinect sensors and a power metre. This approach uses a 3D skeletal positions obtained from the Kinect depth sensors and the power usage of the electrical appliances from the power metre as parameters. The data obtained from the Kinect sensor is used to identify a posture of the person. To identify the activities involving the use of electrical appliances, the data from the Kinect sensors as well as power metre is required.

The sensors are used to detect all range of actions including very small and complex actions. Data obtained from Kinect sensor is used to extract the depth map attributes which are labelled to different epochs according to both time and location, and then rules modelling frequent behaviour patterns for each epoch are generated using a fuzzy association rule mining algorithm. Unusual behaviours are detected in subsequent data by looking for patterns which differed from the learned normal behaviour.

The major drawback of this system is that it can be used to monitor only a single subject at a time and also it does not adapt to any changes made in the room layout. Though Kinect sensors are cheap, power metres are expensive. Various machine learning techniques were applied on the datasets obtained to analyse the ADLS where k-means gave an average accuracy of 89.4%, fuzzy $C$-means gave an average accuracy of 74%, mean shift gave 98% accuracy, GMM gave 80% accuracy and DBSCAN gave an accuracy of 66.7%.

## 2.3 Social Data Shoes for Gait Monitoring of Elderly People in Smart Home [5]

A data shoe system is developed for gait monitoring. Consequently, the summary of the gait data obtained is showcased on social media websites like Facebook for the relatives. Zigbee technology is used to transfer wireless data from the sensor suite to receiver which is USB connected to the computer at cheaper price.

The sensor suite installed on the insole of the shoe consists of five force sensitive resistors (FSRs) in order to observe the gait behaviours. The system was focused on six common postures patterns which included walking, sitting, standing, heel walking, front walking and standby (no shoes) were targeted to be classified. Features like amplitude and frequency of all the five force sensitive resistors are used to classify normal working or abnormal walking.

Principal component analysis (PCA) pattern recognition of the experimental data obtained has proved that the system can classify normal and abnormal walking patterns. Different activities are classified based on the difference in pressure among the sensors. The system was tested by taking the walking data of 25 people aged over 50 and the system was able to classify the people into 6 groups as mentioned. Also, overall this is not very costly as even the Zigbee technology requires less power and is less expensive.

## 2.4 A Development of System that Monitors ADL by Microwave Doppler Sensor Classification by the Presence/Absence System Incorporating Respirational Signals [6]

The proposed system makes use of respirational signal as a classification factor to detect the presence of person who is monitored, which is then classified using support vector machine algorithm.

The proposed system consists of the following three steps: (a) Demodulation of microwave doppler sensor, (b) Calculation of feature vector and (c) Calculation of the presence or absence of person being monitored. The output of the sensor consists of two components: (a) quadrature signal and (b) inphase signal. These are quantized and demodulated, and further they are calculated in two types of feature: (a) vector time and (b) frequency domain which give six feature vectors: time (3) and frequency (3) which are classified using SVM.

The training dataset included 1134 data points and 126 data points are selected for testing dataset. Further, the system was trained 1000 times to reduce the arbitrariness. On testing 21 subjects in an area of $2 * 2 \text{ m}^2$, the system achieved 99.68% accuracy rate. This system gives great accuracy when deployed in small areas; however, it needs to be checked for the real-time environment. Also, it recognizes object efficiently only when the object being monitored is within conical space. Also, to compare the accuracy of results from SVM, naive Bayes classifier was also used. Both achieved 100% precision with SVM giving 99.66% accuracy and naive Bayes 98.79%.

## 2.5 Daily Living Activity Recognition with Echonet Lite Appliances and Motion Sensors [7]

The proposed system makes use of Echonet Lite ready appliances and motion sensors in place of power metres and indoor positioning system. It also works in the case when there is a change in furniture layout. However, all the appliances are not echonet ready.

The proposed system consists of appliance property information obtained from the echonet, motion data sensor and activity labelling collection system. The system records on/off state of appliances, open/close door for fridge, air dirt level content from ac, information from motion sensors and activity labelling button given to the subject. There is a software which records the timestamp of a specific activity when the subject presses the activity labelling button at the start and end of the activity. Evaluation of system was tested on training data of 13 days and 14th-day data was kept as a test case data. Also, for testing the effect of layout change, for first 9 days, normal data was used, and for day 10–14 data corresponding to layout change was used.

The proposed method achieved an accuracy rate of 68% when performed on classification of nine activities. However, the system is only able to do the classification when only a subject is monitored.

## 2.6 Fall Detection Analysis with Wearable MEMS-Based Sensors [8]

The sensor module comprising of three-axis accelerometer, three-axis gyroscope and three-axis magnetometer is used to detect the fall. The system focuses on detecting true falls from the false (quickly sitting down and to jump) by introducing attitude angles by static postures and dynamic transition test. Three attitude angle pitch, roll and yaw are used for fall detection in this, whereas the accelerometer and gyroscope methods make use of the parameter nom. It also makes a comparison between the approach based on attitude angles and the accelerometer- and gyroscope-based approach.

Sensors are placed either on chest, thigh or waist. Chest is a preferred position since in case of fall due to more centre of gravity, the change in position is appreciable. Three tests—ADL, fall activities and fall-like activities—are carried out by fixing wearable electronic model on chest. Though the attitude angles-based approach is complex, it is able to identify the fall direction and is able to distinguish between the fall-like and the fall activities as compared to the approach using the accelerometer and gyroscope. The Kalman filter algorithm was used for analysing the data obtained from the sensors.

The drawback of the accelerometer- and gyroscope-based approach inability to distinguish between fall and fall-like activities and thus attitude-based approach proves to be better approach than the approach using accelerometer and gyroscope.

## 2.7 Human Activity Recognition Using Body Pose Features and Support Vector Machine [9]

Results are obtained by capturing the videos from Microsoft Kinect depth sensor. The system is tested on 10 individuals for recognizing 13 different activities (drinking, walking, waving, writing, clapping, etc.) in indoor and outdoor environment. Microsoft PrimeSense extracts the skeletal representation of an individual; sequence of tracked human skeleton joints are captured from RGBD images and used as a feature. Joint positions and joint distances are recorded and also the motion and the velocity information frame over specified time is observed. Each video consists of 6–7 sequences of actions per category performed by different people so to as to accommodate the variations in the way people perform an activity in a varied way.

For indoor activities, hands are involved and thus the joint position and joint distance of hand with respect to head and torso is calculated, whereas in outdoor activities select 10 frames over a specified time to get information about the joint rotation and joint angles. Also, the velocity of particular joint that remains static in all the frames observed also called as a reference joint is useful during activities involving motion.

Total 650 videos of 50 activities are training samples. SVM classifiers are used, and the activities observed have many features in common; thus, the classifier creates a hyperplane to distinguish between the activities after mapping the features into the respective feature space. The approach correctly identifies single-person one activity, two-person interaction and single-person performing two activities.

## 3 Comparison of Methods

See Tables 2 and 3.

**Table 2** Comparison of methods monitoring ADLs of elderly

| No | Method used | Advantage | Limitations |
|---|---|---|---|
| 1 | Wearable textronic system | The system is part of garment, monitoring system, is portable and easy to use | Safety issues because of the battery placed within the garment |
| 2 | Unsupervised monitoring using Kinect sensors and a power metre | Algorithms used for classification with mean shift giving 98% accuracy | Not adaptive to layout changes. Only monitors single subject at a time |
| 3 | Social data shoes for gait monitoring | Classifies normal and abnormal walking Calculates the amount of bad gait behaviour | Appropriate infrastructures like Wi-Fi and RFID required |
| 4 | Use of microwave doppler system to classify presence/absence of person | 99.68% accuracy obtained on a training data of 21 subjects | The Doppler sensor can only detect the object within the conic space |
| 5 | Echonet lite appliances and motion sensor based approach | 68% accuracy rate in classifying nine ADL activities | More features need to be added for recognizing ADLs of multiple residents at time. Also, all appliances are not echonet ready |
| 6 | Use of MEMS-based sensors | Attitude angle method identifies between fall and fall-like activity and fall direction | Accelerometer- and gyroscope-based methods cannot distinguish fall direction. Attitude-based approach is complex |
| 7 | SVM and body pose features based approach | Correctly classifies single-person performing one activity as well as two activities, two-person interaction | Accuracy not 100% and large size of data from multiple views required, overall accuracy rate of 89% achieved |

**Table 3** Overview of algorithms

| Method | Machine learning algorithms used |
|---|---|
| Method 2-Kinect sensors and motion sensors | Comparison of accuracies obtained by $k$-means, Fuzzy $C$, GMM, DBSCAN, mean shift |
| Method 3-force sensitive resistors | Principal component analysis |
| Method 4-microwave doppler sensor | SVM and naive Bayes classifier |
| Method 6-sensor module comprising accelerometer, gyroscope | Kalman filter algorithm |
| Method 7-Kinect sensors | One-against-one and one-against-all SVM classifier |

## 4 Proposed Model

The proposed model will overcome the limitations of the methods such as inconsistencies incorporated due to layout change, failure to monitor more than a subject using Kinect sensors and echonet lite appliances that are safe to use. The proposed model will overcome the limitation of not adaptive to layout change using echonet lite appliances and motion sensors (Fig. 2).

Kinect sensors will play a major role in the model as the information obtained from the depth maps can be used to extract the joint coordinates and to determine the joint angle and rotation during the activities involving motion [9], thus it can be used in identifying the physical activities of the elderly. The Kinect sensors use infrared camera and should be placed at specific angular positions where it will not be blocked by the obstacles like wall, thus preventing from capturing information of the subject being monitored. Also, Kinect sensor is a low-cost technology and thus can be incorporated in the model [10, 11].

In order to identify activities involving the use of the appliances, the echonet lite protocol can be used as it keeps the record of the activities by monitoring the various states of the appliance because of the compatibility with the Internet

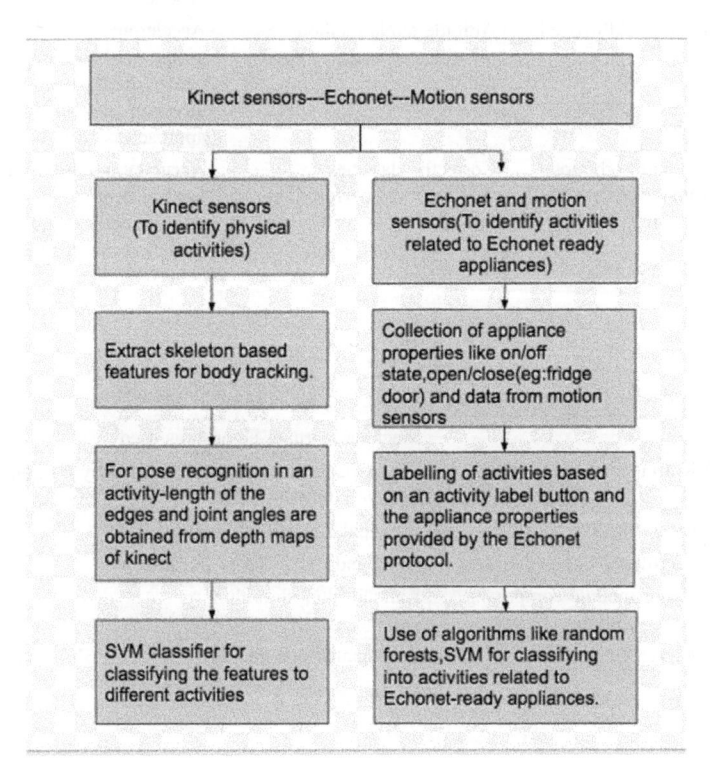

**Fig. 2** Proposed model

protocol. Using the data obtained from Kinect sensor and the echonet appliance protocol, the model can be made efficient by applying support vector machine learning algorithm where the data obtained from these sensors can be classified and mapped according to the feature space. Then, the data can be fed into the system to train it for various training samples. In case of detection of any abnormal behaviour, an alert message can be sent to the relatives of the elderly person being monitored through an application. Similarly, in order to accommodate the changes in the layout, the model can be trained using machine learning for different scenarios with layout modification.

## 5 Conclusion

This paper put forward a comparative analysis of various methods for monitoring the activities of daily living of the elderly people. Thus, this can be used for designing more efficient systems for the monitoring of the ADLs by providing solutions for the limitations identified. The paper proposes an effective approach that takes into account drawbacks of the several systems analysed and feasibility as well as the accuracy rates achieved. The future scope of this will be implementing the proposed model and comparing it with the above methods analysed.

**Acknowledgements** We, co-authors sincerely thank our Research Mentor Prof. Dr. Dhananjay R. Kalbande for providing us an opportunity to work under his guidance. Beside his support in his area of expertise, he took the daunting task of going beyond his area of expertise contributing to social cause, thus making it an enriching and learning experience for all of us.

## References

1. Borah, H., Shukla, P., Jain, K., Kumar, S.P., Prakash, C., Gajrana, K.R.: Elderly in India—Profiles and Programmes 2016. Central Statistics Office Ministry of Statistics and Programme Implementation Government of India (2016)
2. http://www.investopedia.com/terms/a/adl.asp
3. Frydrysiak, M., Tesiorowski, L.: Wearable textronic system for protecting elderly people. In: IEEE Instrumentation and Measurement Society (2016)
4. Pazhoumanndar, H.: Unsupervised Monitoring of ADL's of Elderly Person Using Kinect Sensors and a Power Meter. Edith Cowan University, School of Science (2016)
5. Nilpanapan, T., Kerdcharoen, T.: Social data shoes for Gait monitoring of elderly people in smart home. In: The 2016 Biomedical Engineering International Conference (2016)
6. Shiba, K., Kaburagi, T., Ozaki, K., Nakamura, T., Kurihara, Y.: A development of system monitors ADL by microwave doppler sensor—classification of presence/absence system incorporating respiration signals. In: SICE Annual Conference, Japan, 20–23 Sept 2016
7. Moriya, K., Nakagawa, E., Fujimoto, M., Suwa, H., Arakawa, Y., Kimura, A., Miki, S., Yasumoto, K.: Daily living activity recognition with ECHONET lite appliances and motion sensors. In: First International Workshop on Mobile and Pervasive Internet of Things (2017)

8. Yuan, X., Yu, S., Dan, Q., Wan, G., Liu, S.: Fall detection analysis with wearable MEMS-based sensors. In: 16th International Conference on Electronic Packaging Technology (2015)
9. Bengalur, M.D.: Human Activity recognition using body pose features and support vector machine. In: 2013 International Conference on Advances in Computing, Communication and Informatics (2013)
10. https://msdn.microsoft.com/en-us/library/hh438998.aspx
11. United Nations, Department of Economic and Social Affairs, Population Division. *World Population Ageing 2015* (ST/ESA/SER.A/390) (2015)

# Smart Farming System: Crop Yield Prediction Using Regression Techniques

Ayush Shah, Akash Dubey, Vishesh Hemnani, Divye Gala and D. R. Kalbande

**Abstract** Due to ever increasing global population, there is an ever increase in demand for food; hence, new methods need to be devised to increase the crop yield. This paper proposes an intelligent way to predict crop yield and suggest the optimal climatic factors to maximize crop yield. With the advancement in technology, the focus has now shifted to using machines and control systems to automate the processes and optimize productivity. The paper uses multivariate polynomial regression, support vector machine regression and random forest models to predict the crop yield per acre. The proposed method uses yield and weather data collected from United States Department of Agriculture. The various parameters included in the dataset are humidity, yield, temperature and rainfall. This prediction will help the farmers choose the most suitable temperature and moisture content at which the crop yield will be optimal. The paper uses RMSE, MAE, median absolute error, and *R*-square values to compare between multivariate polynomial regression, support vector machine regression and random forest.

**Keywords** Machine learning · Multivariate polynomial regression
Crop yield · Automation · Dataset

---

A. Shah (✉) · A. Dubey · V. Hemnani · D. Gala · D. R. Kalbande
Computer Engineering, University of Mumbai, Mumbai, India
e-mail: 1595ayush@gmail.com

A. Dubey
e-mail: aakashdubey1995@gmail.com

V. Hemnani
e-mail: visheshhemnani@gmail.com

D. Gala
e-mail: divyegala@gmail.com

D. R. Kalbande
e-mail: drkalbande@spit.ac.in

# 1 Introduction

Agriculture is the science and practice of farming, including cultivation of the soil for the growing of crops and the rearing of animals [1]. Although farmers being skilled in cultivation, there lies a huge gap between scientific and technological knowledge available to them in rural areas.

In order to enhance the production, it is necessary to incorporate technology in agriculture. It will not only boost the quantity of production but will also help improve the quality of crops.

A lot of research is being conducted on the effects of temperature, humidity, pH and moisture in the soil on yields of different crops. However, farmers rarely have any knowledge about such type of scientific developments, which hinders the scientific developments in the agricultural sector [1].

Apart from all these problems, there is also a great deal of uncertainty regarding the yield of crops. The paper aims to establish a method which will not only automate the agricultural practices but also help the farmers get a rough estimate of the crop yields. The principal concept used in this research paper is finding relationship between yield (dependent variable) and other independent variables like temperature and rainfall. Once the relationship has been established, it will assist the farmers to get an estimate about the yield for the crop. Also, the prediction results might help the farmers decide which crop is best suited for a particular piece of land and the environmental conditions in that region. Thus, the system will also help quantify the existing yields which can be used for future predictions.

# 2 Literature Survey

The work done in previous published papers to predict the yield of various crops and the selection of the best crop is surveyed in this section.

The comparative study conducted by Shastry et al. shows the implementation of linear regression, fuzzy logic and anfis. The data was collected from the APSim and the input parameters used were biomass, esw, radiation, rain and wheat yield. The accuracy of the models was determined by comparing the RMSE values and the anfis model showed the best accuracy [2].

But as pointed out in Kumar et al., anfis is suitable only for those models that have a smaller number of input parameters. This paper does a comparative study of machine learning algorithms like artificial neural networks, *k*-nearest neighbours, random forests, support vector machines and decision trees. After the analysis of these models, the paper presents its own crop selection algorithm to choose the sequence of crops to be planted over a season [3].

In the paper by Sellam et al., a linear regression model is computed to analyse the relationship between annual rainfall, area under cultivation, food price index and yield. Least-square fit is used as it can fit linear as well as polynomial regression. The paper highlights four steps; computing a linear regression model, computing residual values, computing residual sum of squares to obtain $R^2$ and implementation. The measure of $R^2$ is used to evaluate the accuracy of their model which comes to about 0.7 [4].

A region-specific crop yield analysis is examined in the paper by Ramesh. It employs multiple linear regression and density-based clustering estimate yield as a function of rainfall, area of sowing and fertilizers. A density-based clustering technique for six cluster approximations of sample data is used in the research paper. The results obtained using multiple regression are verified using density-based clustering [5].

The paper prediction of crop yield using big data proposes a solid architecture for managing big data in agriculture. It consists of three primary components. A MapReduce weather data processing component is used to calculate big datasets on a group of computers. The second component involves finding similar years using nearest neighbours and the last component involves building ARMA model based on the nearest year and then obtaining the prediction number. The experimental evaluation of the above-mentioned paper suggests a good performance of the employed nearest neighbour method, thereby suggesting that crop yield relates closely to weather patterns [6].

The paper by Sujatha titled crop yield forecasting using classification techniques employs naive Bayes, J48, random forest, artificial neural network, decision tree and support vector machine. The yield calculation is done for legume and is calculated in metric tones per hectare. Thus, the above paper gives a comprehensive description of the method to select the best crop for farmer, to plant depending on the weather situation and also help determine suitable season to assist in agriculture [7].

The paper by Kaur stresses the importance of adopting machine learning techniques in agriculture. The paper specifies the areas in which machine learning can be used to enhance the yield and automate agriculture process. The areas where machine can be used as mentioned in the paper are crop selection and yield prediction, weather forecasting, smart irrigation system, crop disease detection and determining the minimum support price. The author also suggests the type of algorithms that can be used for each application [8].

In the paper by Stas, Van Orshoven, low-resolution spot vegetation imagery is used to calculate Normalized Difference Vegetation Indices (NDVI); NDVI is used to construct models based on Boosted Regression Trees (BRT) and Support Vector Machines (SVM). SO BRT and SVM are used to select features with high relevance for predicting yield. So the above-mentioned machine learning techniques were applied to the subset of selected features for yield forecasting. Once forecasting was done, RMSE values were calculated for comparing the BRT and SVM methods. Although both machine learning algorithms had low RMSE values, BRT outperformed SVM [9].

In order to obtain the influence of climatic parameters on crop production, the paper by Veenadhari suggests developing a website to depict the influence of various parameters. The study is carried out in some districts of Madhya Pradesh and the selection of districts depends on the area under the particular crop in consideration. For 20 years, the yield of the crop has been tabulated along with rainfall, maximum and minimum temperature, potential evapotranspiration, cloud cover, and wetday frequency. C 4.5 algorithm is used to find out the most influencing climatic parameter on the crop yields of selected crops. The accuracy of prediction obtained after forecasting is 75% in the above-mentioned paper [10].

The paper by Gandhi discusses the experimental results obtained by applying SMO classifier using the WEKA tool on districts of Maharashtra state, India. The parameters considered for study were precipitation, minimum temperature average temperature, maximum temperature and reference crop evapotranspiration, area production and yield for the Kharif season (June–November) for the years 1998–2002. Analysis of results was done using Mean Absolute Error (MAE), Root Mean Squared Error (RMSE), Relative Absolute Error (RAE) and Root Relative Squared Error (RRSE). The results depicted that classifiers such as naive Bayes, BayesNet and multilayer perceptron performed better by achieving the highest accuracy, sensitivity and specificity as compared to SMO which had the lowest accuracy [11].

## 3 Methodology

### 3.1 Datasets

The datasets were obtained from the USDA website for the state of Iowa. The yield and weather datasets contained 423 observations of harvested corn yield and monthly normals of temperature and precipitation data.

### 3.2 Units

The proposed system is supposed to fit a regression model which would help us predict yield of a crop as function of monthly temperature and rainfall. The system will make use of regression because we need to predict yield based on historical data of yield versus environmental parameters. Since the yield is dependent on multiple parameters, it becomes necessary to make use of multivariate polynomial regression. Also, under extreme environmental conditions, the yield will tend to decrease, which rules out linear regression. Hence, multivariate polynomial regression will serve as the best technique for accurately fitting the appropriate model.

Once the model has been defined, the next intuitive step is to check the accuracy of the model, i.e. how well does the model fit the given data. This is basically done by computing residual values. Residuals can be basically defined as leftovers from computed model fit [5]. The residual ($e$) is the difference between the observed value ($y$) of dependent variable and expected value ($\hat{y}$) of dependent variable [5].

$$e = y - \hat{y}$$

Then, the residual sum of squares was calculated. It depicts the error between data and fitted model. Its value lies between 0 and 1 [5]. The $R$-squared value found upon applying this model to our dataset was found to be 0.89.

## 3.3 Support Vector Machine Regression

Multivariate polynomial regression gave an RMSE value of 9.4. In order to improve the accuracy of system, support vector machine regression was used. So, on the given dataset, the support vector regression attempts to fit a hyperplane. The hyperplane is fitted in such a way that model provides the best fit for future prediction. Also, for our system, the vectors defined for hyperplanes are nonlinear.

For fine-tuning of the model, a cost parameter of 4 is used to control the influence of each individual support vector. The gamma ($\gamma$) value is used to determine the variance and influence of the support vectors. The $\gamma$ value is chosen as

$$\gamma = 1/(\text{no of predictors}).$$

The model uses 20 predictors consisting of combinations of monthly rainfall and temperature. Therefore, $\gamma = 0.05$

The value of epsilon defines a margin of tolerance where no penalty is given to errors. The larger the epsilon ($\varepsilon$), the greater the number of errors permitted by the model. Different values of $\varepsilon$ were tested, and it was seen that the model performs well when the value of $\varepsilon$ is 0.05

## 3.4 Random Forest Algorithm

Random forests or random decision forests are an ensemble learning method for classification, regression, and other tasks, that operate by constructing a multitude of decision trees at training time and outputting the class that is the mode of the classes (classification) or mean prediction (regression) of the individual trees. Random decision forests correct for decision trees' habit of overfitting to their training set.

For this research, we have selected the number of decision trees to be 50 and amount of data to be sampled as 70% of the dataset. Bagging meta-algorithm is used so that every tree generated is independent of the previously generated trees. This is achieved by bootstrapping the samples. Once the model has been trained, we use the remaining 30% of the dataset to test our model.

## 4 Regression

After analysis and research of all machine learning algorithms, we have shortlisted three algorithms.

1. Support Vector Machines (SVM)
2. Multivariate Polynomial Regression
3. Random Forest Algorithm

Model files created at the training phase along with the input values are used for testing.

## 5 Result and Analysis

The above-shortlisted algorithms were used for predicting the yield of the crops. The algorithms used are compared on basis of predicted yield, RMSE, MAE, median absolute error, and $R$-squared values (Fig. 1).

This graph depicts the RMSE, MAE, and MdAE (Median Absolute Error) values calculated over a dataset of 426 observations. MAE does not give as much penalty to outliers as RMSE, so it is a better metric if outliers are few. MdAE is also quite robust to outliers and will ignore outliers completely as it chooses only the median as compared to the mean in MAE. A lower value in RMSE, MAE, and MdAE indicates better performance. Here, SVM performs better than the other two models in all three metrics (Fig. 2).

This graph depicts the $R$-squared value of the three models. $R$-squared values are usually between 0 and 1 (they can be negative if the model fits worse than a horizontal line). A value closer to 1 indicates that the regression line is fitting the data very well. SVM performs better than the other models based on the $R$-squared value too.

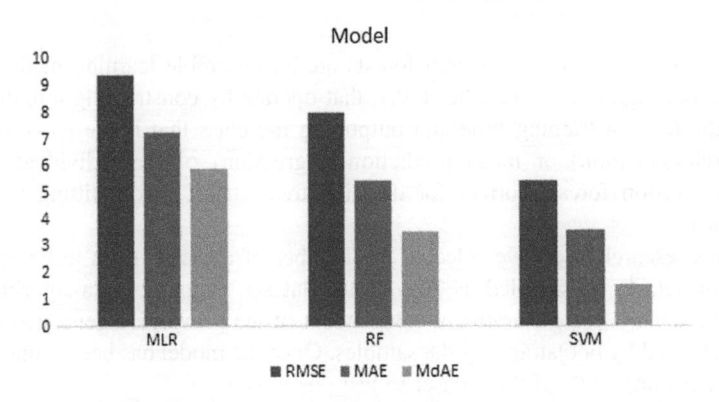

**Fig. 1** RMSE, MAE and MdAE value of three models

R-Square

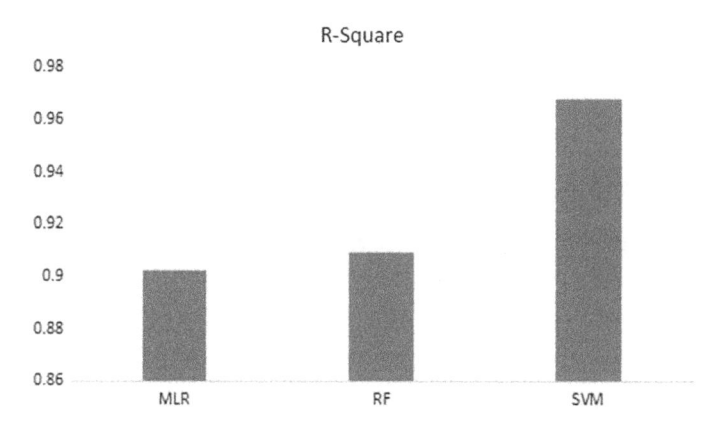

**Fig. 2** $R$-squared value of the three models

## 6 Future Scope

The future scope of this paper involves using hardware interface to gather real-time data from Indian farms, and applying this model to that data. Apart from using temperature and rainfall as our primary predictors, more features will be added to improve the prediction of yield.

## 7 Conclusion

We used three regression-based algorithms to predict the crop yield. These were multivariate polynomial regression, support vector machine regression and random forest. When the dataset was tested, the minimum RMSE, MAE and median absolute error value was found to be obtained by support vector machine regression. The RMSE, MAE and median absolute error values obtained were 5.48, 3.57 and 1.58, respectively, for State crop production. The maximum $R$-squared value was also obtained for support vector machine regression, which was 0.968. Thus, in our approach, we have used the support vector machine regression to obtain the best possible results for predicting the crop yield.

**Acknowledgements** We would like to thank the open-source community on the internet, whose help was instrumental in our learning and execution process.

# References

1. Shashwathi, N., Borkotoky, P., Suhas, K.: Smart farming: a step towards techno-savvy agriculture. Int. J. Comput. Appl. **57**(18) (0975-8887) (2012)
2. Shastry, A., Sanjay, H.A., Hegde, M.: A parameter based ANFIS model for crop yield prediction. In: 2015 IEEE International Advance Computing Conference (IACC), Bangalore (2015)
3. Kumar, R., Singh, M.P., Kumar, P., Singh, J.P.: Crop selection method to maximize crop yield rate using machine learning technique. In: 2015 International Conference on Smart Technologies and Management for Computing, Communication, Controls, Energy and Materials (ICSTM), Chennai (2015)
4. Sellam, V., Poovammal, E.: Prediction of crop yield using regression analysis. Indian J. Sci. Technol. **9**(38) (2016). https://doi.org/10.17485/ijst/2016/v9i38/91714
5. Ramesh, D., Vishnu Vardhan, B.: Analysis of crop yield prediction using data mining techniques. IJRET Int. J. Res. Eng. Technol. (2015)
6. Fan, W., Chong, C., Xiaoling, G., Hua, Y., Juyun, W.: Prediction of crop yield using big data. In: 2015 8th International Symposium on Computational Intelligence and Design (ISCID), Hangzhou, pp. 255–260 (2015)
7. Sujatha, R., Isakki, P.: A study on crop yield forecasting using classification techniques. In: 2016 International Conference on Computing Technologies and Intelligent Data Engineering (ICCTIDE'16), Kovilpatti pp. 1–4 (2016)
8. Kaur, K.: Machine learning: applications in Indian agriculture. Int. J. Adv. Res. Comput. Commun. Eng. (IJARCCE), **5**(4) (2016)
9. Stas, M., Van Orshoven, J., Dong, Q., Heremans, S., Zhang, B.: A comparison of machine learning algorithms for regional wheat yield production using NDVI time series of SPOT-VGT. In: 2016 Fifth International Conference on Agro-Geoinformatics, Tianjin, China (2016)
10. Veenadhari, S., Misra, B., Singh, C.D.: Machine learning approach for forecasting crop yield based on climatic parameters. In: 2014 International Conference on Computer Communication and Informatics, Coimbatore, India (2014)
11. Gandhi, N., Armstrong, L.J., Petkar, O.: Rice crop prediction in India using support vector machines. In: 2016 13th International Joint Conference on Computer Science and Software Engineering (JCSSE), Khon Kaen, Thailand (2016)

# Part II
# Antenna Design

# Elliptical Planar Dipole Antenna with Harmonic Rejection

P. Jishnu, Arnab Pattanayak, Siddhartha P. Duttagupta and Amit A. Deshmukh

**Abstract** This paper focuses on the design of an elliptical planar dipole antenna with excellent harmonic suppression for wireless power transfer applications (WPT). The proposed antenna operates in the GSM 900 frequency bands. The simulation results show 10 dB bandwidth of 133 MHz stretching from 843 to 976 MHz. The fundamental frequency is at 900 MHz with return loss 32 dB. The antenna performance for variation in slots' location and dimension is simulated and studied using CST Microwave Studio.

**Keywords** Elliptical · Harmonic suppression · Return loss · WPT

## 1 Introduction

In the ever growing field of wireless communication, radiation is now ubiquitous, viz., cell phones, Wi-Fi, Bluetooth, etc. These ambient RF energies can be converted into DC power for operating or charging low-power electrical devices. For such wireless power transfer applications, proper design of the antenna is extremely essential for the overall system performance. The presence of harmonic signals due to the higher order resonant frequency of the antenna adversely affects the reception and conversion of RF to DC power. The RF power received at the antenna flows to the rectifier diode, which being a nonlinear device produces DC output current as well as harmonics of fundamental frequency of the received signal. These

---

P. Jishnu (✉) · S. P. Duttagupta
Electrical Engineering Department, IIT Bombay, Mumbai 400076, India
e-mail: jishnuchelsea99@gmail.com

A. Pattanayak (✉)
CRNTS, IIT Bombay, Mumbai 400076, India
e-mail: arnab1134@gmail.com

A. A. Deshmukh
Department of Electronics and Telecommunication Engineering,
SVKM's, D J Sanghvi College of Engineering, Mumbai, India

harmonics leak through the antenna at its higher order resonant modes, contributing to loss of energy, thereby resulting in poor power conversion efficiency [1].

There are numerous methods tried and tested with regards to suppressing these unwanted harmonics. Using low-pass filter stages after the receiving, antenna is the most conventional method [2]. Even though it does remove the harmonics, addition of extra filter components introduces something called insertion loss which further deteriorates the objective.

The more preferred option would be to design an antenna which itself possesses the ability to reject harmonics, thereby eliminating the need for additional components. Various studies have shown to perform the task by modifying the basic metallic structure by cutting out slots [3], using stubs [4], photonic band structure (PBS) [5], defected ground structure (DGS) [6], shorting pins [7], etc. The required complexities vary depending upon the applications.

In this paper, we propose a simple structure by carving out elliptical slots to suppress all the unwanted frequency components. An elliptical planar dipole antenna is designed to operate in GSM 900 uplink (890–915 MHz) and downlink (935–960 MHz) frequency bands. The harmonics at 1.8, 2.7, 3.6, 4.5 GHz, and so on are suppressed using elliptical-shaped slots. The dimension and location of these slots are varied and analyzed for the optimum performance.

## 2 Antenna Geometry

The dipole antenna is designed for the fundamental frequency to be at 900 MHz. Low-cost FR4 lossy material having dielectric constant, $r = 4.3$ with a thickness of $h = 0.16$ mm, is used as the substrate.

### 2.1 Without Slots

Planar elliptical structure is used to construct the dipole antenna as shown in Fig. 1. The dimensions of the antenna are tabulated in Table 1.

We would observe later that the odd harmonics are significant for the basic configuration and therefore, will result in poor conversion efficiency after the rectifier stage in a WPT system.

### 2.2 With Slots

The aim is to maximize the power transfer at the fundamental frequency and suppress every other undesired signals. For this, we cut elliptical slots in the antenna metal surface as shown in Fig. 2. The dimensions of the modified antenna are tabulated in Table 2.

# Elliptical Planar Dipole Antenna with Harmonic Rejection

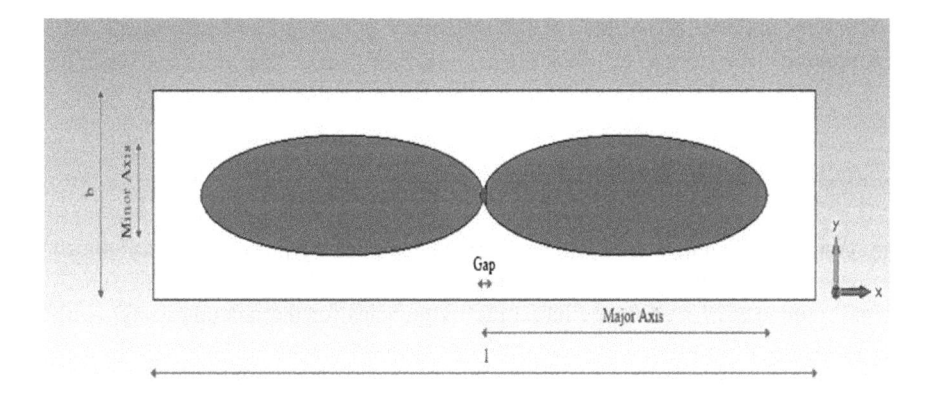

**Fig. 1** Elliptical planar dipole antenna

**Table 1** Antenna design without slots

| Parameter | Values (mm) |
| --- | --- |
| Semi-major axis | 37 |
| Semi-minor axis | 10 |
| Thickness ($t$) | 0.017 |
| Gap ($g$) | 0.5 |
| Substrate dimension ($l, b, h$) | 173, 35, 0.16 |

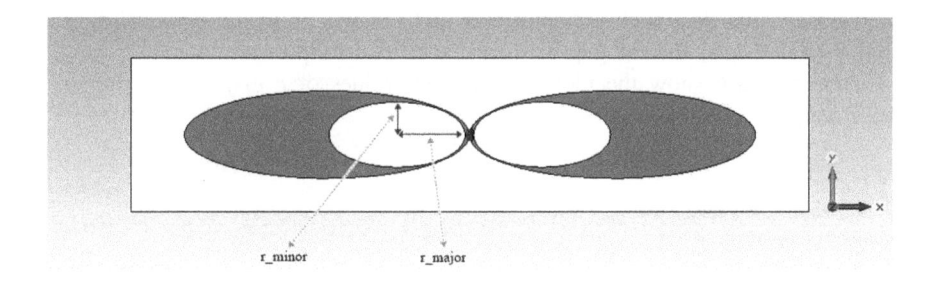

**Fig. 2** Elliptical planar dipole antenna with elliptical slots

**Table 2** Antenna design with slots

| Parameter | Values (mm) |
| --- | --- |
| Semi-major axis | 33.5 |
| Semi-minor axis | 10 |
| Thickness ($t$) | 0.017 |
| Gap ($g$) | 0.5 |
| $r$_major | 16 |
| $r$_minor | 7.5 |
| Substrate dimension ($l, b, h$) | 159, 35, 0.16 |

The semi-major axis of elliptical slot is denoted as r_major and semi-minor axis as r_minor.

## 3 Analyzing the Return Loss

The return loss or reflection coefficient or $S11$ parameter represents the amount of power reflected from the antenna [8]. The comparison of return loss at the fundamental frequency as well as the harmonic frequencies for the antenna design with and without slots is plotted in Figs. 3 and 4, respectively.

We can observe from Fig. 3 that the return loss at the fundamental frequency for elliptical antenna without slots is 18.917 dB, whereas it jumps to 32.543 dB for the design with the slots. Figure 4 shows that the harmonics are suppressed effectively in the modified structure. The even harmonics at 1.8, 3.6, and 5.4 GHz are reduced by roughly 1–2 dB, while the odd harmonics at 2.7, 4.5, and 6.3 GHz which are significant in the original slotless structure undergo much higher suppression as near as 4.5–6 dB.

### 3.1 Parametric Variation and Results

The return loss versus frequency plots is computed for different scenarios by varying the parameters such as major axis and minor axis of the elliptical slots, location of the slots, and the gap between the elliptical dipole sections.

Figures 5 and 6 show the plots for different values of r_major and r_minor, respectively.

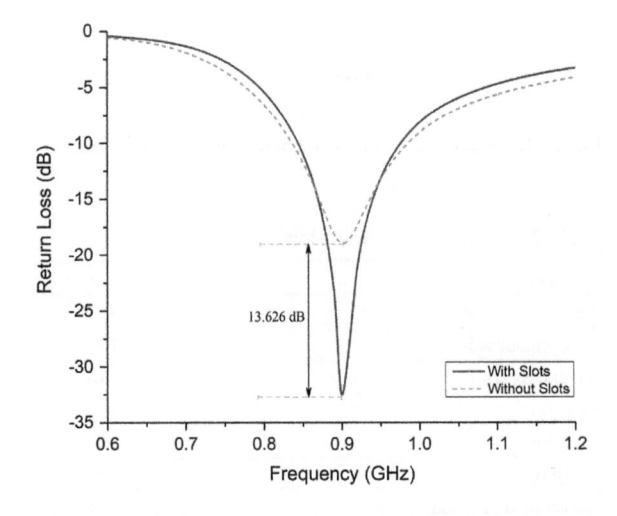

**Fig. 3** Return loss versus frequency comparison of original and modified structures at fundamental frequency

# Elliptical Planar Dipole Antenna with Harmonic Rejection 63

**Fig. 4** Return loss versus frequency comparison of original and modified structures at harmonic frequency

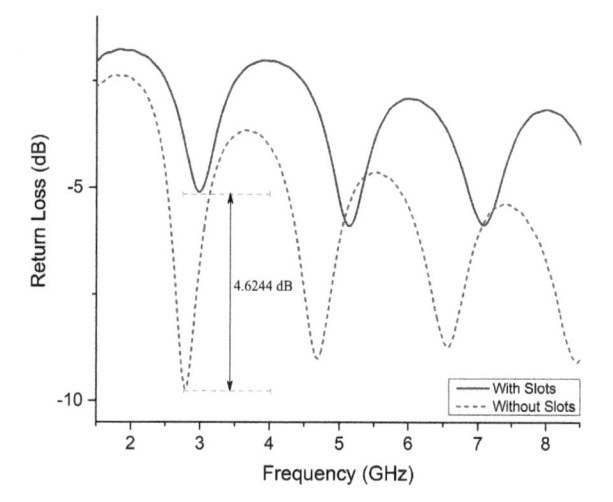

**Fig. 5** Comparison of $S11$ versus frequency for variation in semi-major axis of elliptical slots

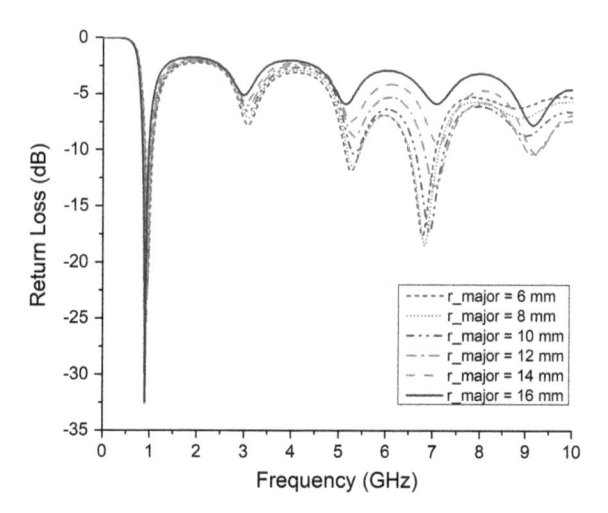

From Fig. 5, we can observe that the optimum system performance is when the semi-major axis of the slot is 16 mm. The suppression level reduces as length of the major axis decreases.

The optimum performance is observed to be at $r\_minor$ = 7:5 mm from Fig. 6. Similar to the earlier case, harmonic suppression deteriorates as length of the minor axis decreases.

Figure 7 shows the plot with respect to variation in the location of center of elliptical slots from the origin. The best-case scenario is when the center is 17 mm from the origin.

**Fig. 6** Comparison of $S11$ versus frequency for variation in semi-minor axis of elliptical slots

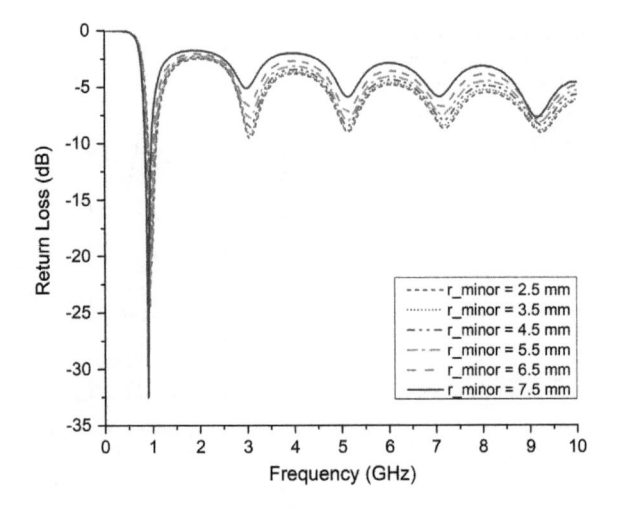

**Fig. 7** Comparison of $S11$ versus frequency for variation in location of elliptical slots

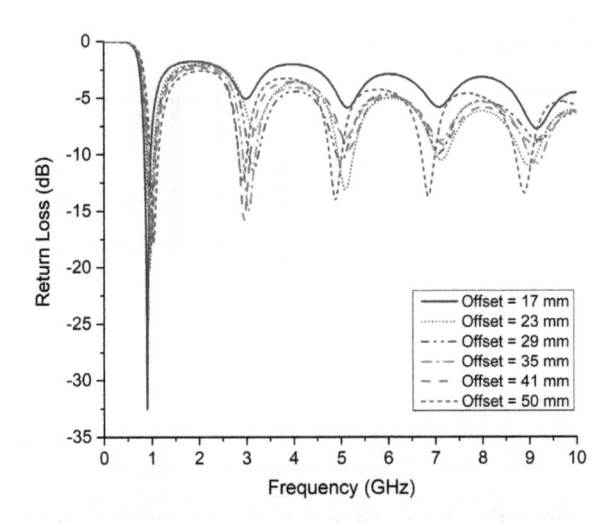

Figure 8 shows the plot for variation in the gap between the two dipole sections. We observe that a gap of 0:5 mm gives the best return loss value at fundamental frequency, but with respect to harmonic suppression, 5 mm and/or 9 mm seems to be the better choice.

# Elliptical Planar Dipole Antenna with Harmonic Rejection

**Fig. 8** Comparison of $S11$ versus frequency for variation in gap between the dipole sections

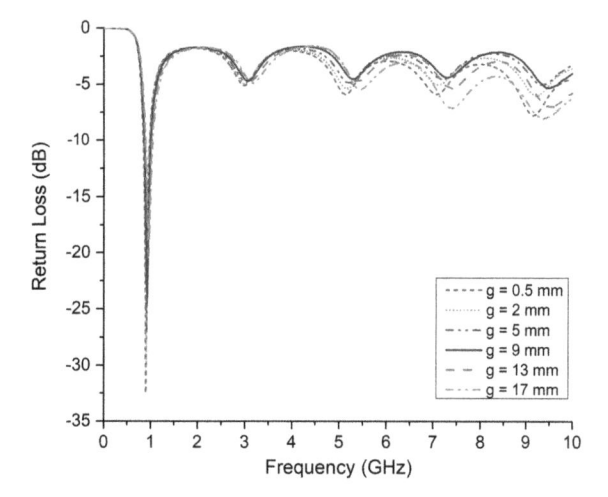

## 4 Radiation Pattern

Figure 9 gives the comparison of the radiation patterns of the two antenna designs at 900 MHz. It can be easily observed that cutting slots do not alter the radiation properties at the fundamental frequency.

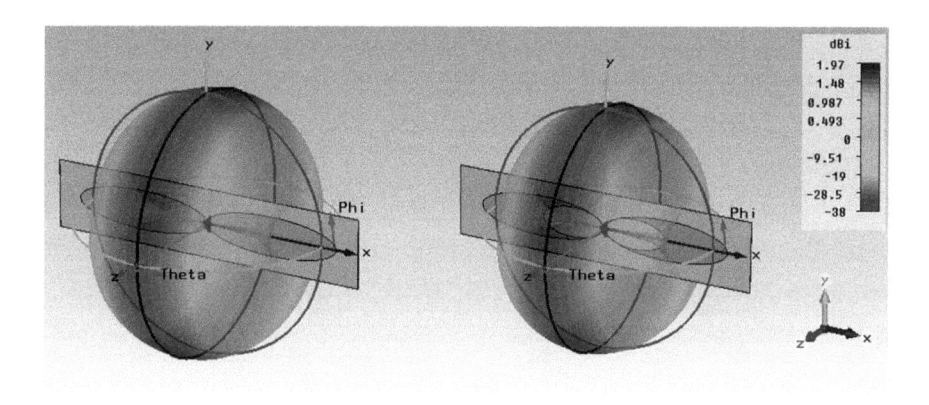

**Fig. 9** Antenna radiation pattern: without slots (left), with slots (right)

## 5 Conclusion

In this paper, we have simulated the design of an elliptical planar dipole antenna with good harmonic rejection capabilities, finding it potentially useful as the RF front-end device for wireless power transfer applications. The proposed antenna is designed for GSM 900 bands. The parametric variation of slot properties and the corresponding effects in the return loss are also observed.

## References

1. Huang, F.-J., et al.: Design of circular polarization antenna with harmonic suppression for rectenna application. IEEE Antennas Wirel. Propag. Lett. **11**, 592–595 (2012)
2. Chang, K.: Microwave Solid-State Circuits and Applications. Wiley (1994)
3. Zainol, N., et al.: A review of antenna designs with harmonic suppression for wireless power transfer. ARPN J. Eng. Appl. Sci. **10**(11) (2015)
4. Tu, W.-H., Chang, K.: Compact second harmonic-suppressed bandstop and bandpass filters using open stubs. IEEE Trans. Microw. Theory Tech. **54**(6), 2497–2502 (2006)
5. Kim, T., Seo, C.: A novel photonic bandgap structure for low-pass filter of wide stopband. IEEE Microw. Guided Wave Lett. **10**(1), 13–15 (2000)
6. Ahn, D., et al.: A design of the low-pass filter using the novel microstrip defected ground structure. IEEE Trans. Microw. Theory Tech. **49**(1), 86–93 (2001)
7. Kwon, S., Yoon, H.K., Yoon, Y.J.: Harmonic tuning antennas using slots and short-pins. In: Antennas and Propagation Society International Symposium, 2001, vol. 1. IEEE (2001)
8. Welcome to Antenna-Theory.com! http://www.antenna-theory.com/

# Resonance Frequency Estimation for Equilateral Triangular Microstrip Antennas Using Artificial Neural Network Model

Amit A. Deshmukh, Megh Shukla, Stuti Patel, Saurabh Labde and A. P. C. Venkata

**Abstract** Neural Network modeling as computational tool has been widely used in many practical problems in the field of communication. This paper proposes the use of neural network modeling to microstrip antenna. The same has been developed for equilateral triangular microstrip antenna which is used on suspended air as well as glass epoxy suspended substrates. Using proposed model, frequencies calculated over wide range and for increasing substrate thickness give closer agreement with the simulated data obtained using IE3D simulations. At some frequencies and substrate thickness, measurements have also been carried out to validate the proposed model. A good agreement between simulated, measured, and calculated results is observed with % error of less than 2%.

**Keywords** Suspended equilateral triangular microstrip antenna
Resonance frequency · Artificial neural network model

## 1 Introduction

Ever since the invention in 1969 by Deschamps, microstrip antenna (MSA) finds many applications in the design of modern day communication systems [1–3]. In MSA, it is always preferred to use regular patch shapes for the radiating element as from mathematical modeling point of view, they are simpler to analyze and to mathematically model. Also, they offer better radiating characteristics as compared with the modified shape MSAs [1–3]. To enhance the gain and bandwidth (BW) in MSA, suspended configurations are used [1–3]. In suspended design, thicker substrates are used which have lower dielectric constants. The radiating patch is either suspended in air above the ground plane or it is fabricated on the microwave substrate which is suspended above ground plane using finite air gap [1–3].

---

A. A. Deshmukh (✉) · M. Shukla · S. Patel · S. Labde · A. P. C. Venkata
Department of Electronics and Telecommunication Engineering,
SVKM's, D J Sanghvi College of Engineering, Mumbai, India
e-mail: amit.deshmukh@djsce.ac.in

© Springer Nature Singapore Pte Ltd. 2018
H. Vasudevan et al. (eds.), *Proceedings of International Conference on Wireless Communication*, Lecture Notes on Data Engineering and Communications Technologies 19, https://doi.org/10.1007/978-981-10-8339-6_8

The fringing field's enhancement happens in suspended thicker substrate which improves the gain. Further, the suspended designs improve upon the reliability of antenna parameters against substrate parameter variations [4]. However from design point of view, direct close form expressions to calculate patch dimension parameters on thicker suspended substrates are not readily available. The artificial neural network (ANN) is an effective computational tool and it finds wide range of applications in wired and wireless communication problems. The ANN model has been reported in analyzing problems related to MSA field. The ANN model for rectangular MSA, equilateral triangular MSA and shorted compact MSAs have been reported which gives closer prediction of patch dimensions for different substrate thicknesses for a wide range of frequencies [5–8]. However, the work reported in [5–8] was only limited to air dielectrics and accuracy of predicting, the patch dimension for given resonance frequency was in the range of 3–5%, with respect to simulated frequency.

In this paper, ANN model for suspended designs of ETMSA is proposed. Two configurations namely, air suspended ETMSA and ETMSA fabricated on glass epoxy substrate ($\varepsilon_r$ = 4.3, $h$ = 0.16 cm, tan $\delta$ = 0.02) which is suspended above ground plane using finite air gap are discussed. To train the ANN model, first in proposed work, ANN network is trained using data sets which contain data like, frequency, substrate thickness in terms of operating wavelength, substrate thickness, simulated patch side length for the given frequency and the half wavelength value at frequency under consideration. The training data sets are taken at every 400 MHz of frequency intervals across 800–6000 MHz frequency spectrum. The network code is prepared using Python programing language. Further using this trained ANN model, side length of ETMSA is predicted for different substrate thickness over 600–6000 MHz of frequency spectrum. The ETMSA when simulated using this length yields nearly the same resonance frequency for ETMSA with % error less than 2% for air as well as suspended dielectric substrate. Thus, proposed ANN model will be helpful to estimate side length of ETMSA on air as well as suspended dielectric configurations that are preferred in broadband high gain variations of ETMSA designs.

## 2 ANN Model for ETMSAs

The suspended variations of ETMSA are shown in Fig. 1a–c. In first variation, ETMSA is suspended above ground plane using air gap of thickness "$h$" cm, whereas in second variation, ETMSA fabricated on glass epoxy substrate which is suspended above ground plane using air gap "ha". In each of the two variations, first training data sets to train ANN model were developed. The ANN model for same is shown in Fig. 1d. ANN architecture is proposed in "Multilayer Perceptron". This is a feedforward network consisting of input layer, hidden layers, and output layers. Further back propagation algorithm is employed in learning algorithm which is a variation of the steepest gradient method that calculates the weights across network. The back propagation algorithm is a supervised learning algorithm which reduces error by changing the weights along direction of steepest gradient. The weights closest to the output layer are altered followed by the weights closer to the input layer. The proposed neural network model is implemented to estimate the patch side length "$S$" for the air suspended and dielectric fabricated air suspended equilateral triangular patch side length. Predicted patch side length includes the effect of fringing fields which are present around patch edges. Inputs to the ANN network are, resonant frequency "$f_r$", Dielectric constant "$\varepsilon_r$", half-wavelength value at given frequency ($\lambda/2$), and the substrate height ($h$). The network has two hidden layers followed by a one neuron output layer. The output of the network is the predicted side length. To evaluate the optimum number of hidden layers and the hidden neurons in each layer, the network was simulated for various configurations. Mean square errors for each configuration were calculated at different epoch values. A comparison of these errors is shown in Fig 1f. It is observed that the error for a single layer hidden neuron is greater than the error generated by a network consisting of two hidden layers. It is also observed that the performance of the network converges to a finite nonzero value as the number of epoch increases. Inferring from the plot, the appropriate configuration is a network consisting of 8, 4 neurons in each hidden layer. Other configurations such as 12, 6 were rejected as they increased network complexity while yielding no appreciable performance improvement.

While training the ANN model, at different frequencies across 600–6000 MHz frequency band, frequency using IE3D simulations is calculated for total substrate thickness increasing from 0.04 to $0.1\lambda_0$. Each of those readings is used to feed to ANN model using which the network coefficients are calculated. For suspended dielectric substrate, the effective dielectric constant for every height is calculated by using effective capacitance formulae [1]. Using the network, side length is predicted at different frequencies across 600–6000 MHz frequency range. These predicted frequencies are different than those used in training the ANN model. Using predicted side length, ETMSA was simulated for different substrate thickness and results for simulated data against predicted one are shown in Tables 1, 2, 3, 4, 5, 6, 7, 8, 9, 10, 11 and 12. Here, Tables 1–6 contain results for ETMSA on-air dielectric whereas Tables 7–12 contains results for ETMSA on suspended glass epoxy substrate.

For varying substrate thickness and different frequencies, prediction of patch side length with error less than 2% is obtained. Although tables are provided at specific frequencies but across entire range (i.e., till 6000 MHz), closer approximation for frequency is obtained. As the close formed expression to estimate side length of ETMSA on varying substrate thickness and frequencies is not available, proposed ANN model here will be helpful to estimate the same. The knowledge of side length is essential to design broadband variations of ETMSAs for higher frequencies and thicker substrates.

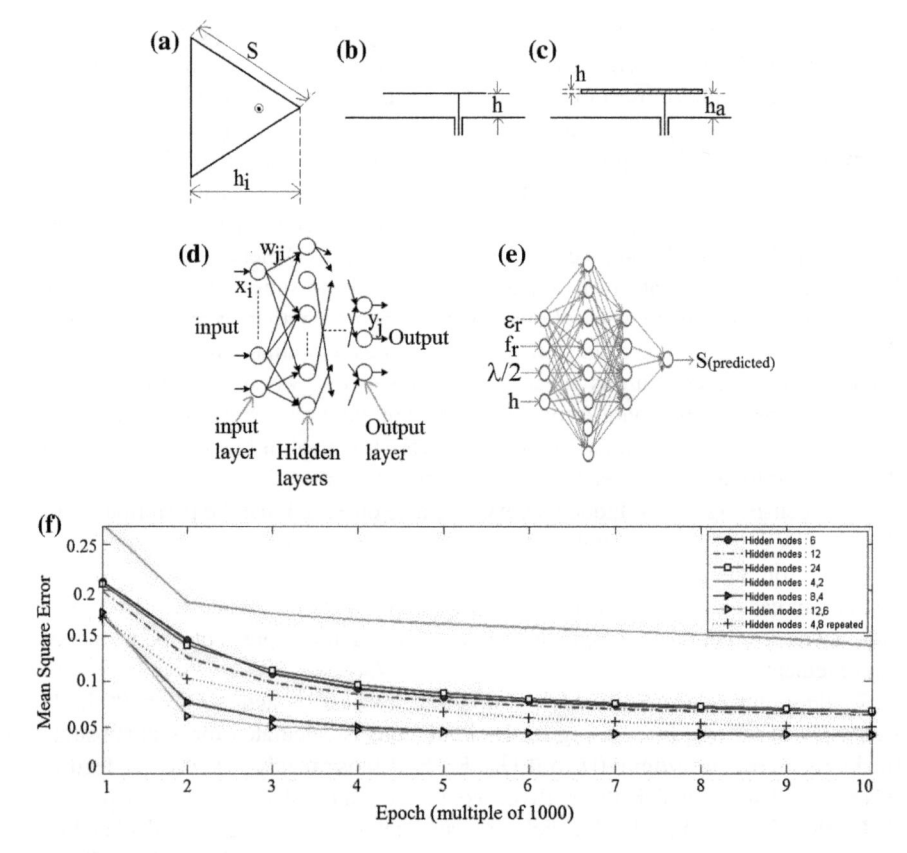

**Fig. 1** **a–c** Coaxially fed suspended ETMSAs, its **d**, **e** ANN model and its **f** variations in mean square error against hidden nodes

**Table 1** Frequency and % error plots obtained using ANN model at 800 MHz

| S. No. | $h$ (mm) | $h/\lambda_0$ | $f_{ie3d}$ (MHz) | $f_{ANN}$ (MHz) | $S$ (predicted) (mm) | % Error |
|---|---|---|---|---|---|---|
| 1 | 15.00 | 0.04 | 798.4 | 800 | 208 | 0.2 |
| 2 | 18.75 | 0.05 | 795.2 | 800 | 203 | 0.6 |
| 3 | 22.50 | 0.06 | 795.2 | 800 | 197 | 0.6 |
| 4 | 26.25 | 0.07 | 793.6 | 800 | 192 | 0.8 |
| 5 | 30.00 | 0.08 | 795.2 | 800 | 186 | 0.6 |
| 6 | 33.75 | 0.09 | 800 | 800 | 180 | 0 |
| 7 | 37.50 | 0.1 | 803.2 | 800 | 174 | 0.4 |

**Table 2** Frequency and % error plots obtained using ANN model at 1200 MHz

| S. No. | $h$ (mm) | $h/\lambda_0$ | $f_{ie3d}$ (MHz) | $f_{ANN}$ (MHz) | $S$ (predicted) (mm) | % Error |
|---|---|---|---|---|---|---|
| 1 | 10.0 | 0.04 | 1202.4 | 1200 | 139 | 0.2 |
| 2 | 12.5 | 0.05 | 1193.6 | 1200 | 134 | 0.53333 |
| 3 | 15.0 | 0.06 | 1193.6 | 1200 | 130 | 0.53333 |
| 4 | 17.5 | 0.07 | 1196 | 1200 | 126 | 0.33333 |
| 5 | 20.0 | 0.08 | 1200 | 1200 | 122 | 0 |
| 6 | 22.5 | 0.09 | 1200.8 | 1200 | 118 | 0.06667 |
| 7 | 25.0 | 0.1 | 1208 | 1200 | 113 | 0.66667 |

**Table 3** Frequency and % error plots obtained using ANN model at 2000 MHz

| S. No. | $h$ (mm) | $h/\lambda_0$ | $f_{ie3d}$ (MHz) | $f_{ANN}$ (MHz) | $S$ (predicted) (mm) | % Error |
|---|---|---|---|---|---|---|
| 1 | 6.00 | 0.04 | 2020.8 | 2000 | 81 | 1.04 |
| 2 | 7.5 | 0.05 | 2000 | 2000 | 79 | 0 |
| 3 | 9.00 | 0.06 | 2000 | 2000 | 76 | 0 |
| 4 | 10.50 | 0.07 | 2003.2 | 2000 | 74 | 0.10667 |
| 5 | 12.00 | 0.08 | 2014.4 | 2000 | 71 | 0.72 |
| 6 | 13.50 | 0.09 | 2000 | 2000 | 68 | 0 |
| 7 | 15.00 | 0.1 | 2004.8 | 2000 | 66 | 0.24 |

**Table 4** Frequency and % error plots obtained using ANN model at 2600 MHz

| S. No. | $h$ (mm) | $h/\lambda_0$ | $f_{ie3d}$ (MHz) | $f_{ANN}$ (MHz) | $S$ (predicted) (mm) | % Error |
|---|---|---|---|---|---|---|
| 1 | 4.61 | 0.04 | 2647.2 | 2600 | 62 | 1.8153 |
| 2 | 5.76 | 0.05 | 2612.8 | 2600 | 60 | 0.4923 |
| 3 | 6.92 | 0.06 | 2600.8 | 2600 | 58 | 0.0307 |
| 4 | 8.07 | 0.07 | 2667.2 | 2600 | 56 | 2.5846 |
| 5 | 9.23 | 0.08 | 2600 | 2600 | 54 | 0 |
| 6 | 10.38 | 0.09 | 2600 | 2600 | 52 | 0 |
| 7 | 12.50 | 0.1 | 2562.4 | 2600 | 49 | 1.4461 |

**Table 5** Frequency and % error plots obtained using ANN model at 3200 MHz

| S. No. | $h$ (mm) | $h/\lambda_0$ | $f_{ie3d}$ (MHz) | $f_{ANN}$ (MHz) | $S$ (predicted) (mm) | % Error |
|---|---|---|---|---|---|---|
| 1 | 3.75 | 0.04 | 3280 | 3200 | 50 | 2.5 |
| 2 | 4.68 | 0.05 | 3222.4 | 3200 | 48 | 0.7 |
| 3 | 5.62 | 0.06 | 3200 | 3200 | 47 | 0 |
| 4 | 6.56 | 0.07 | 3184 | 3200 | 45 | 0.5 |
| 5 | 7.50 | 0.08 | 3174.4 | 3200 | 44 | 0.8 |
| 6 | 8.43 | 0.09 | 3160 | 3200 | 42 | 1.25 |
| 7 | 9.37 | 0.1 | 3112 | 3200 | 41 | 2.75 |

**Table 6** Frequency and % error plots obtained using ANN model at 3800 MHz

| S. No. | $h$ (mm) | $h/\lambda_0$ | $f_{ie3d}$ (MHz) | $f_{ANN}$ (MHz) | $S$ (predicted) (mm) | % Error |
|---|---|---|---|---|---|---|
| 1 | 3.16 | 0.04 | 3928 | 3800 | 41 | 3.3684 |
| 2 | 3.94 | 0.05 | 3838.4 | 3800 | 40 | 1.0105 |
| 3 | 4.73 | 0.06 | 3804.8 | 3800 | 39 | 0.1263 |
| 4 | 5.83 | 0.07 | 3760 | 3800 | 37 | 1.0526 |
| 5 | 6.31 | 0.08 | 3742.4 | 3800 | 36 | 1.5158 |
| 6 | 7.10 | 0.09 | 3715.2 | 3800 | 35 | 2.2316 |
| 7 | 7.89 | 0.1 | 3644.8 | 3800 | 34 | 4.0842 |

**Table 7** Frequency and % error plots obtained using ANN model at 1400 MHz

| S. No. | $h + ha$ (mm) | $h/\lambda_0$ | $f_{ie3d}$ (MHz) | $f_{ANN}$ (MHz) | $\varepsilon_{re}$ | $S$ (mm) predicted | % Error |
|---|---|---|---|---|---|---|---|
| 1 | 8.57 | 0.04 | 1400 | 1400 | 1.167242 | 102 | 0 |
| 2 | 10.71 | 0.05 | 1401.6 | 1400 | 1.129497 | 100 | 0.1142 |
| 3 | 12.85 | 0.06 | 1400 | 1400 | 1.105653 | 97 | 0 |
| 4 | 15.00 | 0.07 | 1400 | 1400 | 1.08916 | 95 | 0 |
| 5 | 17.14 | 0.08 | 1400 | 1400 | 1.077168 | 92 | 0 |
| 6 | 19.28 | 0.09 | 1402.4 | 1400 | 1.0680202 | 89 | 0.1714 |
| 7 | 21.43 | 0.1 | 1404 | 1400 | 1.060781 | 87 | 0.2857 |

**Table 8** Frequency and % error plots obtained using ANN model at 1800 MHz

| S. No. | $h + ha$ (mm) | $h/\lambda_0$ | $f_{ie3d}$ (MHz) | $f_{ANN}$ (MHz) | $\varepsilon_{re}$ | $S$ (mm) predicted | % Error |
|---|---|---|---|---|---|---|---|
| 1 | 6.66 | 0.04 | 1802.4 | 1800 | 1.22605 | 77 | 0.13333 |
| 2 | 8.33 | 0.05 | 1802.4 | 1800 | 1.172894 | 75 | 0.13333 |
| 3 | 10.00 | 0.06 | 1802.4 | 1800 | 1.139979 | 73 | 0.13333 |
| 4 | 11.66 | 0.07 | 1800.8 | 1800 | 1.117705 | 71 | 0.04444 |
| 5 | 13.33 | 0.08 | 1800 | 1800 | 1.1014624 | 69 | 0 |
| 6 | 15.00 | 0.09 | 1799.2 | 1800 | 1.089159 | 67 | 0.44444 |
| 7 | 18.75 | 0.1 | 1795.2 | 1800 | 1.070078 | 63 | 0.26666 |

**Table 9** Frequency and % error plots obtained using ANN model at 2600 MHz

| S. No. | $h + ha$ (mm) | $h/\lambda_0$ | $f_{ie3d}$ (MHz) | $f_{ANN}$ (MHz) | $\varepsilon_{re}$ | $S$ (mm) predicted | % Error |
|---|---|---|---|---|---|---|---|
| 1 | 4.61 | 0.04 | 2604.8 | 2600 | 1.363061 | 50 | 0.18461 |
| 2 | 5.76 | 0.05 | 2615.2 | 2600 | 1.270936 | 48 | 0.58461 |
| 3 | 6.92 | 0.06 | 2613.6 | 2600 | 1.215721 | 47 | 0.52307 |
| 4 | 8.07 | 0.07 | 2602.4 | 2600 | 1.179464 | 46 | 0.09230 |
| 5 | 9.23 | 0.08 | 2594.4 | 2600 | 1.153448 | 45 | 0.21538 |
| 6 | 10.38 | 0.09 | 2581.6 | 2600 | 1.134167 | 44 | 0.70769 |
| 7 | 12.50 | 0.1 | 2555.2 | 2600 | 1.108933 | 42 | 1.72307 |

**Table 10** Frequency and % error plots obtained using ANN model at 3200 MHz

| S. No. | h + ha (mm) | h/λ₀ | f_ie3d (MHz) | f_ANN (MHz) | $\varepsilon_{re}$ | S (mm) predicted | % Error |
|---|---|---|---|---|---|---|---|
| 1 | 3.75 | 0.04 | 3192 | 3200 | 1.48686 | 39 | 0.25 |
| 2 | 4.68 | 0.05 | 3220 | 3200 | 1.3557 | 38 | 0.625 |
| 3 | 5.62 | 0.06 | 3218.4 | 3200 | 1.27957 | 37 | 0.575 |
| 4 | 6.56 | 0.07 | 3200 | 3200 | 1.230286 | 36 | 0 |
| 5 | 7.50 | 0.08 | 3184 | 3200 | 1.19577 | 35.5 | 0.5 |
| 6 | 8.43 | 0.09 | 3160 | 3200 | 1.1704931 | 34.5 | 1.25 |
| 7 | 9.37 | 0.1 | 3134.4 | 3200 | 1.15081 | 33.5 | 2.05 |

**Table 11** Frequency and % error plots obtained using ANN model at 3800 MHz

| S. No. | h + ha (mm) | h/λ₀ | f_ie3d (MHz) | f_ANN (MHz) | $\varepsilon_{re}$ | S (mm) predicted | % Error |
|---|---|---|---|---|---|---|---|
| 1 | 3.16 | 0.04 | 3761.6 | 3800 | 1.63553 | 32 | 1.0105 |
| 2 | 3.94 | 0.05 | 3822.4 | 3800 | 1.45275 | 31 | 0.5895 |
| 3 | 4.73 | 0.06 | 3820.8 | 3800 | 1.35062 | 30 | 0.5474 |
| 4 | 5.83 | 0.07 | 3785.6 | 3800 | 1.26681 | 29 | 0.3789 |
| 5 | 6.31 | 0.08 | 3768 | 3800 | 1.24161 | 28.5 | 0.8421 |
| 6 | 7.10 | 0.09 | 3729.6 | 3800 | 1.209109 | 28 | 1.8526 |
| 7 | 7.89 | 0.1 | 3691.2 | 3800 | 1.184312 | 28 | 2.86316 |

**Table 12** Frequency and % error plots obtained using ANN model at 4800 MHz

| S. No. | h + ha (mm) | h/λ₀ | f_ie3d (MHz) | f_ANN (MHz) | E (ref) | S (mm) predicted | % Error |
|---|---|---|---|---|---|---|---|
| 1 | 2.50 | 0.04 | 4662.4 | 4800 | 1.96527 | 24 | 2.86667 |
| 2 | 3.125 | 0.05 | 4806.4 | 4800 | 1.64726 | 23 | 0.1333 |
| 3 | 3.75 | 0.06 | 4830.4 | 4800 | 1.48686 | 22.5 | 0.6333 |
| 4 | 4.37 | 0.07 | 4800 | 4800 | 1.39079 | 22 | 0 |
| 5 | 5.00 | 0.08 | 4747.2 | 4800 | 1.32552 | 21.5 | 1.1 |
| 6 | 5.62 | 0.09 | 4688 | 4800 | 1.27957 | 21 | 2.3333 |
| 7 | 6.25 | 0.1 | 4627.2 | 4800 | 1.2445 | 21 | 3.6 |

## 3 Conclusions

Antennas for wider BW and gain are realized by using suspended variations in which thicker substrates are used. To evaluate patch parameters for such suspended variations as direct formulations are not available, here in proposed work, ANN model for suspended variations is proposed. The ANN models for MSAs, namely for air suspended and suspended microwave substrate with finite air gap above ground plane, are selected for neural network modeling. Using back propagation algorithm and parametric study for hidden nodes by using python programing language, algorithm was developed. The side length predicted by using proposed model yields simulated frequencies that closely match the desired predicted frequency with error less than 2%. Thus, proposed model can be used to accurately calculate side length of equilateral triangular patch.

# References

1. Kumar, G., Ray, K.P.: Broadband Microstrip Antennas, 1st edn. Artech House, USA (2003)
2. Wong, K.L.: Compact and Broadband Microstrip Antennas, 1st edn. Wiley, New York, USA (2002)
3. Garg, R., Bhartia, P., Bahl, I., Ittipiboon, A.: Microstrip Antenna Design Handbook. Artech House, USA (2001)
4. Deshmukh, A.A., Kumar, G.: Reliability of suspended rectangular microstrip antenna. In: Proceedings of International Conference on Quality, Reliability and Control, Bombay, India, 26th–28th Dec 2001
5. Deshmukh, A.A., Kulkarni, S.D., Venkata A.P.C., Phatak, N.V.: Artificial neural network model for suspended rectangular microstrip antennas. In: Proceedings of ICAC3—2015, Mumbai, India, 1st–2nd Apr 2015
6. Deshmukh, A.A., Kulkarni, S.D., Venkata A.P.C., Nagarbowdi, S.B.: Artificial neural network model for suspended equilateral triangular microstrip antennas. In: Proceedings of ICCICT—2015, Mumbai, India, 15th–17th Jan 2015
7. Deshmukh, A.A., Kulkarni, S.D., Venkata A.P.C.: Artificial neural network model for suspended shorted rectangular microstrip antennas. In: Proceedings of ICCT—2015, Mumbai, India, 25th–26th Sept 2015
8. Deshmukh, A.A., Kulkarni, S.D., Venkata, A.P.C.: Artificial neural network model for suspended shorted 90° sectoral microstrip antennas. In: Extended Version of Proceedings of ICCT-2015, CAE J. (2016)

# Analysis of Circular Microstrip Antenna with Single Shorting Post for 50 Ω Microstrip-Line Feed

## S. M. Rathod, R. N. Awale and K. P. Ray

**Abstract** In this paper, a technique using a single shorting post has been used to match the peripheral impedance of the circular microstrip antenna (CMSA) with that of the 50 Ω-microstrip (MS)-line feed. The shorting posts perturb the current distribution on the patch, altering the input impedance at the periphery. Thus, by selecting shorting post positions, a wide range of impedance has been adjusted without changing the patch geometry. A systematic analysis of the performance of the CMSA at a lower order mode and fundamental mode frequencies with a loading of a single shorting post has been presented. Simulated results of the directly 50 Ω-MS-line-fed CMSA are experimentally validated with good agreement.

**Keywords** Fundamental mode · Lower mode · Microstrip-line feed Peripheral-fed circular microstrip antenna · Shorting post

## 1 Introduction

In past few years, the concept of a compact antenna has attracted the increasing attention. A theory of miniaturization using a shorting post and various impedance matching techniques for CMSA has been reported [1, 2]. However, CMSA cannot be fed directly with 50 Ω-MS-line, without an inset or an impedance transformer, as all around the periphery, the impedance is equally higher. The feeding technique

---

S. M. Rathod (✉) · R. N. Awale
Electronics Engineering Department, Veermata Jijabai Technological Institute,
Matunga (East), Mumbai 400019, India
e-mail: rathod.shivraj@gmail.com

R. N. Awale
e-mail: rnawale@vjti.org.in

K. P. Ray
Department of Electronics Engineering, Defence Institute of Advance Technology (DIAT), Government of India, Girinagar, Pune, Maharashtra 411025, India
e-mail: kpray@rediffmail.com

such as inset fed or quarter wave transformer makes the antenna geometry asymmetrical [1]. The method of loading the patch by either shorting post (pin) or shorting plate has been used extensively to realize compactness [1–5].

To make the electronic systems compact, large size integration on the same plane is desirable, wherein MSAs are connected directly to other active devices and microwave integrated circuits through MS-line [1, 2]. A rectangular MSA (RMSA) provides flexibility for directly feeding through 50 $\Omega$-MS-line along the non-radiating edges. There is no flexibility to feed CMSA directly with 50 $\Omega$-MS-line as there are no non-radiating edges along the periphery. The peripheral input impedance of a CMSA typically varies between 400 and 500 $\Omega$, necessitating the use of either inset or impedance transformer for MS-line feed. Both these feeding methods make the antenna configuration asymmetrical along the $E$-plane, which increases the cross-polarization level in the $H$-plane. Additionally, inset feed becomes unrealizable for some cases at higher range of microwave frequency.

Recently, RMSA configurations directly fed with 110 $\Omega$ MS-line have been reported [6, 7]. The position of a single shorting post inside the patch has been selected to reduce the impedance to 110 $\Omega$ at the radiating edge, to feed it directly with MS-line [6]. This shorted RMSA because of asymmetrical field distribution along the $E$-plane has a higher cross-polarization level in the $H$-plane. To overcome this problem, a pair of symmetrical shorting posts was used along with a center line of the RMSA, which reduced the cross-polarization level besides facilitating the direct feed with MS-line [7]. Shorting posts in RMSA slightly increase the resonant frequency of the modified fundamental mode. In addition, a lower resonant mode is also introduced, whose resonant frequency is a function of the position of the shorting posts. This mode with lowest resonant frequency has not been presented in either of these reported MS-line fed RMSA configurations. Further, these RMSAs are fed with MS-line of 110 $\Omega$, which is not a standard characteristics impedance. Thus, it requires additional TRL calibration kit to de-embed the input impedance. The detail resonant mode analysis of shorted square/rectangular MSA with frequency formulation has been reported in [8, 9]. It is preferred, at least in lower microwave frequency range, to feed MSA with 50 $\Omega$-MS-line. Moreover, the detailed analysis of a directly fed CMSA with MS-line feed is not available in the literature.

In this paper, a detailed analysis of a shorted CMSA with 50 $\Omega$-MS-line-feed using a single shorting post has been presented. Analysis of the lower order mode and the modified fundamental $TM_{11}$ mode of a CMSA with respect to the positions of shorting posts has been carried out. Initially in this paper, a model analysis of a basic coaxial fed CMSA with and without center shorting posts are presented for lower order and fundamental mode frequencies. Further, systematic study of all the characteristics mainly input impedance, surface current distribution, directivity, and radiation pattern of a proposed 50 $\Omega$-MS-line-feed CMSA configuration has been conducted. Proposed shorted CMSA antennas are fabricated and measurements have been carried out to validate simulated results with good agreement.

## 2 Analysis of a Single Shorted CMSA

### 2.1 Modal Analysis of a Basic Coaxial Fed CMSA

A CMSA, as depicted in Fig. 1a, is the preferred radiator, as besides being a perfectly symmetric geometry, it suppresses the harmonics generated by transmitters/oscillators, which are in multiple of fundamental mode frequency. An accurate formula of the resonant frequency of a CMSA is given in [1]. Resonant characteristics of different modes of CMSA using both high and low permittivity substrates, different substrate thickness, and ground plane radius are investigated in detailed [1, 10, 11]. In this section, a CMSA with diameter $D$ has been designed at the fundamental mode frequency $F_F = 1.8$ GHz with dielectric substrate Arlon having a relative permittivity of $\varepsilon_r = 2.5$, a dielectric thickness of $h = 1.59$ mm, and the loss tangent $\tan \delta = 0.003$.

The radius of the circle $a = 30.01$ mm and the 50 $\Omega$ coaxial feed position $x = 8.5$ mm are calculated using formulations given in [1]. The coaxial fed CMSA has been analyzed using IE3D software [12]. The modal plots, showing real ($R_{in}$) and imaginary ($X_{in}$) parts of the input impedance for the fundamental and next four higher order modes, are shown in Fig. 1b. The resonant frequencies of various modes are calculated using formulation available in [1], which are $F_{TM11} = 1.805$ GHz, $F_{TM21} = 2.986$ GHz, $F_{TM02} = 3.746$ GHz, $F_{TM31} = 4.107$ GHz, and $F_{TM12} = 5.112$ GHz. For the feed point $x = 8.5$ mm, it shows matched 50 $\Omega$ impedance for fundamental mode ($TM_{11}$) frequency $F_F$, whereas, the impedances are different for higher order modes. Now, a shorting post/via of radius $r = 0.5$ mm is placed at the center of the patch. The modal plot of the shorted configuration is compared with that of the un-shorted one in Fig. 1b.

For the fundamental $TM_{11}$ mode, the center of the patch has zero potential, thus, a short at the center does not alter the fundamental mode frequency. So is the case for $TM_{21}$ and $TM_{12}$ modes, whereas, for $TM_{02}$ mode there is no zero potential at the center; it gets suppressed with center short. For the present feed location $x$ and

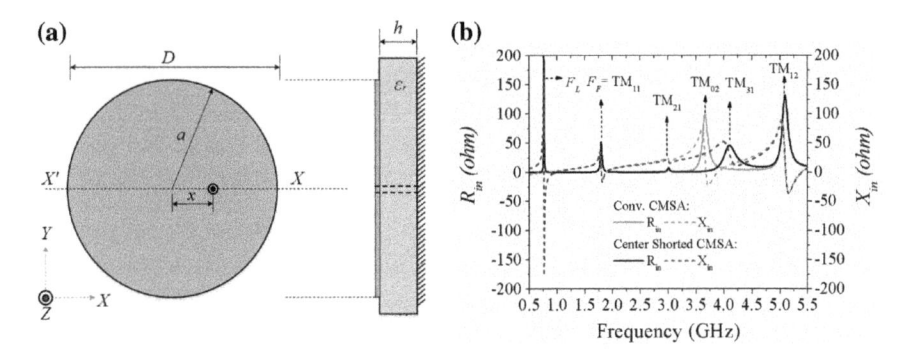

**Fig. 1** a Geometry of a coaxial fed CMSA, b comparison of input resistance and input reactance for conventional coaxial fed CMSA with and without center shorting post

center shorting post, the $TM_{31}$ mode gets properly excited, which was a very small kink in the modal plot for un-shorted CMSA. Additionally, center short introduces a new lower order mode with resonant frequency $F_L$. Lower frequency $F_L$ corresponds to the condition that the longest distance from the shorting point to the opposite open end on the periphery of a CMSA is the approximate effective radius.

## 2.2 Investigation of $F_L$ and $F_F$ with Shorting Position

To find out the effect of the position of the shorting post on $F_L$ and $F_F$, the investigation has been carried out on a CMSA by varying the shorting post position ($d/D$). A geometry of the coaxial end fed CMSA with the shorting post is shown in Fig. 2a. For each position of the shorting post, the value of $F_L$ and $F_F$ are determined. Figure 2b depicts the variation of $F_L$ and $F_F$ of the shorted CMSA with different ($d/D$). It is observed that with an increase in $d/D$ from 0 to 0.5, $F_L$ increases while $F_F$ decreases. The maximum reduction in the $F_L$ is obtained when the shorting post is placed close to the periphery of the CMSA [1, 5]. For the fundamental $TM_{11}$ mode of the CMSA, the radius of the patch is effectively a distance between zero potential point (center) and the open circuit (periphery). Using the same analogy, the resonant frequency $F_L$ of the peripheral shorted CMSA is calculated and discussed in [1].

$F_L$ increases from 0.592 to 0.766 GHz, when the shorting post is moved from the periphery to the center along the center line of the CMSA. The shorting post, when it is not at the center, modifies the field distribution of the fundamental mode, which does not remain $TM_{11}$. In terms of the surface current density, it is very high at the shorting post and becomes minimum at the open end periphery. This modification leads to increase in $F_F$ from 1.816 GHz for $d/D = 0.5$ to 2.123 GHz when the shorting post is close to the periphery, i.e., $d/D = 0.05$. Figure 2b depicts the same effect graphically.

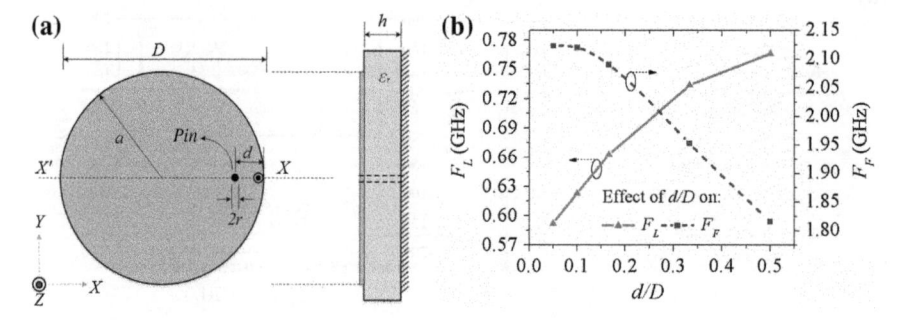

**Fig. 2** a Geometry of a coaxial end fed shorted CMSA, b effect of position $d/D$ of shorting post on $F_L$ and $F_F$

Analysis of Circular Microstrip Antenna … 79

## 3 Single Shorting Post CMSA with 50 Ω-MS-Line-Feed

A shorting post on the patch modifies the field distribution, leading to change of resonant frequency of different modes and their input impedance at the periphery. In order to match the impedance of a CMSA at periphery with 50 Ω-MS-line-feed, a single shorting post is used. The analysis of a single shorted CMSA, as designed below, with 50 Ω-MS-line-feed has been carried out. Figure 3a depicts the CMSA with a single shorting post, which is fed by the 50 Ω-MS-line with width $W_s$ of 4.538 mm and length of 10 mm. Use of MS-line reduces the various resonant frequencies slightly because of a marginal increase in effective dimension.

### 3.1 Input Impedance

A single shorting post of radius $r = 0.5$ mm is introduced along the center line $XX'$ at position $d$ from the peripheral feed point. Figure 3b depict the simulated result of input resistance ($R_{in}$) and reactance ($X_{in}$) of 50 Ω-MS-line-feed shorted CMSA with different $d/D$ ratio. It is observed that as $d/D$ ratio decreases (pin moves from the center to the periphery), i.e., $d/D = 0.5$ to 0.05, the input resistance at the periphery for $F_F$ and $F_L$ reduces from 504 to 50 Ω and 710 to 42 Ω, respectively. For the same variation of $d/D$, the resonant frequency $F_F$ increases from 1.740 to 2.095 GHz while $F_L$ reduces from 0.750 to 0.588 GHz. Hence, by choosing the proper position of a shorting post, a wide range of input impedance of MS-line fed CMSA can be adjusted for a various range of frequency without using other impedance matching technique. Thus, for 50 Ω-MS-line-fed CMSA, when shorting post is kept at $d/D = 0.05$, the input matching is obtained at both the frequencies modified $F_F = 2.095$ GHz and $F_L = 0.588$ GHz. For the requirement of the compact antenna configuration of 50 Ω-MS-line-fed CMSA, it can be operated at resonant frequency $F_L = 0.588$ GHz, whereas for the frequency closer to the

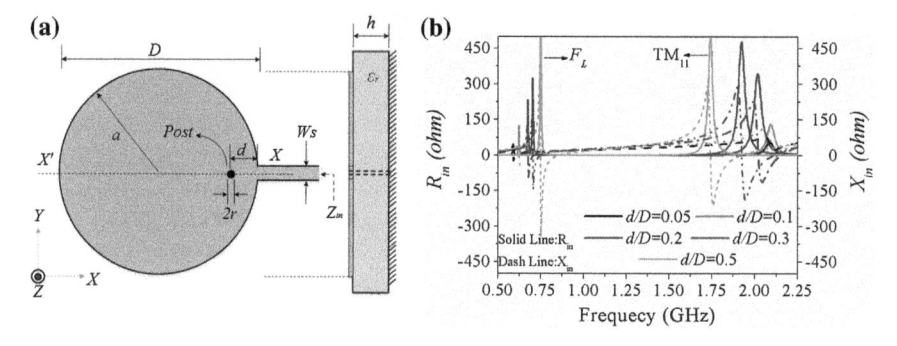

**Fig. 3** Single shorting post CMSA with 50 Ω-MS-line-fed **a** geometry, and **b** simulated input resistance and input reactance for a different shorting position at $F_L$ and $F_F$

fundamental mode of the CMSA, $F_F$ is chosen. Further on, investigations have been carried out only for the modified mode with frequency $F_F$ for the peripheral fed single shorting post CMSA in the following sections.

## 3.2 Surface Current Distribution and Radiation Characteristics

Figure 4 depicts the surface current distributions for three cases of a single shorting post CMSA at corresponding frequencies $F_F$. For the center shorting post, the current distribution is symmetrical on the patch, which is the same as that of $TM_{11}$ mode. When the shorting post is shifted toward periphery, the current distribution becomes increasingly asymmetrical and the $TM_{11}$ mode gets modified. Figure 5 depicts the comparison of simulated co-polarization levels in $E$-plane and corresponding cross-polarization levels in $H$-plane of the 50 $\Omega$-MS-line-feed CMSA with different $d/D$ ratio at corresponding modified $F_F$. It is noticed from Fig. 5a that as the shorting post moves from the center to the periphery of the CMSA, the cross-polarization level in the off broadside direction in the $H$-plane increased from −26 to −5 dBi, while the co-polarization levels in the $E$-plane are almost the same. The performance in terms of the modified resonant frequency $F_F$, directivity and the input resistance of the peripheral fed single short CMSA with respect to the position of the shorting post is summarized in Fig. 5b. When the frequency increases, the directivity increases from 7.26 to 7.97 dBi. The proposed 50 $\Omega$-MS-line-feed CMSA eliminates the need of a separate TRL calibration kit used in [6, 7] for de-embedding the input impedance.

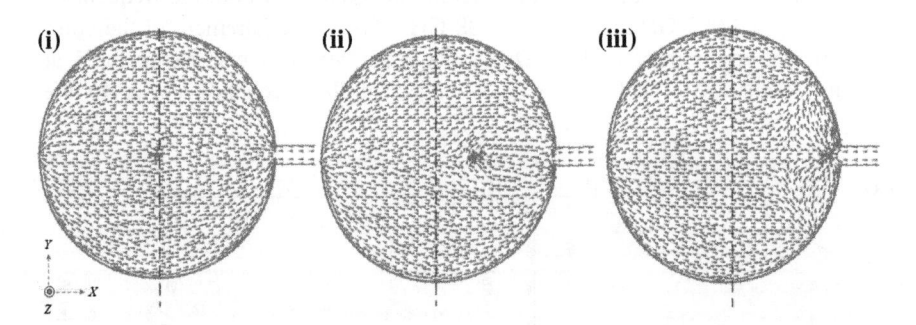

**Fig. 4** Surface current distribution with different shorting positions (i) $d/D = 0.5$ (equivalent to a conventional CMSA), (ii) $d/D = 0.3$, and (iii) $d/D = 0.05$ at corresponding $F_F$

Analysis of Circular Microstrip Antenna ...

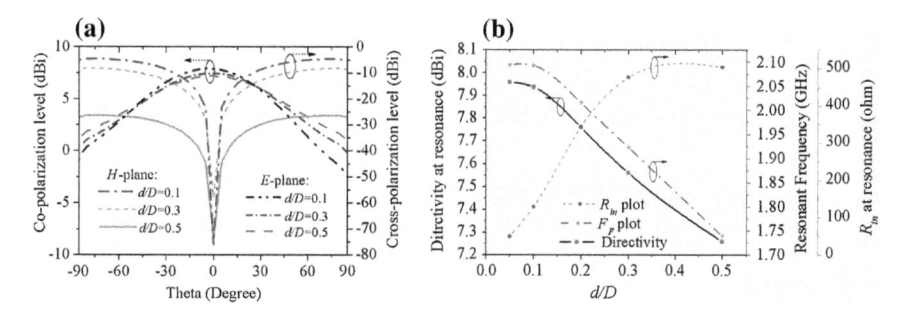

**Fig. 5** **a** Simulated co-polarization in $E$-plane and cross-polarization in $H$-plane with different $d/D$ ratio at corresponding modified $F_F$, and **b** effect of a single shorting post position ($d/D$) on resonant frequency, directivity and resonant resistance of the 50 $\Omega$-MS-line-fed CMSA

## 4 Experimental Validations and Discussions

For verification of the proposed design of MS-line-fed configurations, two 50 $\Omega$-MS-line-fed CMSA with a single and without shorting posts are fabricated on Arlon dielectric substrate with $\varepsilon_r = 2.5$, $h = 1.59$ mm and $\tan \delta = 0.003$. The shorting post with radius 0.5 mm is located at $d/D = 0.05$. CMSAs are fed with 50 $\Omega$-MS-line of width $W_s$ of 4.538 mm and length of 10 mm. The ground plane dimension is 80 × 80 mm². Photographs of proposed CMSA are shown in Fig. 6a. Measurements have been carried out using Agilent-Fieldfox made Vector Network Analyzer. Figure 6b compares simulated and measured VSWR for fundamental and modified fundamental modes (at $F_F$) for the 50 $\Omega$-MS-line-feed CMSA without

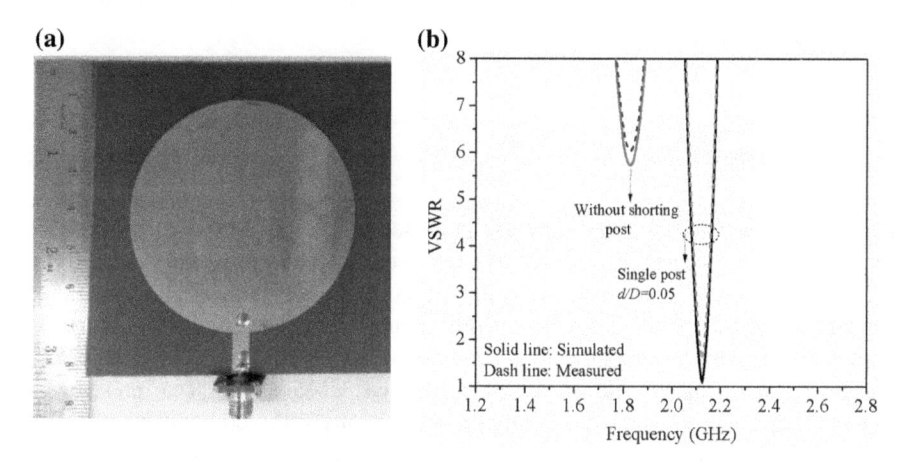

**Fig. 6** **a** Photographs of proposed configuration with a single shorting post at $d/D = 0.05$, and **b** comparison of simulated and measured VSWR for without shorting post and a single shorting post 50 $\Omega$-MS-line-feed CMSAs

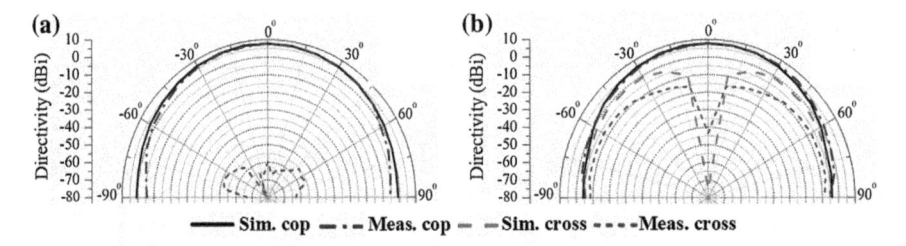

**Fig. 7** Comparison of simulated and measured radiation pattern of a single shorting post CMSA in **a** $E$-plane, and **b** $H$-plane

shorting post, and with a single shorting post. The simulated and measured results of radiation pattern are compared in Fig. 7a, b, which are in good agreement.

The CMSA with 50 Ω-MS-line-feed without shorting post has high resonant input resistance, thus does not get matched with the MS-line feed. The simulated and measured resonant frequencies for this case are 1.740 and 1.747 GHz, respectively. With a single shorting post, input impedance is close to 50 Ω, leading to good input matching with the MS-line-feed. For a single shorting post CMSA, the simulated and measured modified $F_F$ are 2.095 and 2.099 GHz, respectively with respective percentage bandwidth (BW) of 1.6 and 1.4. Figure 7a, b give comparisons of simulated and measured radiation patterns in $E$-plane and $H$-plane for a single shorting post CMSA with 50 Ω-MS-line-feed, respectively at corresponding $F_F$. The maximum radiation is in the broadside direction for the proposed antenna configurations. In Fig. 7a, for the case of $E$-plane of a single shorting post CMSA, the simulated and measured half-power beam widths (HPBWs) are 84.06° and 82.12°, respectively. For this case, the simulated cross-polarization levels are less than −70 dB, whereas the corresponding measured level is less than −50 dB. With reference to Fig. 7b in the $H$-plane, the simulated and measured HPBWs are 76.61° and 74.21° respectively.

## 5 Conclusions

In this paper, a 50 Ω-MS-line has been proposed to feed a CMSA using a single shorting post. This method of direct MS-line feed overcomes the disadvantage of using quarter wave transformer line and inset feed, which makes the antenna geometry asymmetrical. The peripheral impedance of the CMSA can be adjusted for a wide range by selecting the proper position of the shorting post. An analysis of the lower and fundamental mode of a shorted CMSA has been presented. Simulation results are validated through measurements with good agreement. The proposed impedance matching technique can be useful for series feed antenna array with CMSA elements.

**Acknowledgements** The authors would like to thank Shri. S. S. Kakatkar and Shri. Jagdish Prajapatee from RFMS Division, SAMEER, Mumbai for their kind help in fabrication and valuable input during experimental measurement.

# References

1. Kumar, G., Ray, K.P.: Broadband Microstrip Antennas, 1st edn. Norwood, MA, Artech House, USA (2003)
2. Wong, K.L.: Compact and Broadband Microstrip Antennas, 1st edn. Wiley, New York (2002)
3. Waterhouse, R.: Small microstrip patch antenna. Electron. Lett. **31**(8), 604–605 (1995)
4. Satpathy, S.K., Ray, K.P., Kumar, G.: Compact shorted variations of circular microstrip antennas. Electron. Lett. **34**(2), 137–138 (1998)
5. Ray, K.P.: Broadband, dual frequency and compact microstrip antennas. Ph.D. Thesis, Indian Institute of Technology, Bombay, India (1999)
6. Zhang, X., Zhu, L.: An impedance-agile microstrip patch antenna with loading of a shorting pin. In Asia-Pacific Microwave Conference (APMC), pp. 1–3 (2015)
7. Zhang, X., Zhu, L.: Patch antennas with loading of a pair of shorting pins toward flexible impedance matching and low cross polarization. IEEE Trans. Antennas Propag. **64**(4), 1226–1233 (2016)
8. Rathod, S.M., Awale, R.N., Ray, K.P.: Analysis of a single shorted rectangular microstrip antenna for 50 Ω microstrip line feed. In: 15th Biennial International Symposium on Antennas & Propagation (APSYM), pp. 1–4, CUSAT-Cochin (2016)
9. Deshmukh, A.A., Pawar, S., Kadam, P., Odhekar, A., Ray, K.P.: Analysis of single shorted square microstrip antenna. In: 2017 International Conference on Emerging Trends & Innovation in ICT (ICEI), pp. 123–128, Pune (2017)
10. Wood, C.: Analysis of microstrip circular patch antennas. IEE Proc. **128**(2), 69–76 (1981)
11. Shafai, L., Kishk, A.A.: Analysis of circular microstrip antennas. In: Handbook of microstrip antennas, vol. 1, pp. 45–85 (1989)
12. HyperLynx 3D EM Design System: Mentor Graphics Corp, Ver. 15.2, USA (2012)

# Gap-Coupled Designs of Compact F-Shape Microstrip Antennas for Wider Bandwidth

**Amit A. Deshmukh and Shefali Pawar**

**Abstract** F-shape microstrip antenna realized from L-shape microstrip antenna is a compact variation of equivalent rectangular microstrip antenna. The F-shape patch yields triple frequency response with broadside pattern characteristics. In this paper, fundamental and higher order modes of F-shape patch are discussed. After studying current plots at three modes, resonant length formulations at them as function of patch dimensions are proposed. Frequencies obtained using the same at each mode agrees closely with the simulated frequencies. Further, various gap-coupled designs of F-shape patch are discussed. Due to optimum gap-coupling between two patches when coupled along patch longer dimension, wideband response at first two resonant mode frequencies with 3–4% of VSWR bandwidth is obtained. In each frequency band, gap-coupled design yields broadside radiation pattern with similar polarization of radiated signal.

**Keywords** Multiband microstrip antenna · F-shape microstrip antenna
Broadband microstrip antenna · Gap-coupled configuration · Higher order mode

## 1 Introduction

In microstrip antenna (MSA), more commonly used shapes of radiating patch are rectangular, square, circular, equilateral triangular, semicircular, ring shape and their modified variations [1, 2]. The different radiating patch geometries are evolved as one radiating shape offers advantages over the other one in terms gain, harmonic suppression, bandwidth (BW) and cross-polar radiation levels [1–3]. Amongst the various regular shape MSAs, rectangular or square geometry offers better performance in terms the above parameters. The dual-band MSAs are needed when transmission/reception of electromagnetic signals is needed at closely spaced

---

A. A. Deshmukh (✉) · S. Pawar
Department of Electronics and Telecommunication Engineering,
SVKM's, D J Sanghvi College of Engineering, Mumbai, India
e-mail: amit.deshmukh@djsce.ac.in; amitdeshmukh76@gmail.com

© Springer Nature Singapore Pte Ltd. 2018
H. Vasudevan et al. (eds.), *Proceedings of International Conference on Wireless Communication*, Lecture Notes on Data Engineering and Communications Technologies 19, https://doi.org/10.1007/978-981-10-8339-6_10

frequencies. Although with respect to fundamental and higher order modes in regular shape MSAs dual and multiband response can be obtained, but polarization of transmitted signal varies, and thus, they are not suitable from practical applications [1–4]. Using techniques like placement of stub or cutting the slots in regular shape MSA variations, multiband response with single and dual polarization is achieved [1–8]. While realizing dual-polarized response, modes introduced due to modifications in patch are orthogonal to that of fundamental resonant mode. In these dual-polarized MSAs, change of polarization with respect to stub or slot dimensions for any of their combinations is not realizable. Which means that once polarization is realized (either dual or single), then it remains fixed for all the slot/stub dimensions. By suitably modifying rectangular MSA (RMSA), a compact L-shape MSA was realized [9]. Due to larger length of excited surface currents in fundamental mode, it offers a larger reduction in resonance frequency. By placing open circuit stub on the edge of L-shape MSA, new compact F-shape MSA is realized [9]. For varying stub length, F-shape MSA offers dual-polarized multiband response as well as single-polarized multiband response with broadside pattern at each of the frequencies. In [9], path lengths for the various resonant modes were defined; however, resonant length formulation at different modes was not presented. These formulations are needed since it helps in the redesigning of F-shape antenna in other frequency bands.

In this paper, first design of multiband F-shape MSA fabricated on low-cost FR4 substrate having parameters, $h = 0.16$ cm, $\varepsilon_r = 4.3$, tan $\delta = 0.02$, at frequency of 1000 MHz, is discussed. In F-shape geometry, multiband single- or dual-polarized response is the result of coupling between patch's first three resonating modes, namely, $TM_{10}$, $TM_{20}$ and $TM_{30}$. Further, by studying modal current plots at three modes, their resonant length formulation is proposed. Calculated frequencies using the same closely agree with simulated frequency. Therefore, these formulations of lengths can be helpful to design F-shape patch at different frequency ranges. To enhance the BW at respective resonant modes, various gap-coupled variations of two F-shape patches are proposed. Amongst all the gap-coupled designs, a configuration with F-shape patches gap-coupled along the larger patch dimension yields optimum results. At first two frequency bands (i.e. patch $TM_{10}$ and $TM_{20}$ modes), it yields BW of 3–4%. Across the BW, gap-coupled antenna exhibits pattern maximum in the bore-sight direction. Thus, the proposed work here explains functioning of polarization tunable F-shape patch antenna and provides its resonant length formulations and also discusses its wideband gap-coupled designs. In comparison, equivalent RMSA which has total patch area as that of the gap-coupled F-shape MSAs does not provide dual- or single-polarized wideband response. Therefore, novelty in proposed work lies in providing dual wideband response in closely spaced frequency bands with possible two polarizations in a compact design. The antenna proposed in this paper is first studied using simulations using IE3D that is followed by measurements. MSAs were fed using SMA connector in which to realize 50 $\Omega$ impedance, while using Teflon substrate in the coaxial cable, 1.2 mm of inner wire diameter is used. Ground plane size in the measurements is selected to be more than six times substrate thickness in all four directions with reference to edges of the patch. This ground plane size helps into realize infinite ground plane effect that has been used in simulations [1]. Further in experimentation, input impedance at antenna port and radiation pattern in far field distance were measured using high frequency instruments like, 'ZVH–8', 'SMB 100A' and 'FSC–6', inside the Antenna lab.

## 2 Resonant Length Formulation for F-Shape MSA

An optimum design of F-shape MSA is shown in Fig. 1. It yields simulated triple frequencies and BW's as 721, 1376 and 1636 MHz and 12, 23 and 29 MHz, respectively [9]. Whereas measured three frequencies and BW's are 730, 1360 and 1620 MHz and 11, 25 and 32 MHz, respectively [9]. At three frequencies pattern exhibits maximum in bore-sight direction showing E-plane directed along $\Phi = 0°$ for first two resonant modes. It is along $\Phi = 90°$ at the third resonant mode.

In [9], although current path lengths at modes were shown at three modes but formulations at them were not given. Here, by studying surface current distributions at $TM_{10}$, $TM_{20}$ and $TM_{30}$ resonant modes, their resonant length formulation is realized as shown in Fig. 2a, c and e and also as given in Eqs. (1)–(3).

At $TM_{10}$ mode,

$$L_{e1} = L + \left[ \left( \frac{l_s}{2} + \frac{w}{2} \right) \times \sin\left(\frac{\pi x}{2L}\right) \right] + 2(\Delta L) \qquad (1)$$

At $TM_{20}$ mode,

$$L_{e2} = L + \left[ \left( \frac{l_s}{2} + \frac{w}{2} \right) \times \sin\left(\frac{\pi x}{L_h}\right) \right] + 2(\Delta L) \qquad (2)$$

At $TM_{30}$ mode,

$$L_{e3} = L + \left[ (2l_s + w) \times \sin\left(\frac{\pi x}{L_h}\right) \right] + 2(\Delta L) \qquad (3)$$

**Fig. 1** F-shape MSA

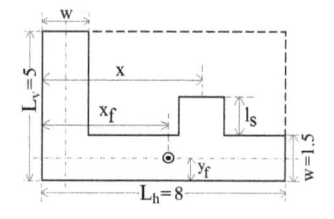

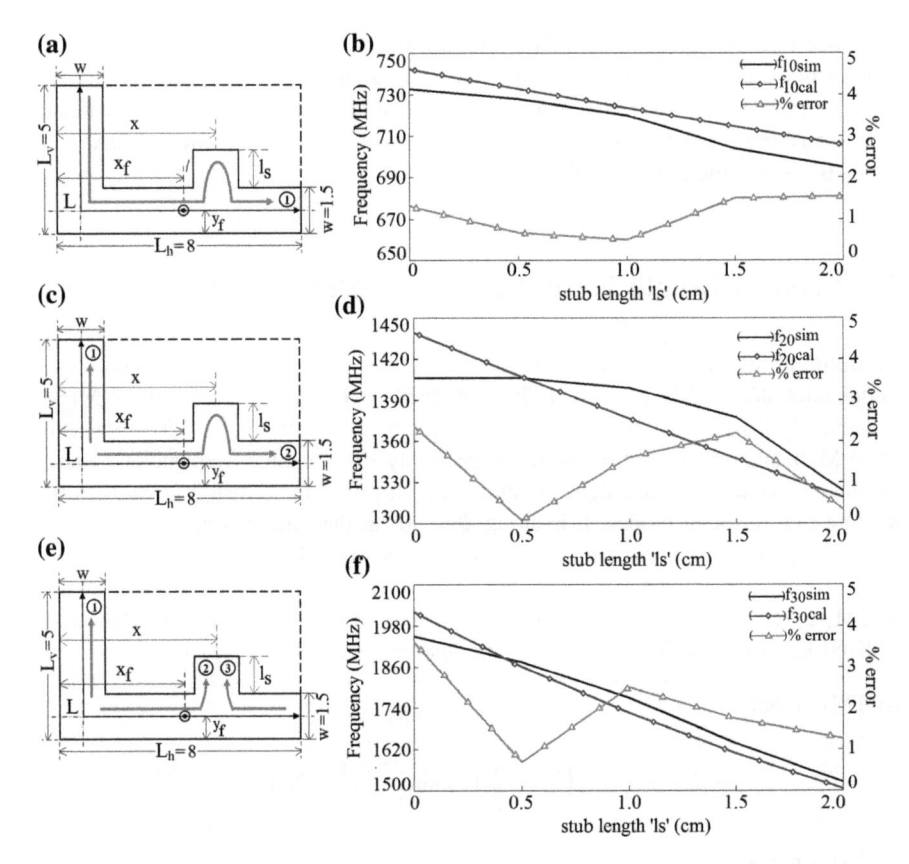

**Fig. 2** Resonant length formulations and frequencies and % error plots at **a, b** TM$_{10}$, **c, d** TM$_{20}$ and **e, f** TM$_{30}$ modes for F-shape MSA

$$\%\text{Error} = \left| \frac{f_{ie3d} - f_{cal}}{f_{ie3d}} \right| \times 100 \quad (4)$$

where $L_v$ = vertical patch length, $L_h$ = horizontal patch length, $w$ = width of the stub, $l_s$ = stub length, $x$ = stub position, $f_{cal}$ = calculated frequency, $f_{sim}$ = simulated frequency, $\Delta L$ = fringing length, $h$ = height of the patch and $\varepsilon_r$ = substrate dielectric constant.

The stub length '$l_s$' provides an alternative path length for modal surface currents and thus effectively modifies patch length '$L_e$' at the same as given in respective equations. For different stub lengths, using above equations frequency is calculated. The % error of calculated frequency with reference to simulated frequency is obtained by using Eq. (4). The plot of frequencies and % error against stub length is given in Fig. 2b, d and f. For all stub lengths variations, an error of less than 5% is realized. Thus for given frequency and substrate parameters, proposed formulations can be used to design F-shape MSA in given frequency band.

Gap-Coupled Designs of Compact F-Shape Microstrip … 89

## 3 Gap-Coupled Variations of F-Shape MSAs

To enhance the BW at respective modal frequencies, various gap-coupled variations of F-shape MSAs were studied as shown below. The air gap between parasitic and fed F-shape MSAs is kept at 0.5 cm in all the configurations. The resonance curve plots for each of the variations are also given below. For the configurations shown in Figs. 3c, e and 4a, c, the coupling between fed and parasitic patch is minimum due to which two separate peaks in the resonance curve at every resonant modes are not

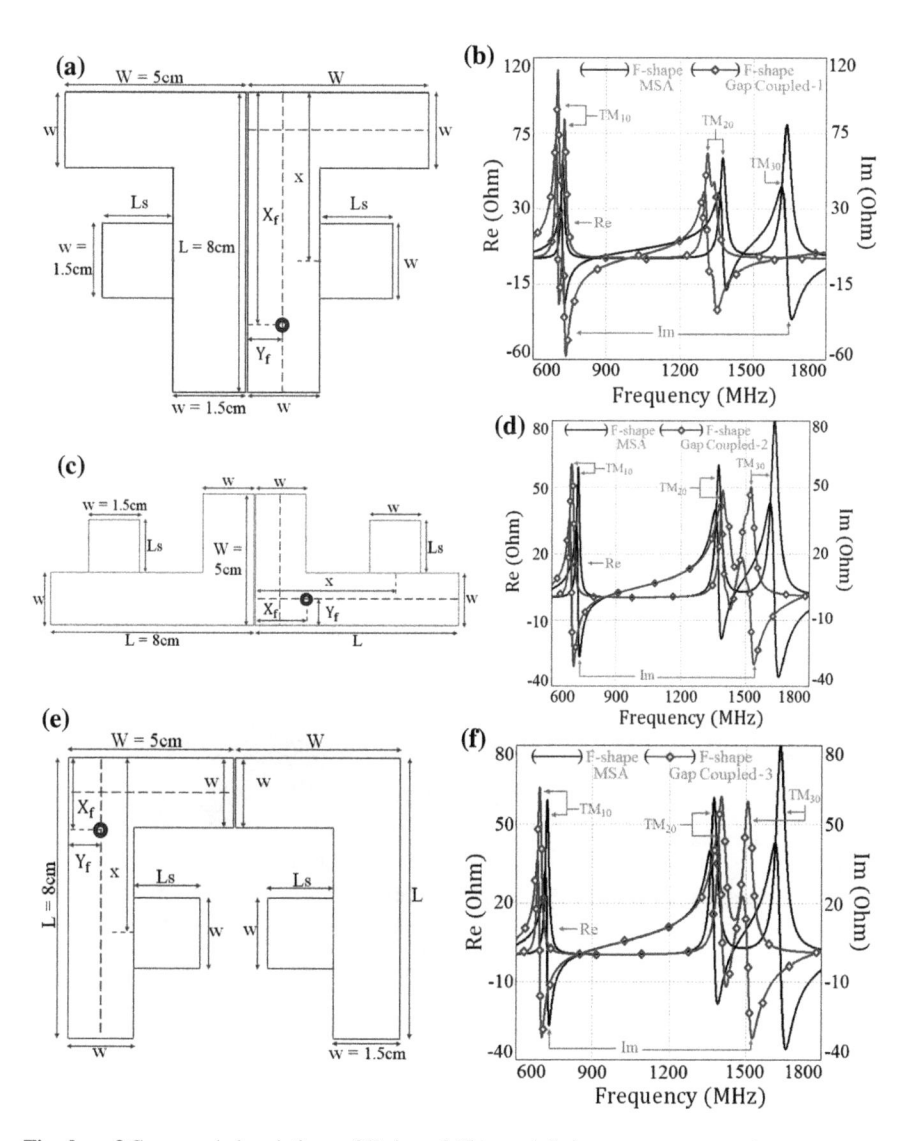

**Fig. 3** a–f Gap-coupled variations of F-shape MSAs and their resonance curve plots

observed. For the configuration shown in Fig. 3a, the optimum coupling is present between fed and parasitic patch that leads to loops in the impedance locus at each mode and two separate peaks near $TM_{10}$ and $TM_{20}$ modes in resonance curve plots as shown in Fig. 3b. For this variation, the coupling is not observed at $TM_{30}$ mode.

Thus, an optimum response with respect to $TM_{10}$ and $TM_{20}$ modes is realized, and its input impedance plots using simulation and measurement are given in

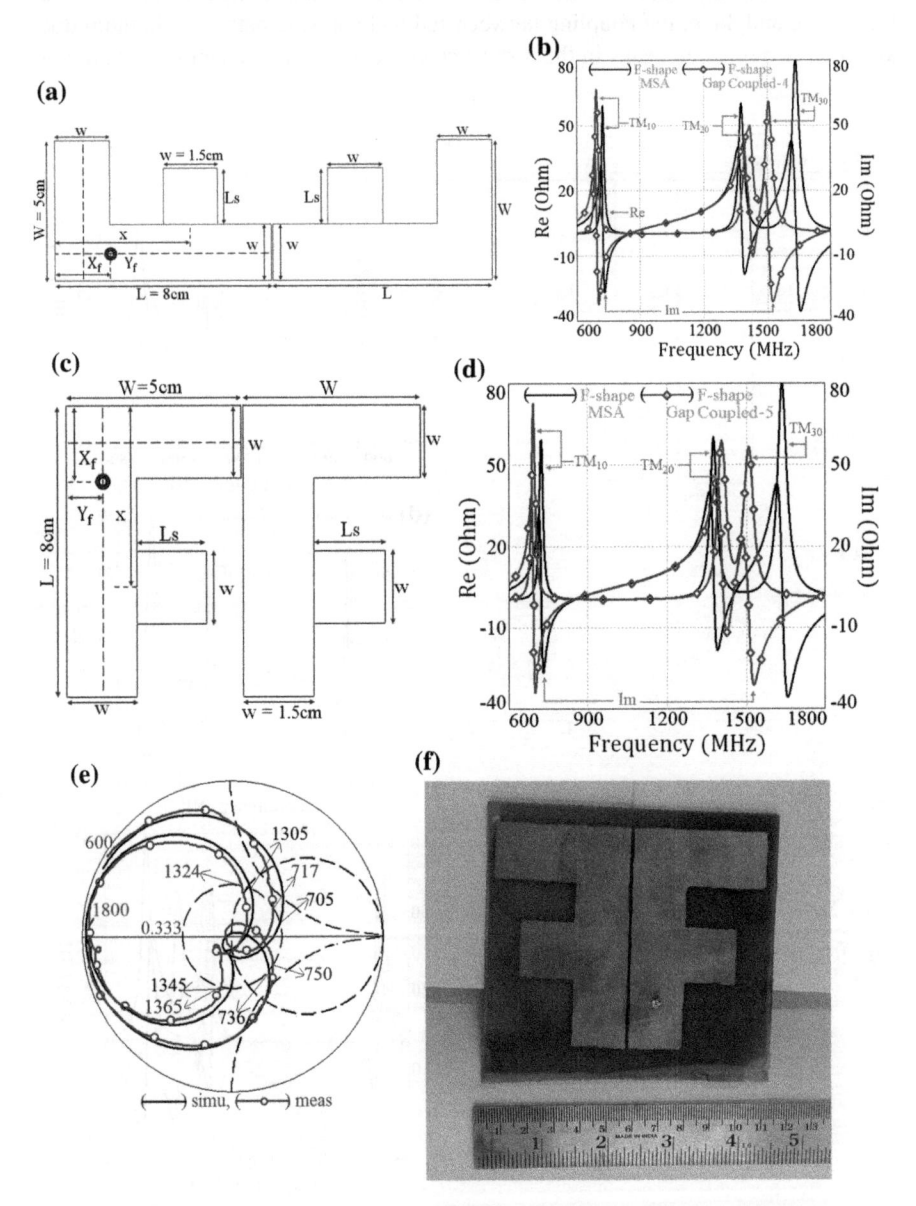

**Fig. 4** a–d Gap-coupled variations of F-shape MSAs and their resonance curve plots, e Smith chart for optimum configuration of gap-coupled F-shape MSAs as shown in Fig. 3a and its f fabricated prototype

Fig. 4e. Simulated frequencies and BW at first two frequency bands are 724 and 1338 MHz and 31 and 40 MHz, respectively. Respective measured values are 728 and 1345 MHz and 33 and 41 MHz, respectively. The fabricated antenna is shown in Fig. 4f.

The simulated and measured patterns at two frequency bands are given in Fig. 5a–d. Pattern at two frequencies shows maxima along bore-sight direction with higher cross-polarization levels is noticed towards higher frequency band. At

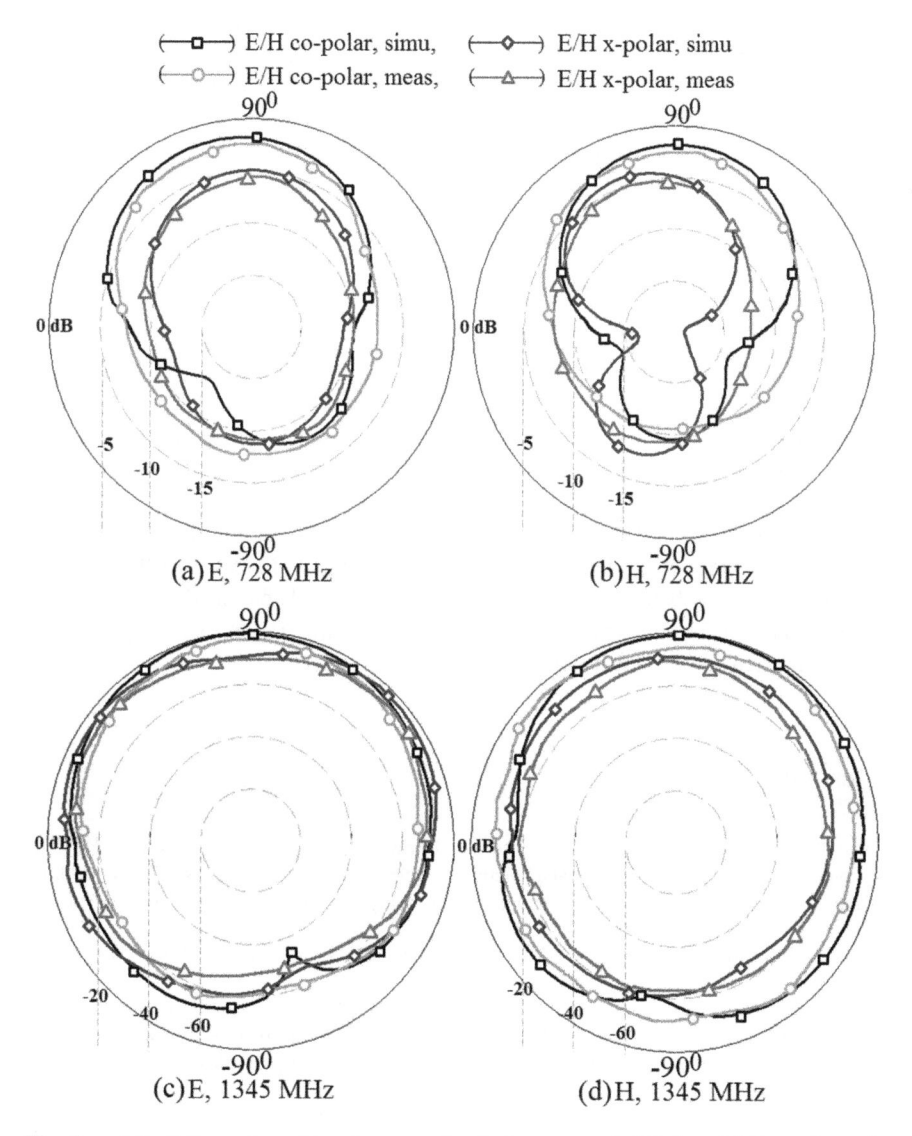

**Fig. 5 a–d** Radiation pattern at two frequency bands for gap-coupled F-shape MSAs as shown in Fig. 3a

second band, current exhibits two half-wavelength variations across patch length. Due to this some contribution of orthogonal currents are present along the overall patch area. This increases cross-polar radiation towards those frequencies.

The total patch area in optimum gap-coupled design of F-shape MSAs is nearly 80 cm$^2$, i.e. length = 10 cm and width of 8 cm each. Here in optimum configuration frequency, ratio between two bands is 1.85. In comparison, RMSA of above equivalent dimension will operate at different resonant mode frequencies such as $TM_{10}$, $TM_{01}$, $TM_{11}$, $TM_{02}$, $TM_{20}$ and $TM_{12}$ at frequencies of 893, 720, 1163, 1436, 1778 and 1716 MHz, respectively. The pattern across these various modes does not remain identical as well as modes with similar kind of patterns may not be excited optimally for same coaxial feed location. Thus, novelty in proposed work lies in presenting new design of polarization tunable F-shape antenna with their resonant length formulations at three modes. Further, design of gap-coupled F-shape MSAs enhances the BW at first two resonant modes.

# 4 Conclusions

The novel design of F-shape MSA for multiband response with polarization tunability is discussed. By observing the variations in surface current components on the F-shape patch, resonant length formulations at three modes are presented. Calculated frequencies obtained using formulations closely agree with simulated frequencies against varying stub length. Further, various gap-coupled variations of F-shape MSA were studied for wideband response. The configuration coupled along patch longer edge yields optimum results with 4–3% of VSWR BW at respective frequency bands. The gap-coupled antenna yields broadside pattern over complete BW. In future scope of work, suspended configuration of F-shape MSA will be analyzed for enhanced gain response.

# References

1. Lee, H.F., Chen, W.: Advances in Microstrip and Printed Antennas. Wiley, New York (1997)
2. Wong, K.L.: Compact and Broadband Microstrip Antennas, 1st edn. Wiley, New York (2002)
3. Garg, R., Bhartia, P., Bahl, I., Ittipiboon, A.: Microstrip Antenna Design Handbook. Artech House, Norwood (2001)
4. Kumar, G., Ray, K.P.: Broadband Microstrip Antennas, 1st edn. Artech House, Norwood (2003)
5. Deshmukh, A.A., Tirodkar, T.A., Jain, A.R., Joshi, A.A., Ray, K.P.: Analysis of multiband rectangular microstrip antennas. Inter. J. Microw. Opt. Technol. **8**(3), 145–154 (2013)
6. Maci, S.: Dual band slot loaded antenna. IEEE Proc. Microw. Antennas Propag. **142**(3), 225–232 (1995)
7. Daniel, A.E., Shevgaonkar, R.K.: Slot-loaded rectangular microstrip antenna for tunable dual-band operation. Microwave Opt. Tech. Lett. **44**(5), 441–444 (2005)

8. Babu, S., Kumar, G.: Parametric study and temperature sensitivity of microstrip antennas using an improved linear transmission line model. IEEE Trans. Antennas Propag. **47**(2), 221–226 (1999)
9. Deshmukh, A.A., Pawar, S., Ray, K.P.: Multi-band configurations of L-shape and F-shape microstrip antennas. Proceedings of ICCUBEA—2017, 17th–18th (2017)

# Naturally Tapered Series Fed Arrays for First Sidelobe Level Reduction

**Bharati Singh, Nisha Sarwade and K. P. Ray**

**Abstract** Series fed arrays with feed at the centre element have been used as a natural tapered array for first sidelobe level (FSLL) reduction. The fact that all elements of the series fed arrays do not receive equal power has been used to realize natural tapering in these arrays. An analysis with respect to number of elements and size of the elements for the series fed arrays used is clearly brought about.

**Keywords** First sidelobe level (FSLL) · Natural tapering · Series fed arrays

## 1 Introduction

The reduction of first sidelobe level (FSLL) of an array is an important parameter for measurement of directivity of the radiating system. Several means are used to reduce the FSLL of an antenna array. A common approach to achieve a reduction in FSLL is to realize amplitude tapering. In order to realize amplitude tapering for an antenna array, there are two basic methods. In the first approach for the corporate fed array, the input to each equi-spaced identical element of the array is controlled. The main beam power to sidelobe power ratio is improved by amplitude manipulation of excitation signals of the array elements [1]. The requirement of a complex feed network, where unequal power dividers and couplers need to be used makes

---

B. Singh (✉) · N. Sarwade
Department of Electronics, VJTI, Matunga, Mumbai, India
e-mail: bhartisingh@somaiya.edu

N. Sarwade
e-mail: nishasarvade@vjti.org.in

B. Singh
KJSCE, Vidyavihar, Mumbai, India

K. P. Ray
Department of Electronics Engineering, Defence Institute of Advance Technology (DIAT), Government of India, Girinagar, Pune, Maharashtra 411025, India
e-mail: kpray@diat.ac.in

the design a very complicated one to realize the amplitude distribution in a given plane [2, 3]. Because of this complicated feed network design, it is difficult to realize the smooth amplitude tapering even for a large number of elements.

The second approach, for the reduction of FSLL, is to use non-identical antenna elements varying the gain of each element according to desired amplitude tapering [4]. The concept of variation of width of the RMSA in series fed array is used in [5–8]. This method simplifies the array feed network but requires the precise design of each element of the array to realize the distribution. None of the work reported so far discusses the possibility of a centrally fed series array with identical elements to realize tapering in the E-plane. In this work, the series fed array is designed as a naturally tapered array resulting in considerable reduction of FSLL. An analysis with respect to number of elements and size of the elements for the series fed arrays used is clearly brought about. Experimental verification at 1.76 GHz validates the analysis and design.

## 2 Design and Analysis of the Naturally Tapered Array

Series fed arrays are well known for their simple design and implementation [1]. Two types of series fed arrays are possible to design as depicted in Fig. 1a, b. The microstrip patches are connected with high impedance microstrip lines. For a broadside beam, the antenna elements have to be in phase. This means that the phase shifts between the patches must be 360° (or it is multiple). The interconnecting lines are thus half-wavelength long (Fig. 1a) and a wavelength long (Fig. 1b).

The input impedance of the RMSA is minimum (zero) along the centre of the length of the RMSA and maximum (around 200 Ω) at its end (at the width) [9]. Each element causes mismatch to the high impedance microstrip lines. The amount

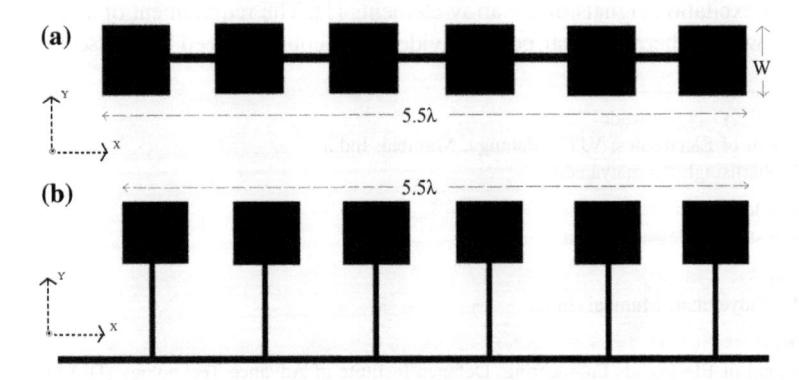

**Fig. 1 a** Layout of the series fed array with half-wavelength high impedance interconnecting microstrip lines and **b** layout of the series fed array with one wavelength high impedance interconnecting microstrip lines

of power an element extracts from the line is related to its impedance relative to that of the line. Hence, due to mismatch, in series fed arrays, power reduces with respect to input to successive elements. This can be used to realize naturally tapered arrays by using identical RMSA elements. In this work, a centrally fed uniform series fed linear array with identical elements is designed for analysis. In corporate fed uniform arrays, the power fed to each antenna elements is identical and as per the Fourier transform far-field radiation pattern, the first sidelobe level is −13.42 dB [1]. When a uniform linear array with identical elements is series fed, the centrally fed element radiates and remaining input power is directed to the next element and so on. The input power reduces at each successive element; thus, far off elements in large arrays may not have sufficient input power [10]. Thus, in a series fed uniform linear array, due to radiation by each element, some obvious reduction of input power to the next successive element causes a natural tapering of power. This results in a far-field pattern with reduced sidelobe levels as compared to a uniform linear array.

Four naturally tapered arrays were designed at the same frequency of 1.76 GHz using two differently sized RMSAs of different widths designed at the same frequency. The size (width * length) of both RMSAs are 4.2 cm * 3.9358 cm and size 5.1 cm * 3.93 cm. Their lengths were optimized so that they resonate at the same frequency. For each of the RMSAs, two series fed uniform linear arrays were designed using five and seven numbers of elements. Glass epoxy substrate with dielectric constant 4.3 with height $h = 1.59$ mm was used for the design. Finite ground plane with $6 * h$ clearance all around the array patches of the PCB is considered which behaves like an infinite ground plane.

Analysis of the four arrays designed using IE3D (simulated) is presented in Table 1.

From Table 1, it is clear that for the naturally tapered arrays, there is an improvement in FSLL from the standard −13.42 dB of a uniform array to about −15 dB for a five-element array. The FSLL further improves with increase in number of elements ($N$). For $N = 7$, the achieved FSLL improves to −16.2 dB for the array with smaller sized elements. The increase in size of the RMSA results in overall gain of the array and a smoother realization of the aperture distribution in the E-plane leading to the improvement of FSLL to −18.39 dB. A reasonably good gain is obtained as the size of RMSA, and hence, antenna array is large enough.

Table 1 Comparison of naturally tapered arrays (uniform elements series fed arrays)

| Case A: Smaller equal elements (size 4.2 cm * 3.9358 cm) ||||
|---|---|---|---|
| N | $S_{11}$ (dB) | $F_r$ (GHz) | FSLL (dB) | GAIN (dBi) |
| 5 | −44 | 1.76 | −15.447 | 7.985 |
| 7 | −21 | 1.76 | −16.2 | 8.74 |
| Case B: Larger equal elements (size 5.1 cm * 3.93 cm) ||||
| N | $S_{11}$ (dB) | $F_r$ (GHz) | FSLL (dB) | GAIN (dBi) |
| 5 | −28 | 1.76 | −15.6052 | 8.83 |
| 7 | −18 | 176 | −18.39 | 9.22 |

If the input power to the central element is $P_{in}$, then the next consecutive element receives power given as in Eq. (1).

$P_{next}$ = Input power − radiated power by the central element, i.e.

$$P_{next} = P_{in} - P_{rad} \qquad (1)$$

where $P_{rad}$ is the power radiated by the central element. This process of reduction of power continues till the last edge element on each side of the central element of the uniform array with identical elements. Thus, all the elements in a centrally series fed array do not receive identical power due to the natural process of radiation by each of the elements. Hence, a natural tapering is realized in the E-plane which in turn improves the FSLL. The increase in number of elements for such an array realizes the improvement of power tapering in the time domain which results in reduction of FSLL in the frequency domain (far-field radiation pattern).

## 3  Results and Experimental Verification of Naturally Tapered Array

The photograph of the naturally tapered array designed at 1.76 GHz is given in Fig. 2. The array is symmetrical with respect to the feed, and hence, the obtained radiation pattern is symmetrical.

For an experimental verification of the proposed concept, a seven-element series centrally fed naturally tapered array is designed and fabricated at 1.76 GHz on glass epoxy substrate with dielectric constant 4.3, tan δ = 0.02 and height of 0.159 cm. Size of the individual elements are 5.1 cm * 3.93 cm (width * length). Size of the fabricated array as depicted in Fig. 2 is 74 cm by 5 cm. Measurements on the fabricated antenna were carried out using ANRITSU Vector Network Analyzer (No. MS2026C) [11]. Figure 3 compares the simulated and measured return loss plots of the nine-element uniform linear tapered antenna array. These two plots are in good agreement. The simulated resonance frequency is 1.76 GHz, whereas, the measured value is 1.756 GHz. Figure 4a, b compares theoretical and measured radiation patterns in the E and the H-plane, respectively.

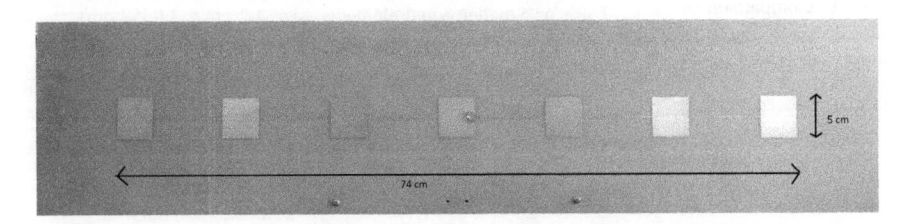

**Fig. 2** Photograph of the naturally tapered fabricated array with identical RMSA elements

**Fig. 3** Simulated and measured return loss of naturally tapered array

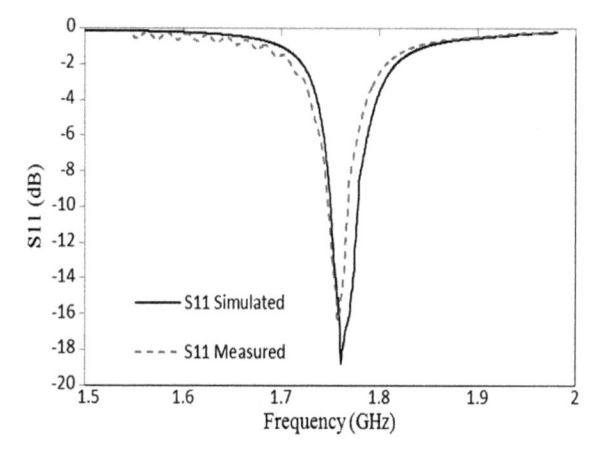

**Fig. 4** Simulated and measured radiation patterns of naturally tapered array: **a** E-plane and **b** H-plane

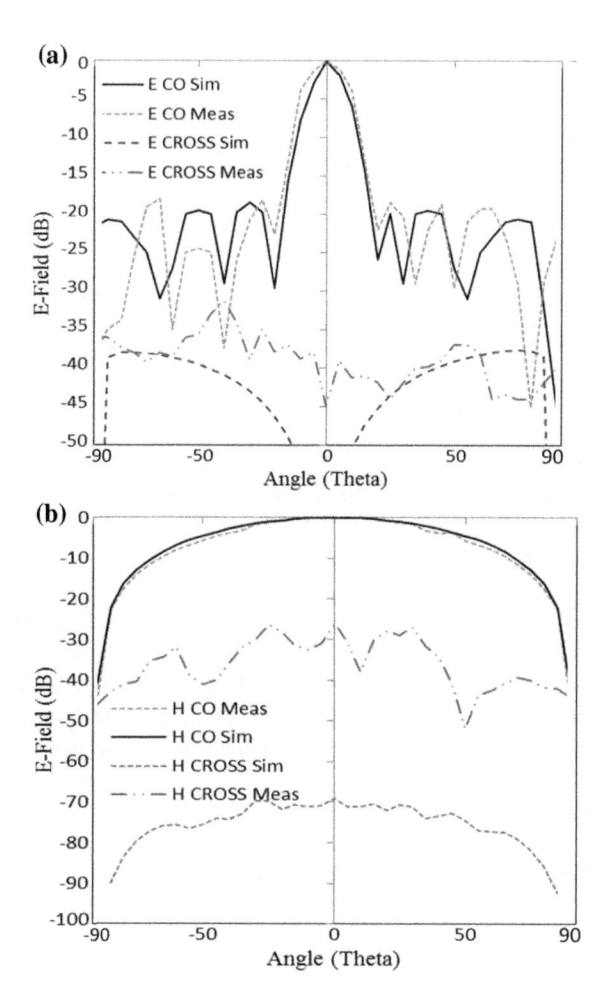

**Table 2** Comparison of simulated and measured parameters of naturally tapered arrays

| Parameters | $S_{11}$ | Half-power beamwidth E-plane | Half-power beamwidth H-plane | FSLL (dB) | VSWR |
|---|---|---|---|---|---|
| Simulated | −18.7738 | 11° | 70° | −18.39 | 1.2603 |
| Measured | −16.206 | 13.5° | 78° | −18.06 | 1.366 |

Table 2 depicts the comparison of the simulated and measured results of the uniform linear tapered array.

Hence, a naturally tapered array has been designed using identical RMSA elements, fabricated and analysed for different numbers of elements. Simulated results have been experimentally verified at 1.76 GHz with a reasonably good agreement with each other. Practically, an improved FSLL of −18.06 dB has been achieved with $N$ equal to 7.

Table 2 shows the reasonable agreement between theoretical and measured results. The measured and theoretical values of FSLL are very close to each other. The half-power simulated and realized beamwidth are also in good agreement. The measured cross-polar in E-plane is less than 40 dB down with respect to main lobe. In the H-plane, the radiation pattern is similar to that of a single RMSA. The measured cross-polar in H-plane is less than 30 dB down with respect to main lobe Thus, it confirms that for seven elements RMSA linear array, natural tapering is realized due to reducing power as inputs to consecutive elements with respect to the central fed element leading to the reduction of FSLL.

## 4 Conclusion

Natural tapering is realized for the series fed arrays and an analysis is carried out with respect to number of elements. An array of seven identical elements is fabricated, and the experimental results validate the proposal. An improved reduced FSLL of −18.06 dB is experimentally achieved, which is lower than 4.86 dB of that of a standard uniform linear array. It is concluded that for even low number of elements in a centrally fed series array, the input power to each element is not identical. Radiation from each element causes less power as input to the next consecutive element which brings about a natural tapering and hence reduction in FSLL. The realized tapering and hence reduction in FSLL depend on the size of the identical elements at a particular frequency. Simulated and experimental results are in good agreement with each other.

**Acknowledgements** The authors acknowledge SAMEER, MeitY, Government of India, Mumbai and VJTI, Mumbai managements for providing facility for experimentation and encouragement.

# References

1. Balanis, C.A.: Antenna Theory Analysis and Design, Third edn. Wiley, New York (2005)
2. Wincza, K., Gruszczynski, S.: Microstrip antenna arrays fed by a series-parallel slot-coupled feeding network. IEEE Antennas Wirel. Propagat. Lett. **10**, 991–994 (2011)
3. Yang, Y., Wang, Y., Fathy, A.E.: Design of compact vivaldi antenna arrays for UWB see through wall applications. Prog. Electromagnet. Res. **82**, 401–418 (2008)
4. Singh, B., Sarwade, N., Ray, K.P.: Non-identical rectangular microstrip antenna arrays with corporate feed for aperture tapering. IETE J. Res. Available: http://dx.doi.org/10.1080/03772063.2017.1356754
5. Yuan, T., Yuan, N., Li, L.W.: A novel series-fed taper antenna array design. IEEE Antennas Wirel. Propagat. Lett. **7**, 362–365 (2008)
6. Otto, S., Rennings, A., Litschke, O., Solbach, K.: A dual-frequency series-fed patch array antenna. In: 3rd European Conference on Antennas and Propagation, pp. 1171–1175, 23–27 Mar 2009
7. Chen, Z., Otto, S.: A taper optimization for pattern synthesis of microstrip series-fed patch array antennas. In: Proceedings of the 2nd European Wireless Technology Conference, pp. 160–163, 28–29 Sept 2009
8. Chong, Y.I., Wenbin, D.O.U.: Microstrip series fed antenna array for millimeter wave automotive radar applications. In: Conference on Microwave Workshop Series on Millimeter Wave Wireless Technology and Applications (IMWS), IEEE MTT-S International, pp. 1–3, 18–20 Sept 2012
9. Kumar, G., Ray, K.P.: Broadband Microstrip Antennas. Artech House, Norwood, USA (2003)
10. Singh, B., Sarwade, N., Ray, K.P.: Compact series fed Tapered Antenna Array using unequal rectangular microstrip antenna elements. Microwave Opt. Tech. Lett. **59**(8), 1856–1861. (2017) https://doi.org/10.1002/mop.30640
11. Chen, Z., Liu, D., Nakano, H., Qing, X., Zwick, Th.: Handbook of Antenna Technologies. Springer Nature Publication (2016)

# Variations of Slot Cut Multiband Isosceles Microstrip Antennas for Dual-Polarized Response

Amit A. Deshmukh, Anish Mishra, Foram Shah, Pooja Patil, Hetvi Shah and Aarti G. Ambekar

**Abstract** A variation of equilateral triangular microstrip antenna is isosceles triangular microstrip antenna. It exhibits identical resonant mode field distributions. In this paper, multiband and dual-polarized designs of isosceles triangular patch antenna for increasing isosceles angle using offset coaxial feed and rectangular slot are proposed. The combination of offset feed and slots yields optimum input impedances at $TM_{10}$, $TM_{01}$, $TM_{11}$, and $TM_{02}$ modes of the patch that yields dual-polarized response with 1–1.5% of bandwidth at each of the frequencies. Here at first two resonant modes, broadside radiation pattern is obtained whereas conical pattern showing pattern maximum along the end-fire direction is obtained at third and fourth resonant modes.

**Keywords** Multi-band microstrip antenna · Isosceles triangular microstrip antenna
Dual polarization · Higher order resonant mode

## 1 Introduction

The radiating patch in microstrip antenna (MSA) can take any arbitrary shape; however from mathematical modeling point of view, regular patch shapes like rectangular/square, circular, and triangular, and their modified variations like semicircular, and half triangular patch, have been used [1–3]. Among these shapes, rectangular geometry offers better performance in terms of gain, bandwidth, and radiation pattern, but circular shape offers better harmonic rejection when input RF signal source outputs frequencies at set desired frequency as well as its harmonic values. Equilateral triangular MSA (ETMSA) is a special case of triangular MSA

---

A. A. Deshmukh (✉) · A. Mishra · F. Shah · P. Patil · H. Shah · A. G. Ambekar
Department of Electronics and Telecommunication Engineering,
SVKM's, D J Sanghvi College of Engineering, Mumbai, India
e-mail: amit.deshmukh@djsce.ac.in

© Springer Nature Singapore Pte Ltd. 2018
H. Vasudevan et al. (eds.), *Proceedings of International Conference on Wireless Communication*, Lecture Notes on Data Engineering and Communications Technologies 19, https://doi.org/10.1007/978-981-10-8339-6_12

(TMSA) for an angle equal to 60°. Further isosceles triangular MSA (ITMSA) for different isosceles angle is an ETMSA variation, and the two exhibits identical modal field distributions. Some practical applications, antenna working in multiple close by frequencies is needed and their dual or multiband antennas are selected [4]. As against all the conventional antennas, MSA offers ease in realization of multiband response by using single radiating element. To realize this, MSAs are embedded with additional slots or loaded with open-ended stubs on patch edges [5–8]. Of the two techniques, slot is preferred since they do not increase overall patch size as well as slot does not lead to spurious radiation, which is observed from stubs that affect the radiation pattern of the patch. When mode introduced by slot is perpendicular to patch mode, then dual-polarized response is obtained [3, 5].

In this paper, first configuration of ETMSA for the fundamental mode ($TM_{10}$) frequency of 900 MHz is discussed. The study is also carried out for offset feed ETMSA where it does not show excitation of any orthogonal resonant modes. Further, various designs of equivalent ITMSA for increasing isosceles angle are investigated for different feed point locations. For angles higher than 60°, along with variation of multiples of half-wavelength variations along patch height, surface currents also exhibit multiples of half-wavelength variations along patch base, which gives multiband dual-polarized response. However as observed from modal current distributions along orthogonal modes, impedance matching at them to yield dual, triple, or multiband response for same feed location cannot be realized. Therefore, ITMSA with higher angles is embedded with rectangular slots which yield impedance matching and help into realize multiband response. In 70° ITMSA, four band responses over a frequency range of 700–1600 MHz were realized, whereas in 80° ITMSA, triple frequency response over the same range was obtained. Four band response in 80° ITMSA was obtained by suitably embedding additional slots in 80° ITMSA that optimizes input impedance at lower order resonant mode. At each frequency, ITMSAs yield 1–1.5% of bandwidth (BW). It shows broadside pattern at first two resonant modes whereas conical pattern (maximum in end-fire direction) at third and fourth mode.

In reported literature, many slot cut configurations that yield similar multiband tunable response are reported. However here in proposed work, first comparison between excitations of orthogonal modes in varying angle ITMSAs is presented. By suitably selecting isosceles angle and coaxial feed location, presented work discusses methods to realize multiband dual-polarized response using ITMSAs. The reported literature does discuss results for ITMSAs but does not provide a detailed modal study which is provided here to yield dual-polarized triple and four band response, and hence, this is the novelty here. The presented ITMSA designs are studied using glass epoxy substrate which has parameters as $\varepsilon_r = 4.3$, $h = 0.16$ cm, and tan $\delta = 0.02$. ITMSAs were first studied using simulations using the method of moment-based IE3D software. An SMA panel type connector was used to feed the antennas, which has 0.12 cm of inner wire diameter that realizes 50 $\Omega$ cable impedance. Antennas were fabricated on substrate which is larger enough to simulate the effects of infinite ground plane, as used in simulations. Further

## 2 ITMSAs Embedded with Slots for Varying Isosceles Angle

The coaxially fed ETMSA is shown in Fig. 1a. Using resonance frequency equation reported in [3], ETMSA side length "$S$" is calculated for $TM_{10}$ mode frequency of 900 MHz. Length is found to be 11.4 cm and when simulated it yields a frequency of 864 MHz. Resonance curve plots for ETMSA, which explain real and imaginary impedance behavior over a wide frequency range, are shown in Fig. 1c. For feed at "$x_f$" = 0 and "$y_f$" = 5.0 cm, it shows excitation of $TM_{01}$, $TM_{11}$, and $TM_{02}$ modes.

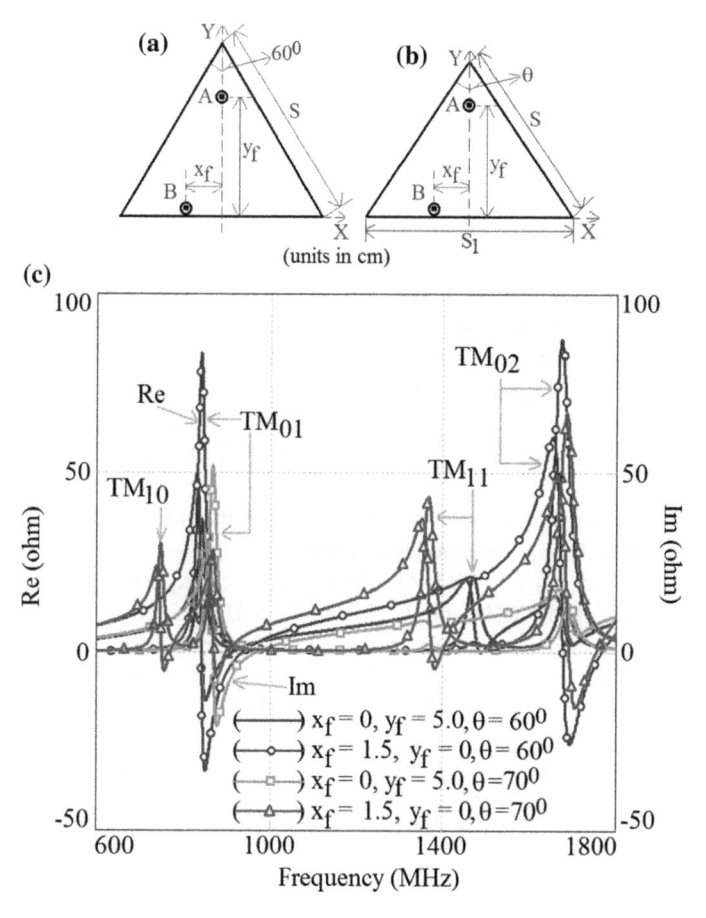

**Fig. 1** a ETMSA, b ITMSA, and their c resonance curve plots

The ETMSA is simulated for offset feed point, i.e., at "$x_f$" = 1.5 and "$y_f$" = 0 and its resonance curve plots are shown in Fig. 1c. At this location, the impedance at respective modes changes but no additional resonant mode was observed. In ETMSA, this is observed because patch is a symmetrical configuration with respect to centroid point [3]. Further, ITMSAs with different angles ($\theta$) were obtained as given in Fig. 1b. In ITMSAs, side length "S" of the patch is unchanged but with changing angle, base length "$S_1$" is varied. Resonance curve plots for $\theta$ increasing from 60° to 90° with offset feed point locations are shown in Figs. 1c and 2a. At observed resonant modes, surface current plots for $\theta$ = 70° are given in Fig. 2b–e.

With an increase in isosceles angle, the frequency of $TM_{01}$ mode increases whereas $TM_{11}$ mode frequency reduces. The $TM_{02}$ mode frequency marginally increases. However, with offset feed position, additional resonant modes are observed for ITMSA with an angle more than 60°. As observed from current distributions for 70° ITMSA, apart from $TM_{01}$, $TM_{11}$, and $TM_{02}$ modes, an orthogonal variation in currents is observed at first resonant mode. Since it shows half-wavelength variation, this distribution is due to $TM_{10}$ mode. Due to symmetry in patch with respect to centroid point, this mode was absent in ETMSA. Therefore with this additional mode, ITMSA will realize dual-polarized multiband response with reference to patch resonant modes. However as noted from impedance curves and as observed from parametric study for varying feed locations, matching of impedances in 25–100 $\Omega$ range at respective resonant modes to achieve dual-polarized multifrequency response was not obtained. Therefore to realize the same, rectangular slot is embedded in varying angle ITMSA as shown in Fig. 3a. The slot is cut along the centroid point, and ITMSA for 70° and 80° was optimized. For 70° ITMSA, multiband response as shown in Fig. 3c was obtained.

Frequencies and respective BWs using simulations are 747, 802, 1345, and 1590 MHz and 8, 11, 23, and 16 MHz. Measurement is carried out, and the four experimental frequencies and their BWs are 730, 784, 1322, and 1560 MHz and 9, 12, 25, and 14 MHz, respectively. For slot cut 80° ITMSA, using single slot, triple frequency response was realized, as matching at lower order mode frequency is absent (i.e., less than 25$\Omega$). Also as observed from its Smith chart as shown in Fig. 4a, impedance BW at two modes is smaller.

Here for three slot cut 80° ITMSAs, simulated three frequencies and their BWs are 834, 1251, and 1661 MHz and 7, 18, and 14 MHz, respectively. Frequencies and BWs obtained using measurements are 834, 1251, and 1661 MHz and 8, 16, and 15 MHz, respectively. For to realize frequency response with four bands, additional slots were cut inside slot cut 80° ITMSA as shown in Fig. 3b. The position of these slots is selected such that it will increase input impedance at first-, second-, and fourth-order resonant modes as well as it will realize impedance matching at first resonant mode. The optimized input impedance plot for multiple slots cut 80° ITMSA is given in Fig. 4b. Here frequencies and BWs using

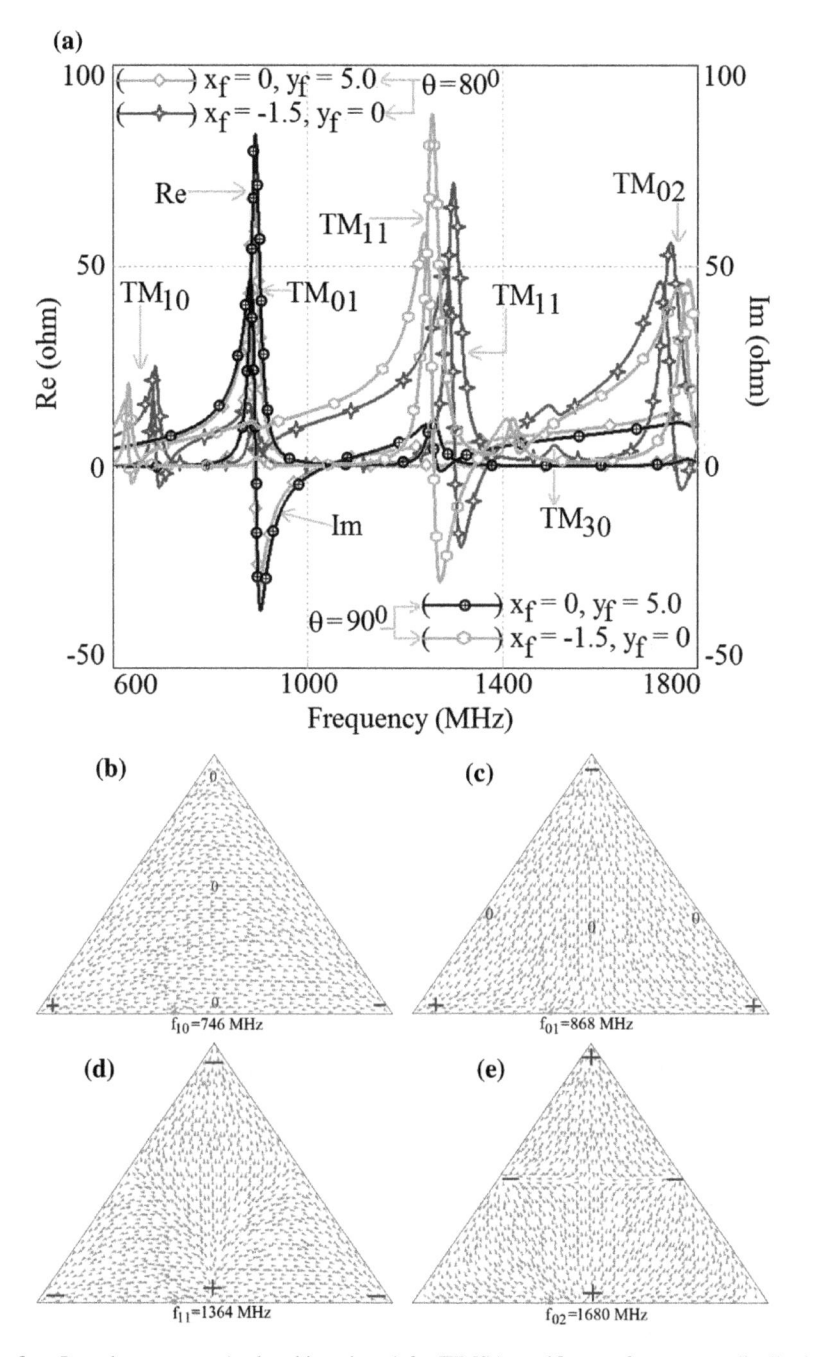

**Fig. 2** a Impedance curves (real and imaginary) for ITMSAs and **b–e** surface current distribution at observed modes for 70° ITMSA

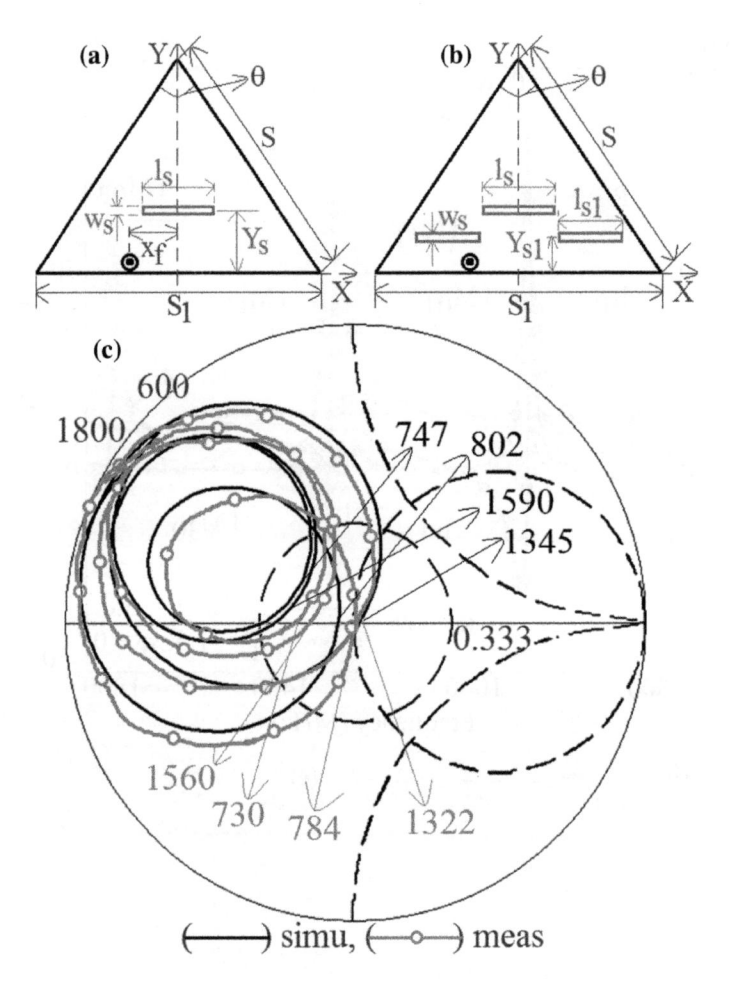

**Fig. 3** **a, b** Slots cut ITMSA variations and **c** input impedance response for slot cut 70° ITMSA

simulation are 685, 789, 1243, and 1481 MHz and 7, 12, 22, and 17 MHz, respectively. Also, frequencies and BWs obtained using measurements are 701, 802, 1256, and 1499 MHz and 7, 11, 23, and 16 MHz, respectively. For slot cut 80° ITMSAs, their fabricated prototypes are shown in Fig. 4c, d. In these multiband ITMSAs, radiation patterns at first two frequencies/modes is broadside, whereas that at third and fourth frequencies/modes, the pattern is conical. The E-plane at first mode is along $\Phi = 0°$, whereas that at next three modes it is along $\Phi = 90°$. Thus in proposed configurations apart from multiband dual-polarized response, pattern scanning from broadside to end-fire directions over four resonant modes is realized.

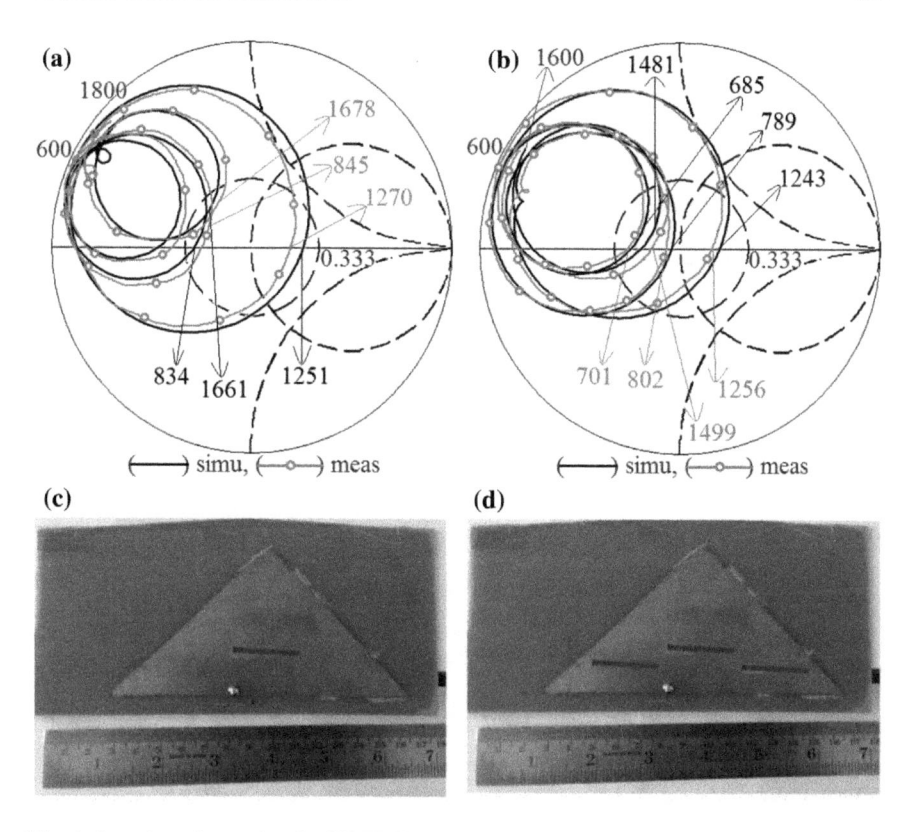

**Fig. 4** Input impedance plots for 80° ITMSA with **a** rectangular slot and **b** multiple slots and their **c, d** fabricated prototype

## 3 Conclusions

Novel designs of slot cut varying angle ITMSAs for dual-polarized multiband frequency response are proposed. Here in the proposed designs, multiband response is obtained using combinations of offset feed and multiple rectangular slots. At each of the modes, 1–1.5% of VSWR BW is obtained with broadside to end-fire radiation pattern scanning across multiple frequencies. The detailed study about modes of ITMSA with varying feed positions and isosceles angles was not reported in the literature to exploit upon possible multiband dual-polarized response. Here, the present detailed work provides the same. In future scope of the work, similar study will be carried out for multiband designs of ITMSA with angle more than 80°. Also, formulations at various resonant modes that yield multiple frequencies will be derived.

# References

1. Lee, H.F., Chen, W.: Advances in Microstrip and Printed Antennas. Wiley, New York (1997)
2. Garg, R., Bhartia, P., Bahl, I., Ittipiboon, A.: Microstrip Antenna Design Handbook. Artech House, Norwood (2001)
3. Kumar, G., Ray, K.P.: Broadband Microstrip Antennas, 1st edn. Artech House, USA (2003)
4. Balanis, C.A.: Antenna Theory: Analysis and Design, 2nd edn. Wiley Ltd, USA
5. Wong, K.L.: Compact and Broadband Microstrip Antennas, 1st edn. Wiley Inc, New York, USA (2002)
6. Daniel, A.E., Shevgaonkar, R.K.: Slot-loaded rectangular microstrip antenna for tunable dual-band operation. Microwave Opti. Tech. Lett. **44**(5), 441–444 (2005)
7. Deshmukh, A.A., Pawar, S., Ray, K.P.: Multi-band Configurations of L-shape and F-shape Microstrip Antennas. Proceedings of ICCUBEA 2017, 17th–18th (2017)
8. Deshmukh, A.A., Tirodkar, T.A., Jain, A.R., Joshi, A.A., Ray, K.P.: Analysis of multiband rectangular microstrip antennas. Int. J. Microwave Opt. Technol. **8**(3), 145–154 (2013)

# Design of a Novel CPW Band Stop Filter Using Asymmetric Meander-Line Defected Ground Structure

Makarand G. Kulkarni, A. N. Cheeran, K. P. Ray and S. S. Kakatkar

**Abstract** In this paper, two configurations of coplanar waveguide (CPW) band stop filter (BSF) using meander-line shaped symmetric-DGS (SDGS) and asymmetric-DGS (ADGS) have been designed and analyzed. For the proposed filters, the equivalent circuit models have been obtained in terms of *R*, *L*, and *C* parameters. Performance assessment of these filters has been carried out based on cutoff frequency, resonant frequency, stopband width, sharpness factor, filter selectivity, etc. The fabricated CPW-ADGS BSF shows two transmission zeros with better than −20 dB attenuation level, stopband width of 660 MHz, and sharpness factor close to unity with filter selectivity of 90 dB/GHz. The measured results of the fabricated CPW-ADGS BSF show a good agreement with its simulated results.

**Keywords** Asymmetric defected ground structure · Band stop filter Coplanar waveguide · Meander-line defected ground structure

---

M. G. Kulkarni (✉)
K. J. Somaiya College of Engineering, Mumbai, India
e-mail: makarandkulkarni@somaiya.edu

M. G. Kulkarni · A. N. Cheeran
Department of Electrical Engineering, VJTI, 400019 Mumbai, India
e-mail: ancheeran@vjti.org.in

K. P. Ray
Department of Electronics Engineering, Defence Institute of Advance Technology (DIAT), Government of India, Girinagar, Pune, Maharashtra 411025, India
e-mail: kpray@rediffmail.com

S. S. Kakatkar
RF and Microwave System (RFMS) Division, SAMEER Mumbai, IITB Campus, Mumbai 400076, India
e-mail: sandeep@sameer.gov.in

# 1 Introduction

Filters play an important role in many microwave applications, such as wireless communications and radar systems. In this paper, two filters have been designed using the coplanar waveguide (CPW) [1] proposed by C. P. Wen in 1969. Compared with microstrip structures, the CPW structures are more attractive because it requires only a single metal level and offers greater design flexibility as well as ease of fabrication. In 1993, Yablonovitch has proposed photonic band gap (PBG) [2] structure (or the term electromagnetic band gap structure (EBG), usually preferred in the microwave community). PBG/EBG structures are large periodic arrays of defects to prevent the propagation of electromagnetic waves in a certain frequency band. In defected ground structure (DGS), the periodicity is usually not implemented and the structures are based on a single defect (or a few defects) [3]. Apart from filters, DGS has prominent advantages in many other applications such as power dividers/couplers, amplifiers, and so on.

Filters can be implemented with shunt stubs or stepped-impedance lines in a microwave circuit, but these techniques require large circuit layout size and provide a narrow passband and spurious passbands in stopband. The band stop filter (BSF) using DGS has a number of attractive features like simple structure, wide and deeper stopband than that of a conventional low-pass filter, low insertion loss, and easy circuit modeling.

Reference [4] proposed EBG-CPW filter with double periodicity with T-shaped capacitive load located at periodic positions with strip width modulation. However, it required larger circuit board area and had a limitation in microwave circuit applications. Also, a few more complex EBG structures have been proposed in [5]. Dumbbell-shaped DGS was explored first time by D. Ahn and applied to design a low-pass filter on microstrip lines [6]. In [7] CPW filters using square dumbbell DGS are proposed but it generates higher radiation losses (31.9%). In the earlier work by the same authors [8], two CPW filter configurations have proposed. One is CPW with "symmetric-DGS (SDGS)", in which upper and lower heads of rectangular dumbbell DGS has been made symmetric in size, and the second is CPW with "asymmetric-DGS (ADGS)", in which lengths of upper and lower heads of DGS have been made asymmetric. However, the simulation as well as the measured results of the fabricated CPW-ADGS filter shows significant radiation loss ($\approx$13%) and also low filter selectivity ($\approx$29 dB/GHz). In the previous work by the same authors [9], the CPW band reject filter has been presented using the cascading arrangement of rectangular dumbbell-shaped DGS, resulting wide rejection bandwidth, less radiation losses with improvements in overall filter performance.

In this paper, design and analysis of two configurations of CPW-BSF using DGS, has been carried out. One is CPW-SDGS (Symmetric-DGS) BSF, in which both, i.e., upper and lower heads of the DGS are of equal area. Second is CPW-ADGS (asymmetric-DGS) BSF, in which upper and lower heads of DGS are of unequal area. The area of upper head is 10% more than that of the head of SDGS, and the area of lower head is 10% less than that of the head of SDGS.

The performance analysis has been presented in terms of −3 dB cutoff frequency ($f_C$), attenuation ($\alpha$) at resonant frequency ($f_0$), and stopband width (*SBW*) at −10 dB. Sharpness factor (*SF*) [10] is the measure of sharp rate of the filter cutoff. The filter selectivity ($\xi$) [11] is the roll-off rate of the filter transfer function between the frequency of the passband and the frequency of the stopband. To achieve higher filter selectivity the difference between the passband and the stopband frequencies must be small. In addition, the equivalent circuit model of parallel *R*, *L*, and *C* circuit has been obtained to estimate radiation losses (% $\eta$) [12] which arise due to DGS. The comparative performance analysis has also been carried out with earlier reported work.

## 2 Design, Fabrication, and Circuit Modeling

CPW of 50 Ω characteristic impedance ($Z_0$) has been designed at 2.5 GHz, with a length of 40.2 mm, strip width of *W* = 6 mm, and gap spacing of *S* = 0.5 mm. The schematics details of the CPW-SDGS BSF and CPW-ADGS BSF, respectively, have been shown in Fig. 1a, b. As shown in Fig. 1a, the area of the meander-line shaped upper as well as lower head of the SDGS is 23.5 mm$^2$ each. The area of the meander-line shaped upper head of the ADGS is designed by increasing 10% of that of head of SDGS, which is equal to 25.85 mm$^2$. And the area of the meander-line shaped lower head of the ADGS is designed by decreasing 10% of that of head of SDGS, which is equal to 21.15 mm$^2$. The overall area of SDGS and ADGS has been kept same which is equal to 71 mm$^2$. The total circuit size is 40.2 mm × 36 mm. The substrate material is Arlon, with thickness of 1.6 mm, $\varepsilon_r$ = 4.5, tan δ = 0.002. Hence, by keeping the DGS area same, the comparative effect of design of ADGS on filter characteristics has been examined in this paper.

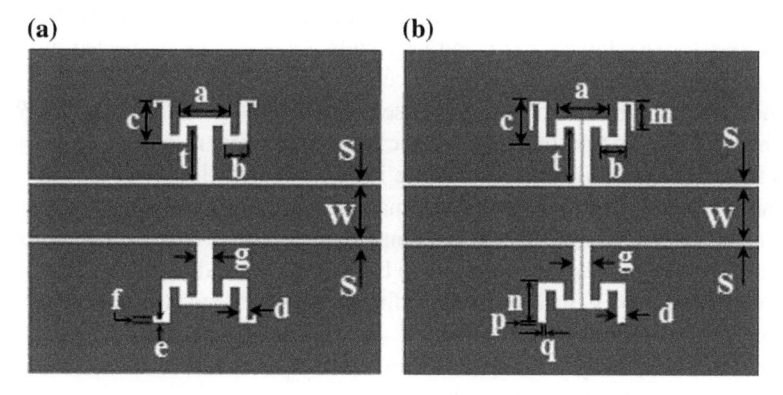

**Fig. 1** Schematic diagrams: **a** CPW-SDGS BSF and **b** CPW-ADGS BSF (where *a* = 6, *b* = 3, *c* = 5, *d* = 1, *e* = 0.5, *f* = 1, *g* = 2, *m* = 3.5, *n* = 4.5, *p* = 0.1, *q* = 0.5, *S* = 0.5, *t* = 6, and *W* = 6; all dimensions are in millimeter)

**Fig. 2** Fabricated CPW-ADGS BSF

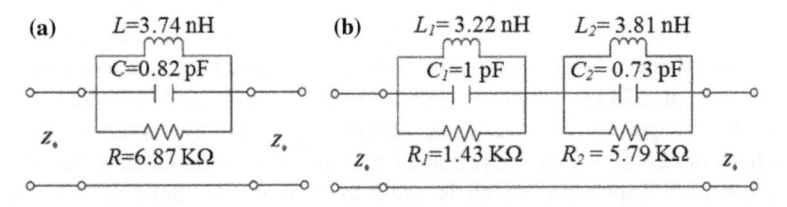

**Fig. 3** Equivalent circuit model: **a** CPW-SDGS BSF and **b** CPW-ADGS BSF

Figure 2 shows the fabricated CPW-ADGS BSF (as per the dimensions given in Fig. 1b). The measurements were carried out on a vector network analyzer.

Figure 3 shows the equivalent circuit model of DGS unit, which consists of the parallel Resistor ($R$), Inductor ($L$), and Capacitor ($C$), elements connected in series with the transmission line and can be calculated using (1-3) as given in [13]. The asymmetric-DGS can be modeled using series combination of two parallel $R$, $L$, $C$ resonators, since its transfer characteristic has two different resonance frequencies at two transmission zeros. The capacitance is due to stored charges at the gap. The inductance comes from the additional magnetic flux flowing through the two apertures, and the radiation effect is explained by resistance. Here, the resistance $R_i$, inductance $L_i$, and the capacitance $C_i$ have been extracted and listed in Table 1, for $i$th resonance frequency, where $i = 1$, for first resonance frequency ($f_{01}$) and $i = 2$, for second resonance frequency ($f_{02}$).

At resonant frequency, the impedance of the any parallel $R$, $L$, and $C$ resonance circuit is equals to $R$ and for Fig. 3b, the equivalent resistance at resonance condition is $R$ equals to ($R_1 + R_2$). The radiation losses [12] due to DGS have been calculated using (4). From (4), when the resistance goes to infinity (lossless), the

Design of a Novel CPW Band Stop Filter Using Asymmetric ...                                    115

**Table 1** Comparative performance assessment

| Parameters | Ref. [12] | This work |  |  |
|---|---|---|---|---|
|  |  | CPW-SDGS BSF (simulated) | CPW-ADGS BSF (simulated) | CPW-ADGS BSF (measured) |
| DGS geometry | Square | Meander-line |  |  |
| $f_{C1}$ (GHz) | 4.1 | 2.06 | 2.11 | 1.98 |
| $f_{C2}$ (GHz) | 7.8 | 3.44 | 3.475 | 3.5 |
| $f_{01}$ (GHz) | 6 | 2.87 | 2.79 | 2.8 |
| $f_{02}$ (GHz) |  |  | 3 | 3.2 |
| SBW (MHz) | 300 | 520 | 600 | 660 |
| Fractional SBW | 5% | 18.11% | 20.72% | 22% |
| $\alpha_{min1}$ (dB) | −14 | −36.87 | −23.75 | −21 |
| $\alpha_{min2}$ (dB) |  |  | −35.41 | −30 |
| $SF_1$ | 0.68 | 0.71 | 0.76 | 0.71 |
| $SF_2$ | 1.3 | 1.19 | 1.15 | 1.09 |
| $\xi_1$ (dB /GHz) | 5.78 | 41.81 | 30.51 | 22 |
| $\xi_2$ (dB /GHz) | 6.11 | 59.42 | 68.23 | 90 |
| % $\eta$ | 31.9 | 2.82 | 2.68 | 4.73 |

radiation loss due to DGS becomes zero. Sharpness factor (SF) [10] is the measure of sharp rate of the filter cutoff has been calculated using (5). As mentioned in the earlier section filter selectivity ($\xi$) [11] depends on passband frequency ($f_p$) and the stopband frequency ($f_s$) and it can be defined as (6). In (6), $\alpha_{max}$ = the attenuation value at 3 dB, $\alpha_{min}$ = attenuation value at resonant frequency $f_0$, $f_S = f_0$, and $f_C$ = 3 dB cutoff frequency.

$$R_i = 2Z_0 \left( \frac{1}{|S_{21}(f_{0i})|} - 1 \right) \quad (1)$$

$$C_i = \frac{f_{C1}}{4\pi Z_0 (f_{0i}^2 - f_{C1}^2)} \quad (2)$$

$$L_i = \frac{1}{4\pi f_{0i}^2 C_i} \quad (3)$$

$$\eta = \frac{1}{\frac{R}{4Z_0} + \frac{Z_0}{R} + 1} \quad (4)$$

$$\mathrm{SF}_i = \frac{f_{Ci}}{f_{0i}} \tag{5}$$

$$\xi_i = \frac{\alpha_{\min} - \alpha_{\max}}{f_{Si} - f_{pi}}. \tag{6}$$

## 3 Results and Discussions

The filter performance has been simulated using IE3D simulation tool and its results have been depicted in Fig. 4. The measured results of the proposed CPW-ADGS BSF have been compared with the simulated results as shown in Fig. 4. And the comparative performance assessment has been given in Table 1.

As per the performance evaluation shown in Table 1, due to ADGS, two transmission zeros occurred at nearby resonance frequencies (2.79 and 3 GHz), the widening effect of *SBW* has been observed. There has been an increase in the fractional *SBW* (20.72%) as compared to that of reported earlier in [12]. For the fabricated CPW-ADGS BSF, due to small shift in the $f_{o2}$ toward higher frequency side, the fractional *SBW* improved to 22%. The SF at lower and higher frequency sides of the stopband, i.e., $SF_1$ and $SF_2$, respectively, have also been improved in ADGS configuration approached the theoretical value of unity. Also, filter selectivity at lower and higher frequency sides of the stopband, i.e., $\xi_1$ and $\xi_2$, respectively, have been enhanced maximum up to 90 dB/GHz in the fabricated CPW-ADGS BSF. Radiation losses (%η) due to DGS also decreased effectively from 31.9% [12] up to 4.73% in the realized CPW-ADGS BSF.

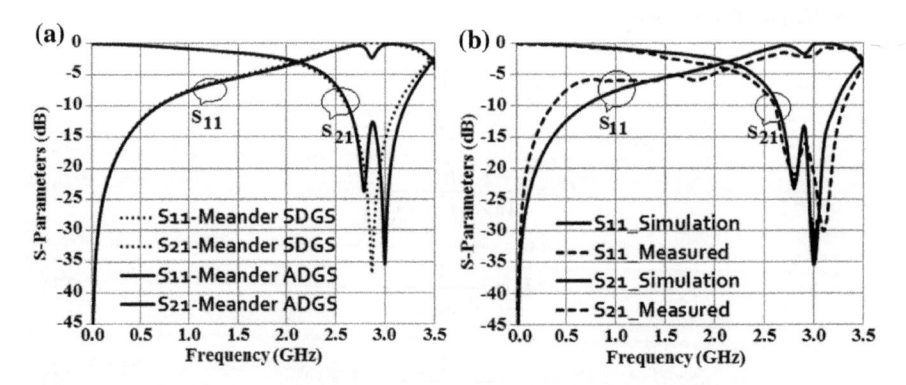

**Fig. 4** a Comparison of resultant S-parameters of CPW-SDGS BSF and CPW-ADGS BSF and b comparison of simulated and measured S-parameters of CPW-ADGS BSF

## 4 Conclusion

Two configurations of CPW-BSF using meander-line shaped DGS have been designed and analyzed in detail in this work. A proposed design of CPW-ADGS BSF is a good alternative to the conventional design of BSF. The CPW-ADGS BSF has been designed just by a slight variation (only 10%) in the area of DGS of the meander-line geometrical heads of CPW-SDGS BSF. The overall performance of these two proposed BSF has also been compared and found better than the earlier reported work. The fabricated CPW-ADGS BSF show improvement in fractional stopband width up to 22%, with sharpness factor close to ideal value of unity, also a good filter selectivity of 90 dB/GHz. The measured filter response on fabricated BSF shows a good agreement with the simulated response.

**Acknowledgements** The authors are thankful to Mr. Jagdish Prajapati and Mr. Prafull Irpache from RFMS Division, SAMEER Mumbai, for their help in fabrication and experimentation.

## References

1. Wen, C.P.: Coplanar waveguide: A surface strip transmission line suitable for nonreciprocal gyro-magnetic device applications. IEEE Trans. Microwave Theor. Tech. **17**(12), 1087–1090 (1969)
2. Yablonovitc, E.: Photonic band-gap structures. J. Opt. Soc. Amer. B, Opt. Phys. **10**, 283–295, (1993)
3. Gupta, K.C.: Microstrip Lines and Slotlines, Third edn. Artech House Microwave Library (2015)
4. Martine, F., Falcone, F., Bonache, J., Lopetegi, T., Laso, M.A.G., Sorolla, M.: Dual electromagnetic band-gap cpw structures for filter applications. IEEE Microwave Wirel. Compon. Lett. **13**, 393–395 (2003)
5. Zoul, Y., Hu, X., He, S., Lin, Z.: Compact coplanar waveguide lowpass filter using a novel electromagnetic bandgap structure. In: 7th International Symposium on Antennas Propagation & EM Theory, pp. 1–4 (2006)
6. Ahn, D., Park, J.-S., Kim, C.-S., Kim, J., Qian, Y., Itoh, T.: A design of the low-pass filter using the novel microstrip defected ground structure. IEEE MTT, **49** (2001)
7. Li, X.: Asymmetric coplanar waveguide filter with defected ground structure. In: XXXIth URSI and General Assembly and Scientific Symposium (URSI GASS), 2014
8. Kulkarni, M.G., Cheeran, A.N., Ray, K.P., Kakatkar, S.S.: Design of a novel CPW filter using asymmetric DGS. In: Proceedings International Symposium on Antennas & Propagation, Kochi, Dec 2016, pp. 13–16
9. Kulkarni, M.G., Cheeran, A.N., Ray, K.P., Kakatkar, S.S.: Coplanar waveguide band reject filter using electromagnetic band gap structure. Prog. Electromagnet. Res. Lett. PIERL-**70**, 53–58 (2017)
10. Smierzchalski, M., Kurgan, P., Kitlinski, M.: Improved selectivity compact band-stop filter with gosper frcatal-shaped defected ground structures. Microwave Opt. Technol. Lett. **52**(1) January (2010)
11. Karmakar N.C.: Improved performance of photonic band-gap microstrip structure with the use of chebyshev distributions. Microwave Opt. Technol. Lett. **33**(1) (2002)

12. Kim, H.M., Lee, B.: Bandgap and slow/fast-wave characteristics of defected ground structures including left-handed features. IEEE Trans. Microw. Theory and Tech. **54**(7), 3113–3120 (2014)
13. Lin, S.Y., Tian, W.Z., Zheng, S.Q., Sun, X.W.: A semicircle DGS with high Q factor for microstrip line and low-pass filter. Proceedings of Asia-Pacific Microwave Conference 2006, pp. 1197–1199

# Multi-Resonant Wide Band Rectangular Microstrip Antenna with U-Shape and Rectangular Slots

Amit A. Deshmukh, Poonam A. Kadam and Akshay Doshi

**Abstract** Wideband design of rectangular microstrip antenna using multiple slots on the radiating patch and ground plane at fundamental mode frequency of 1500 MHz is proposed. The slots tune $TM_{02}$ and $TM_{12}$ mode frequencies on radiating patch and $TM_{02}$ mode frequency of the ground plane, to yield wideband response. Bandwidth of more than 370 MHz (>22%) is realized in multi-resonant configuration. The proposed design gives radiation in bore-sight direction and shows peak gain of more than 8 dBi. Parametric formulations for different antenna parameters for the proposed design are presented. The antenna designed using them at different $TM_{10}$ mode frequencies yields broadband response. Thus, the proposed work clearly explains the functioning of multi-resonant slot cut antenna in terms of patch resonant modes, and formulations will be used to design similar antenna at specific $TM_{10}$ mode frequency.

**Keywords** Broadband microstrip antenna · Slot cut microstrip antenna Defected ground plane microstrip antenna · U-slot · Rectangular slot

## 1 Introduction

The slot cut method has been widely used ever since 1995, when U-slot cut design was reported for the first time to realize wideband response in microstrip antenna (MSA) [1]. For the same, different slots have been used, like U-slot, V-slot, rectangular slot and respective modified variations, like half U-slot, half V-slot, step half U-and V-slots [1–4]. In earlier literature, it was stated that slot introduces additional resonant mode other than the resonant modes governed by patch dimensions [1–4]. The new mode frequency when lies nearer to the patch fundamental mode frequency then it yields wider bandwidth (BW). For this to realize, depending upon the slot

---

A. A. Deshmukh (✉) · P. A. Kadam · A. Doshi
Department of Electronics and Telecommunication Engineering,
SVKM's, D J Sanghvi College of Engineering, Mumbai, India
e-mail: amit.deshmukh@djsce.ac.in; amitdeshmukh76@gmail.com

position, slot length is equated half-wave or quarter-wave in length [1–5]. However, recent more detailed study brings out that introduction of slot does not add to new resonating mode. It rather optimizes placing of higher modes with respect to fundamental mode that realizes wideband response [6]. In broadband slot cut MSAs, as higher order modes are present, surface current distribution over modified patch varies. Orthogonal variations in currents lead to higher cross-polar radiation pattern which also reduces antenna gain. Hence instead of cutting slot on patch, slots have been placed on the ground plane which leads to defected ground plane structure (DGS). By appropriately selecting the slot position on ground plane, the dual and wideband responses have been realized [7, 8]. However, in the reported work on DGS MSAs, an explanation for wideband response in terms of patch resonating modes is not given. It only provides parametric study for optimizing the design. Also reported work lacks in giving resonant length formulations for patch and ground plane modes introduced because of the slots.

In this paper, wideband design of multi-resonant slot cut rectangular MSA (RMSA) using slots on radiating patch and on the ground plane at fundamental frequency of 1500 MHz is proposed. The pair of rectangular slots was cut on RMSA as well as on the ground plane whereas U-slot was only cut on RMSA. These slots modify the resonance frequencies of $TM_{02}$ and $TM_{12}$ modes on RMSA and $TM_{02}$ mode on ground plane, and optimizes there spacing with respect to $TM_{10}$ mode on patch that yields wider BW. Here on substrate thickness of $0.06\lambda_0$, simulated and measured BWs are more than 370 MHz (>20%). As the slots are distributed on ground plane and radiating patch, this slot cut antenna yields broadside radiation pattern with cross-polar level of smaller than 12 dB with reference to co-polarization levels, over most of the BW. Due to this, it yields broadside co-polar gain of more than 7 dBi over an entire BW. Further based on optimum design in 1500 MHz frequency range, parametric formulation for various slot and patch parameters on patch and ground plane are proposed. The multiple slots cut RMSA is designed using these equations at fundamental mode frequency of 2500 MHz. The redesigned antenna also yields wideband response with formations of loops due to modified $TM_{10}$, $TM_{02}$ and $TM_{12}$ modes inside VSWR = 2 circle. Proposed formulations will be helpful to re-design similar multiple slots cut antenna at any given fundamental mode frequency on thicker substrate. The antenna proposed in this paper is first studied using simulations using IE3D software followed by the measurements. The input impedance, radiation pattern and gain measurements were carried out using high-frequency instruments like ZVH—8, SMB 100A and FSC—6, inside an Antenna lab.

## 2 Multiple Slots Cut Multi-Resonant RMSA

The suspended RMSA on air substrate of thickness $0.06\lambda_0$ is shown in Fig. 1a, b. For this thickness, RMSA length '$L$' and width '$W$' are calculated such that its $TM_{10}$ mode frequency is around 1500 MHz. Due to longer probe length, simulated

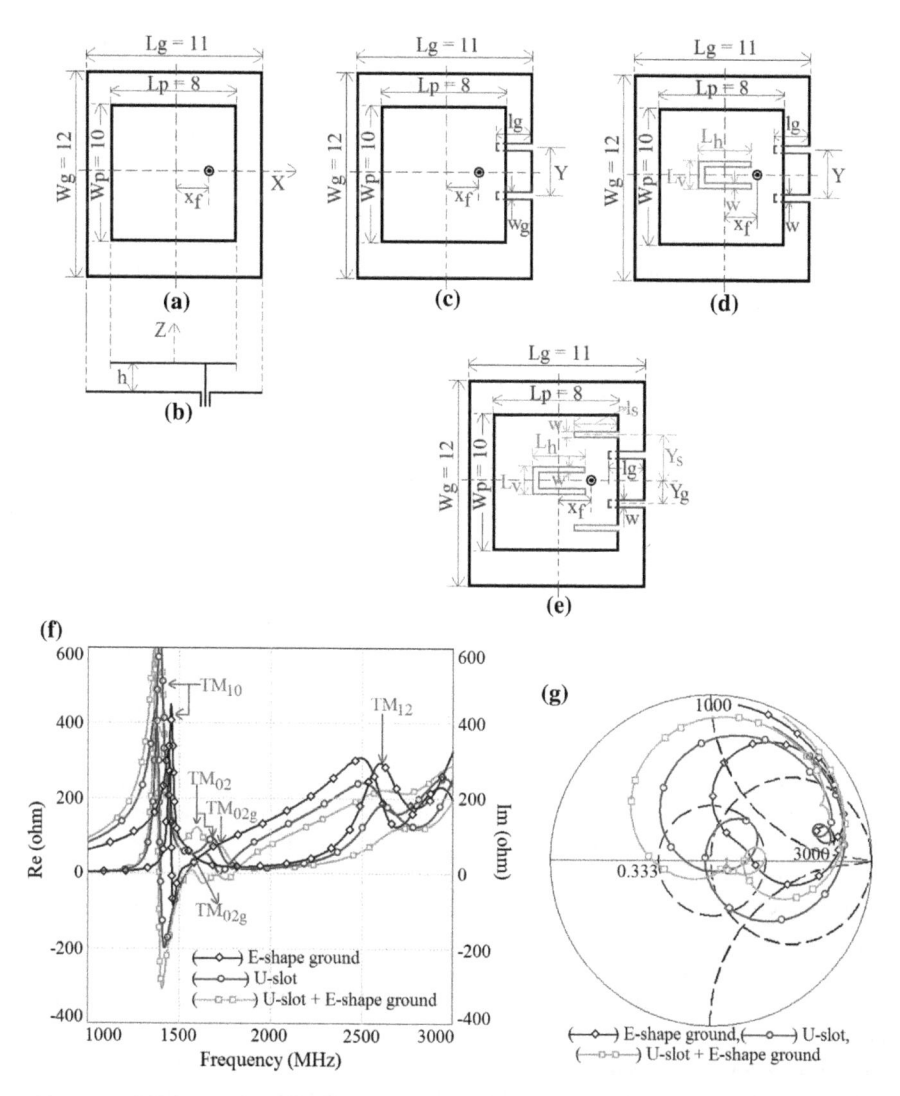

**Fig. 1** **a, b** Coaxially fed RMSA, **c** RMSA backed by E-shape ground plane, **d** U-slot cut RMSA backed by E-shape ground plane, **e** multiple slots cut RMSA backed by E-shape ground plane, and their **f, g** resonance curve plots and Smith charts

and measured BWs in air suspended coaxially fed RMSA are only 76 MHz (4.7%). To further enhance the BW, slot configurations of RMSA using slots are studied as shown in Fig. 1c, d. The RMSA backed by E-shape ground plane (wherein pairs of slots are cut on rectangular ground plane) yields simulated and measured BW of around 110 MHz ($\sim$7.2%). As this BW is lower U-slot cut design with E-shape ground plane is studied. Here first RMSA with U-slot was optimized for wideband characteristics and it gives BW of 250 MHz ($\sim$15%). Further enhancement in its

BW is obtained with embedding slots on ground plane. This design gives BW of around 280 MHz (17%). The simulated resonance curve and Smith charts for these optimum designs are shown in Fig. 1f, g.

In impedance locus for RMSA backed by E-shape ground plane, single loop (smaller in size) inside VSWR = 2 circle is present whereas for RMSA with U-slot, larger size single loop is present. The single loop in Smith chart indicates the presence of two resonant modes. For RMSA backed by E-shape ground plane, wideband response is due to coupling between $TM_{10}$ mode on patch and modified $TM_{02g}$ mode on ground plane. For RMSA with U-slot, wider BW is due to coupling between $TM_{10}$ and modified $TM_{02}$ modes on RMSA. In the impedance locus for RMSA with U-slot which is backed by E-shape ground plane, two loops are present. They are due to $TM_{10}$ and $TM_{02}$ modes on U-slot cut RMSA and $TM_{02g}$ mode on the ground plane. The next resonant peak in these slots cut RMSAs is $TM_{12}$. To further increase the BW of RMSA with U-slot and E-shape ground plane, $TM_{12}$ mode frequency is required to be tuned and this is obtained by cutting pair of slots on RMSA as shown in Fig. 1e. This slot position is selected such that it will realize maximum reduction in $TM_{12}$ mode resonance frequency. The resonance curve plots and Smith charts for increase in slot length '$l_s$', for '$Y_s$' = 3.0 cm and '$w$' = 0.4 cm is given in Fig. 2a–d.

The slot length '$l_s$' is orthogonal to current components at $TM_{12}$ mode; hence, its frequency reduces and also it comes nearer to $TM_{10}$, $TM_{02}$ and $TM_{02g}$ mode frequencies. These lower order modified resonant mode frequencies nearly remains unchanged. An effects of variations in slot position '$Y_s$' in multiple slots cut RMSA is also studied and resonance curve plots and Smith charts for the same with '$l_s$' = 3.5 cm are shown in Fig. 3a, b. The variation in position of slots marginally affects the resonance frequencies of each mode, but it modifies the input impedance at them. This leads to optimization of loop positions formed due to resonant modes nearer to VSWR = circle, to realize enhance BW. The optimum response is obtained for $L_h$ = 4.0, $L_v$ = 2.0, $l_s$ = 4.1, $Y_s$ = 7.8, $x_f$ = 1.3, $l_g$ = 3.0 and $Y$ = 4.8 cm, as shown in Fig. 3c.

Respective simulated and experimental BWs are 371 MHz (22.6%) and 390 MHz (24.3%). Deviation in two values is attributed to the experimental errors. Variation in gain in broadside co-polar direction and pattern at two band edge frequencies are provided in Figs. 3d and 4a–d, respectively. Pattern exhibits maxima in the bore-sight direction which yields gain of above 6 dBi over entire BW. Fabricated antenna configuration is shown in Fig. 4e, f.

Further based upon the optimum design at 1500 MHz, formulations for patch and various slot parameters are realized. For air substrate of thickness ($h$) $0.06\lambda_0$, at a given $TM_{10}$ mode frequency, RMSA length ($L_p$) is calculated by using Eq. (1). For higher MSA gain, RMSA width ($W_p$) is selected as 1.25 times its length. Dimensions of ground plane are selected as '$L_g$' = 1.375 L and '$W_g$' = 1.2 W. This length and width mainly affect $TM_{02}$ mode frequency of grounded rectangular patch. The dimension of ground plane slots are taken as, '$I_g$' = $0.15\lambda_0$, '$Y_g$' = $0.12\lambda_0$. The U-slot dimensions are selected as, '$L_h$' = $0.2\lambda_0$, '$L_v$' = $0.1\lambda_0$,

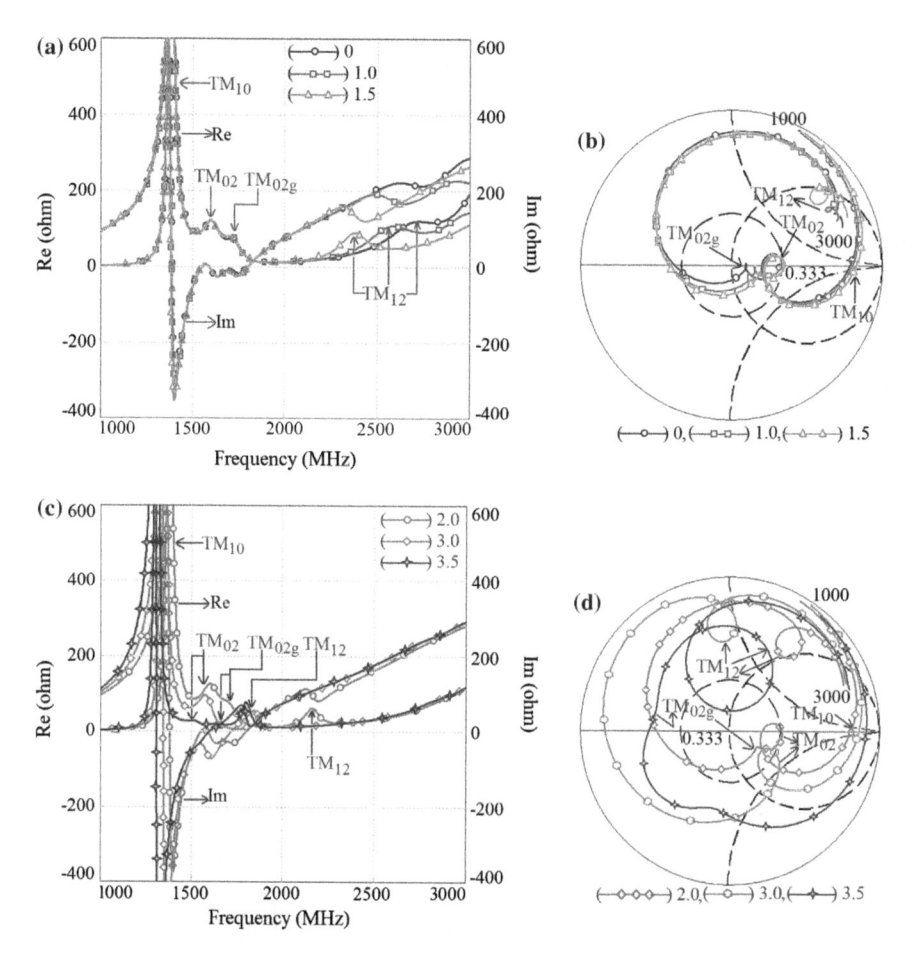

**Fig. 2** a–d Impedance curves and Smith charts with variations in slot length $l_s$' for multiple slots cut RMSA backed by E-shape ground plane

and '$w$' = $0.02\lambda_0$, whereas rectangular slot length and its position on the radiating patch are, '$l_s$' = $0.205\lambda_0$, '$Y_s$' = $0.195\lambda_0$. The coaxial feed is placed at distance of '$x_f$' = $0.065\lambda_0$, from the patch centre.

$$L_p = (30/2f_r) - 2(0.7h) \quad (1)$$

where

'$f_r$' is in GHz,
'$h$' is in cm and
'$L_p$' is in cm.

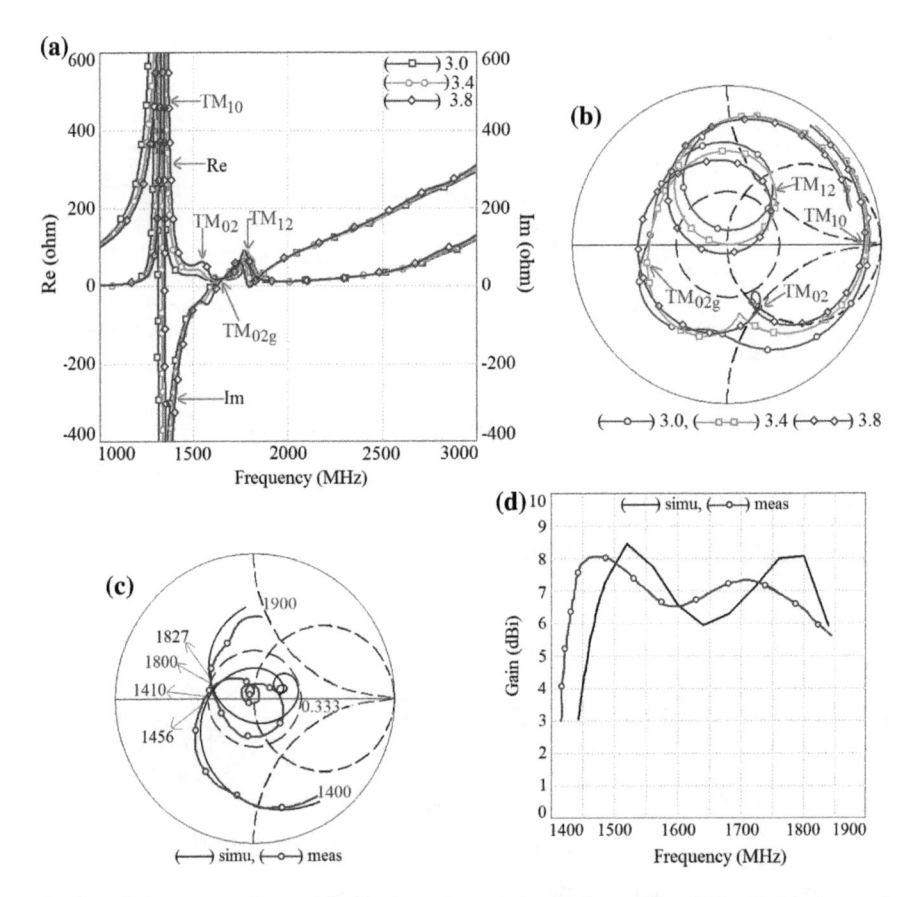

**Fig. 3** **a, b** Resonance plots and Smith charts for variation in slot position '$Y_s$', **c** Smith chart and **d** gain variation over BW for multiple slots cut RMSA backed by E-shape ground plane

The multiple slots cut RMSA is designed using above equations for $TM_{10}$ = 2500 MHz, and the realized configuration giving relevant dimensions and input impedance plots is provided in Fig. 5a, b. On 0.7 cm thicker air substrate, loops formed due to coupling between patch's modified resonating modes, lies completely inside VSWR = 2 circle. For slotted design, respective simulated and experimental BWs are 589 MHz (21.97%) and 618 MHz (23.4%). Redesigned antenna also yields broadside radiation pattern over entire BW with peak gain of more than 8 dBi. Thus, the proposed equation can be used to realize multiple slots cut antenna at given $TM_{10}$ mode frequency on thicker air substrate.

Thus, the proposed work in this paper gives a detailed explanation about effects of multiple slots cut in the rectangular patch that yields broader BW. Similar study explaining the effects of multiple slots (multi-resonant antenna) in defected ground

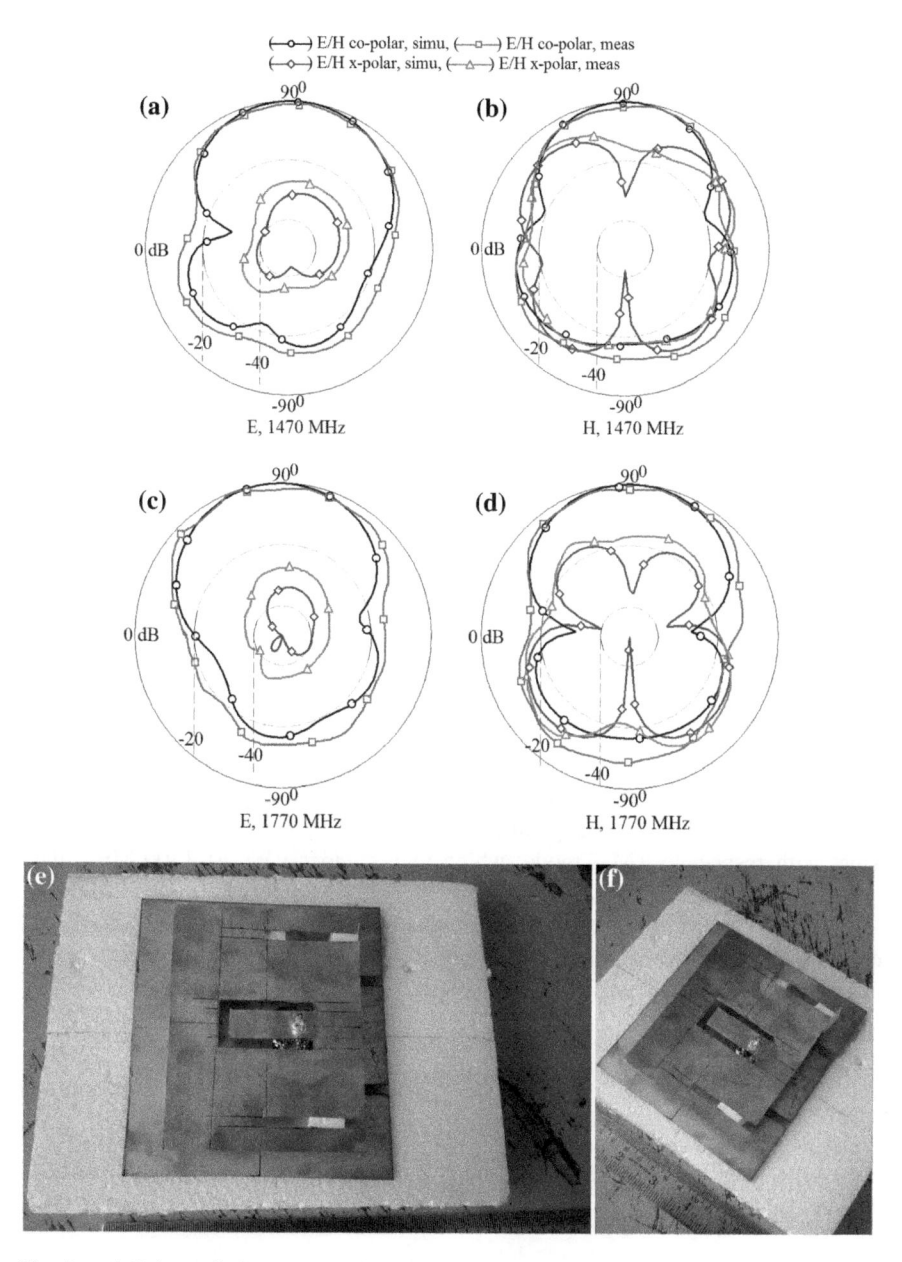

**Fig. 4 a–d** Polar radiation pattern plots at two band edge frequencies, and **e, f** fabricated prototype for multiple slots cut RMSA backed by E-shape ground plane

plane structure MSA is not reported in the available work, and thus, this is the novelty here. Also formulations for patch and slot parameters will be helpful in redesigning of similar antenna at specific $TM_{10}$ mode resonant frequency.

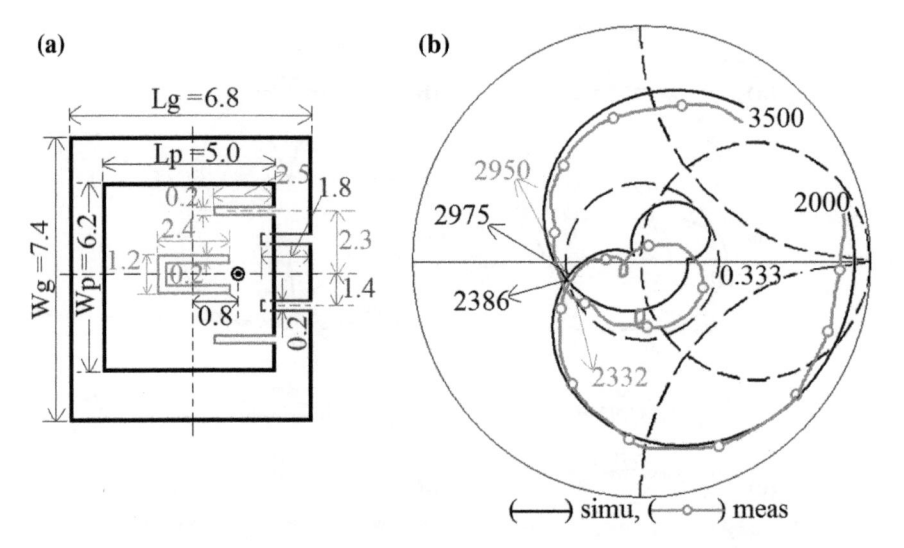

**Fig. 5 a** Multiple slots cut RMSA backed by E-shape ground plane and its **b** input impedance plots at $f_{TM10}$ = 2500 MHz

## 3 Conclusions

Novel design of multiple slots cut RMSA that is backed by modified E-shape ground plane is proposed. The slots tune the resonance frequency of $TM_{02}$ and $TM_{12}$ resonant modes on rectangular patch and $TM_{02}$ mode on rectangular ground plane, with respect to $TM_{10}$ mode, which yields wider BW. At 1500 MHz, the simulated and measured BWs of higher than 350 MHz ($\sim$22%) are realized. Design formulations to realize similar antenna at different frequency are also presented. Using them when antenna is designed on thicker air substrate yields wider BW with formation of loops inside VSWR = 2 circle in the Smith chart.

## References

1. Kumar, G., Ray, K.P.: Broadband Microstrip Antennas, 1st edn. Artech House, Norwood (2003)
2. Wong, K.L.: Compact and Broadband Microstrip Antennas, 1st edn. Wiley Inc, New York (2002)
3. Lee, H.F., Chen, W.: Advances in Microstrip and Printed Antennas. Wiley, New York (1997)
4. Wong, K.L., Hsu, W.H.: Broadband triangular microstrip antenna with U-shaped slot. Electron. Lett. **33**(25), 2085–2087 (1997)
5. Ghalibafan, J., Attari, A.R.: A new dual-band microstrip antenna with U-shaped slot. Progress Electromagn. Res. C **12**, 215–223 (2010)
6. Deshmukh, A.A., Ray, K.P.: Analysis of broadband variations of U-slot cut rectangular microstrip antennas. IEEE Magazine Antennas Propag. **57**(2), 181–193 (2015)

7. Chakraborty, U., Chowdhury, S.K., Bhattacharjee, A.K.: Frequency tuning and miniaturization of square microstrip antenna embedded with "t"-shaped defected ground structure. Microwave Opt. Technol. Lett. **55**(4), 869–872 (2013)
8. Mondal, K., Sarkar, P.P.: M-shaped broadband microstrip patch antenna with modified ground plane. Microwave Opt. Technol. Lett. **57**(6), 1308–1312 (2015)

# Spanner Shape Monopole MIMO Antenna with High Gain for UWB Applications

Vandana Satam, Shikha Nema and Sanjay Singh Thakur

**Abstract** A compact planar two-element spanner shape ultra-wideband multiple input multiple output antenna for UWB applications is proposed. The investigated MIMO antenna comprises two spanner shape radiators separated by distance of 0.25λ with microstrip line feed. The monopole configuration is used to achieve UWB requirement. The antenna prototype is fabricated using FR4 substrate. The antenna is electrically small in size (59.5 mm × 52 mm). The final design of monopole MIMO antenna has return loss less than −10 dB and isolation below −15 dB over the range of 2–12 GHz covering ultra-wideband frequency range. The correlation coefficient between two antenna elements is less than 0.02. The peak gain of antenna is 6.79 dBi for wireless application with efficiency around 90%. The significant improvement in bandwidth, return loss, VSWR, gain, efficiency and envelope correlation coefficient is observed in investigated antenna in comparision with single-element spanner shape antenna. Simulated results are in accordance with measured one.

**Keywords** Coupling · Diversity · Envelope correlation coefficient
MIMO · Monopole · Spanner · UWB

---

V. Satam (✉)
K. J. Somaiya COE, Mumbai, India
e-mail: vandanam@somaiya.edu; vandana.anerao@gmail.com

V. Satam
UMIT, SNDT Women's University, Mumbai, India

S. Nema
Department of Electronics and Communication, UMIT, SNDT Women's University, Mumbai, India

S. S. Thakur
Electronics and Telecommunication Department, VIT, University of Mumbai, Wadala, Mumbai, India

© Springer Nature Singapore Pte Ltd. 2018
H. Vasudevan et al. (eds.), *Proceedings of International Conference on Wireless Communication*, Lecture Notes on Data Engineering and Communications Technologies 19, https://doi.org/10.1007/978-981-10-8339-6_15

# 1 Introduction

In recent wireless systems issues related to bandwidth, capacity, speed, reliability, interference, cost, size, etc. plays a very important role. As antenna is a resonant device, it operates over certain frequency band. Tuning of an antenna according to wireless system to which it is connected must be vital part otherwise reception and/ or transmission will become weak. Higher data rates for mobile users are essential for improvement in multimedia applications. Various wireless standards are used to achieve data rate improvement. MIMO is the advance technique utilized to increase the data rates and throughput for advances in 4G wireless standards. Multiple antennas can be installed on both source and load sides to increase channel capacity. MIMO technology is a new prototype to achieve very high bandwidth, gain and large data rates in modern wireless communications [1]. Based on channel configurations multiple antennas should be placed in nearby environment and can be operated in several modes simultaneously. The planar monopole antenna is a good candidate for portable UWB devices instead of conventional UWB antennas, such as horn, biconical and log periodic antennas because of their bulky size and directional properties. UWB planar antennas are most suited for wireless communication because of their compact size. Due to simple construction and broadband characteristics monopole antennas are good option for planar devices. Ultra-wideband technology operates from 3.1 to 10.6 GHz which finds a large number of applications in the field of wireless communication [2]. UWB technology along with MIMO is a good combination for short-range communications and wireless applications which requires device to transmit very low power [3]. Theoretically, it has been proven that MIMO channel capacity is totally dependent on mutual coupling. As coupling among antenna elements increases, there is reduction in efficiency because of mismatching of impedance. Hence, the mutual coupling reduction becomes very crucial in any MIMO wireless system. Another reason for reduction in radiation efficiency is loss of power from the transmitted antenna to the port of ideal antenna. Therefore, the need of good isolation is important as well. MIMO antenna can be used in different configurations to provide a significant improvement in performance metrics. Hence from last few decades, research is going on different techniques to reduce mutual coupling and providing high isolation between MIMO antenna systems. Use of decoupling and matching networks for reducing mutual coupling and improvement in isolation is highlighted [4]. The prototype consisting of an electromagnetic band gap (EBG) structure consists of an array of metal flange on a flat metal sheet. Improvement in port isolation can be achieved by EBG structures by suppressing surface waves propagating in a frequency band [5]. Miniaturized UWB printed antenna with slot for diversity applications is reported [6]. The antenna consists of staircase-shaped radiating elements for orthogonal radiation patterns with coplanar waveguide (CPW) feeding technique, where a rectangle stub is placed at an angle of 45° between the CPW to make high isolations. A dual-band MIMO antenna along with two back-to-back monopoles is presented, and the author proposed fractal radiating antenna with L shape stub on ground plane suitable for UWB application gives a maximum gain of 5.5 dBi [7]. To enhance the bandwidth of planar antennas, printed monopole antennas have been investigated [8]. In this article, a compact two-element spanner shape MIMO UWB antenna is presented with very low envelope correlation coefficient for wireless application as compared to earlier published configuration [5–9]. In literature, various structures found which includes single band monopole structures with single band, multiband and ultra-wideband antennas [9–11]. Triangular, square, pentagonal and hexagonal radiation structures have already been presented but hexagonal shape has shown better impedance matching with microstrip line feed and hence gives bandwidth enhancement [12, 13]. The partial background plane, microstrip line feed and the lower edge of the radiating patch is forming quarter wavelength transformer therefore hexagonal radiator has shown optimum impedance matching as compared to triangular, rectangular and pentagon radiator [8, 12–14]. Hexagonal shape radiator is modified as spanner shape to attain good isolation and excited by using microstrip line feed. Monopole structure gives ultra-wideband characteristics. The said antenna has a wide range of applications in the field of wireless communication, military applications and short-range RADAR application.

## 2 Proposed Antenna Configuration

Figure 1 shows the configuration of spanner shape MIMO antenna with high isolation.

Rather than resonance frequency, the lower edge frequency is calculated for printed monopole antenna [8, 12–14].

For printed hexagonal monopole antenna, lower edge frequency can be calculated from Eq. (1).

**Fig. 1** Structure of spanner shape monopole MIMO antenna

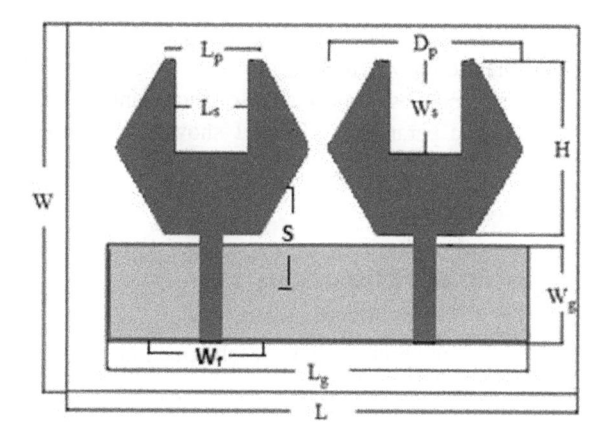

Table 1 Optimized dimensions of proposed antenna

| Parameter | L | W | $L_g$ | $W_g$ | S | $W_s$ |
|---|---|---|---|---|---|---|
| Dimensions (mm) | 59.5 | 52 | 56.5 | 12.5 | 1.5 | 12 |
| Parameter | $W_f$ | $D_p$ | H | $L_p$ | $L_s$ | – |
| Dimensions (mm) | 2.8 | 26 | 22.5 | 13 | 10 | – |

$$f_L = \frac{7.2}{(L+r+p)} \quad (1)$$

$$L = \sqrt{3}H \quad (2)$$

$$r = \frac{3H}{4\pi} \quad (3)$$

p  Length of the 50 Ω feed microstrip line in cm and
H  Side length.

The prototype of the antenna is fabricated on FR4 substrate with dielectric constant 4.4, height of substrate 1.59 mm and loss tangent 0.02. The dimensions of the proposed antenna are 59.5 mm × 52 mm. The hexagonal shape two radiators place parallel to each other and excited by microstrip line of 50 Ω line impedance. The hexagonal shape radiator is modified as spanner shape by etching slot at centre. Current vectors are present on the edges of hexagonal monopole; therefore, central part of radiating patch is removed, and hence, spanner shape is being formed. This reduces actual area of an antenna. Here hexagonal radiator created using empirical formula [14]. The basic characteristics of a linear antenna array for which the element antenna patterns in isolation are identical and similarly oriented. To achieve good isolation characteristics distance between two radiators kept 0.25λ [15].This reduces coupling between elements. Slot of $L_s \times W_s$ created on hexagon which results reduction in patch area and in making antenna low profile. $L_g \times W_g$ partial ground plane with microstrip line feed and the lower part of hexagon are provided to characterize the λ/4 transformer matching network. It increases effective current path and introduces additional resonance bands. IE3D software is used for simulating antenna parameters. Table 1 shows the optimized dimensions for proposed antenna.

## 3  Result and Discussion

The top view and bottom view of fabricated prototype are as shown in Fig. 2. IE3D software is used for optimizing the desired configuration. The fabricated prototype was tested with Agilent N9917A (18 GHz) FieldFox Microwave Analyzer. Furthermore, it is noticed that there is agreement between simulated and measured

**Fig. 2** Top view and bottom view of fabricated prototype of MIMO antenna

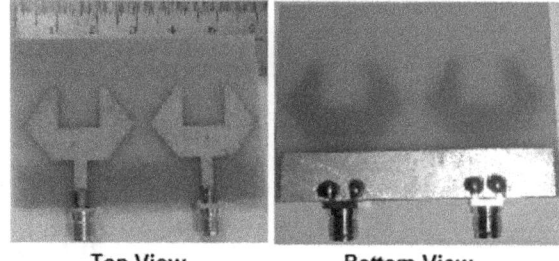

Top View          Bottom View

results. Simulated and measured results of S-parameters are satisfactory. A small variation in result is due to calibration of equipment, connector losses and material properties of fabricated prototype.

## 3.1 Impedance Performance

Figure 3a shows the return loss as a function of frequency. The antenna achieved an impedance bandwidth of 2–12 GHz for $|S_{11}| < -10$ dB. Because of similarity of structure, port 2 also shows similar response $|S_{22}|$ as that of port 1. The other port is terminated with 50 Ω matched load while measuring results. The isolation characteristic within impedance bandwidth is greater than −15 dB as shown in Fig. 3b. The proposed antenna is resonating at different frequencies which show the utilization of designed antenna for various applications like ISM, wireless system, Bluetooth, short-range RADAR, etc. The presented UWB MIMO antenna gives desired impedance characteristics in UWB operating range thus considered to be good candidate for portable UWB MIMO system. The shift between simulated and measured results due to tolerance in substrate material specification, i.e. dielectric constant, losses in cables and connectors.

## 3.2 Current Distribution

Scalar surface current distribution at different frequencies after exciting port 1 is shown in Fig. 4. The current magnitude scale was kept constant for all cases. Most of the energy supplied to monopole couples strongly near the periphery of antenna called active zone. There is very little current flow within the interior part of second radiated antenna element. It has been observed that current distribution is less near non-feeding antenna called neutral zone. In neutral zone, current distribution is very small but non zero. From Fig. 4, it has been seen that the neutral zone for each frequency is found at different locations.

**Fig. 3** **a** Simulated and measured return loss ($S_{11}$), and **b** simulated and measured isolation loss ($S_{12}$)

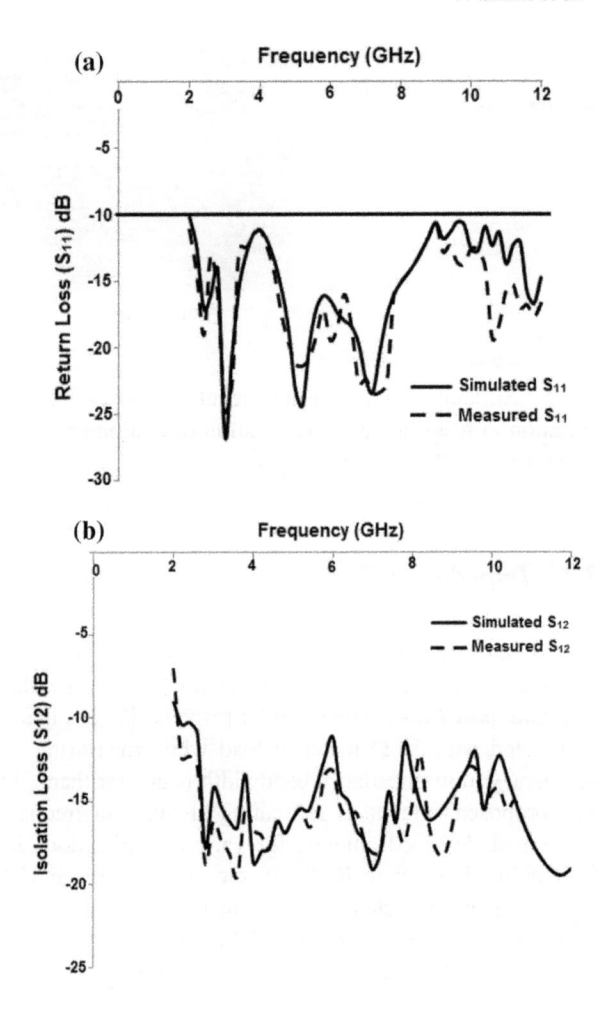

## 3.3 Radiation Performance

Radiation pattern in E-plane and H-plane at different resonance frequencies is shown in Fig. 5. At lower frequencies H-plane radiation patterns are omnidirectional radiation pattern is observed in H-Plane, it is a figure of eight in E-Plane. In addition, at high frequency H-plane pattern shows omnidirectional pattern without any disturbance. Also, E-plane pattern shows good directional properties of antenna at high frequency leading to good pattern diversity. The radiation characteristic shows peak gain of 6.79 dBi at 5.2 GHz frequency in Fig. 6 which is lower band for wireless applications. The peak gain across UWB band varies from 3.65 to 4.79 dBi. The radiation efficiency is above 60% across UWB with maximum value

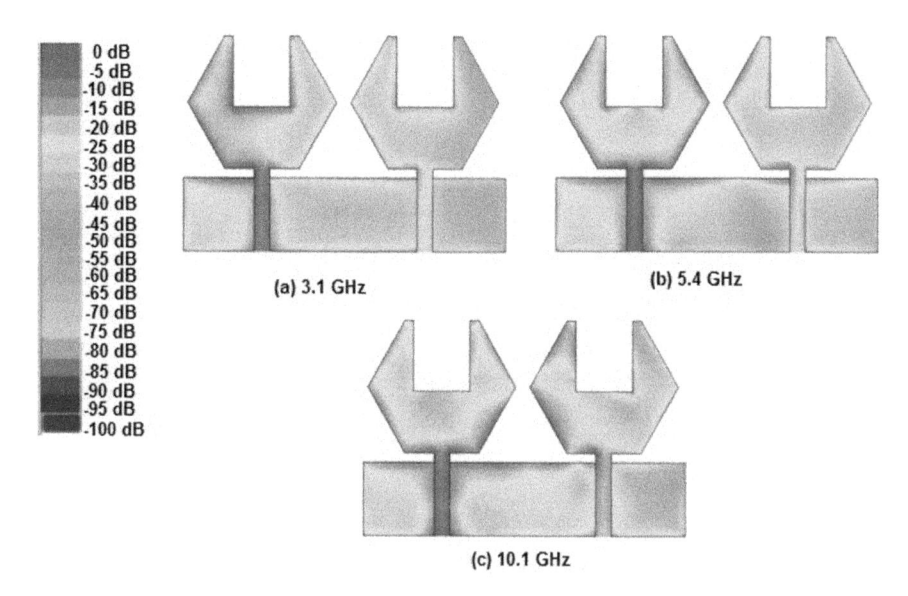

**Fig. 4** Distribution of surface current on structure of MIMO antenna at various frequencies

90% at 5.2 GHz. Simulation study can be made for comparing gain of single-element spanner shape antenna to that of proposed antenna, i.e. MIMO antenna. It is observed from Fig. 6 that gain of MIMO antenna is relatively high compared to single-element antenna.

## 3.4 Diversity Performance

The envelope correlation coefficient (ECC) is an important parameter to estimate diversity performance of MIMO antenna. Preferably, diversity antennas used in wireless systems require zero correlation coefficient but in practical applications correlation coefficient nearer to zero can be achieved. Calculation of ECC of diversified antenna can be done either by radiation pattern measurement in far field or using scattering parameter method. In this paper, ECC is calculated using S-parameters [23].

$$\rho_{21} = \frac{|S_{11}^*S_{21} + S_{21}^*S_{22}|^2}{(1 - |S_{11}|^2 - |S_{21}|^2)(1 - |S_{22}|^2 - |S_{12}|.^2)} \quad (4)$$

Since ECC assumes mutual effects between antenna elements hence it is a better parameter in comparison with reflection coefficient. By using $S$ parameter equation, proposed antenna gives ECC value less than 0.02 over ultra-wideband spectrum at different resonating frequencies. This signifies UWB MIMO antenna has good

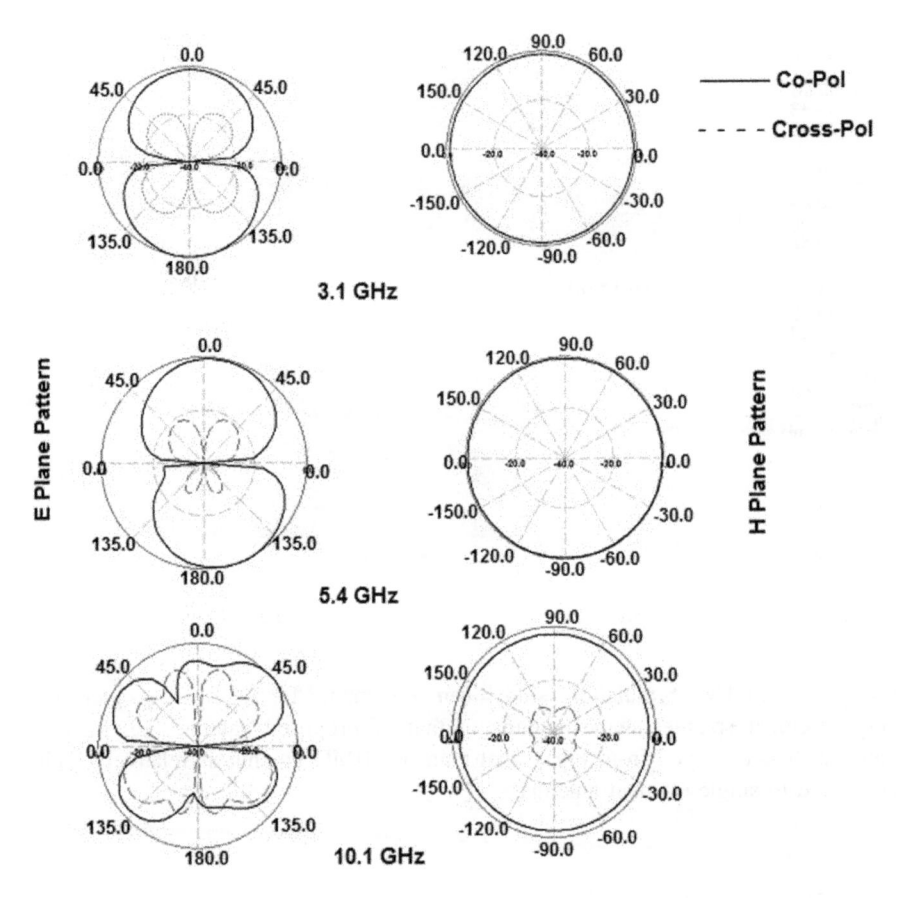

**Fig. 5** Radiation characteristics of MIMO antenna at different resonating frequencies

diversity performance. The diversity gain of the diversity antenna can be calculated approximately using the following formula (5):

$$DG = \sqrt{1 - \rho_e} \tag{5}$$

Since ECC is near zero, hence, diversity gain throughout require band is around 10 dB.

## 3.5 Comparison with the Literature

On comparing gain of proposed antenna with existing in literature [16–23], proposed antenna has better peak gain compare to existing antennas in literature leading to good directional properties.

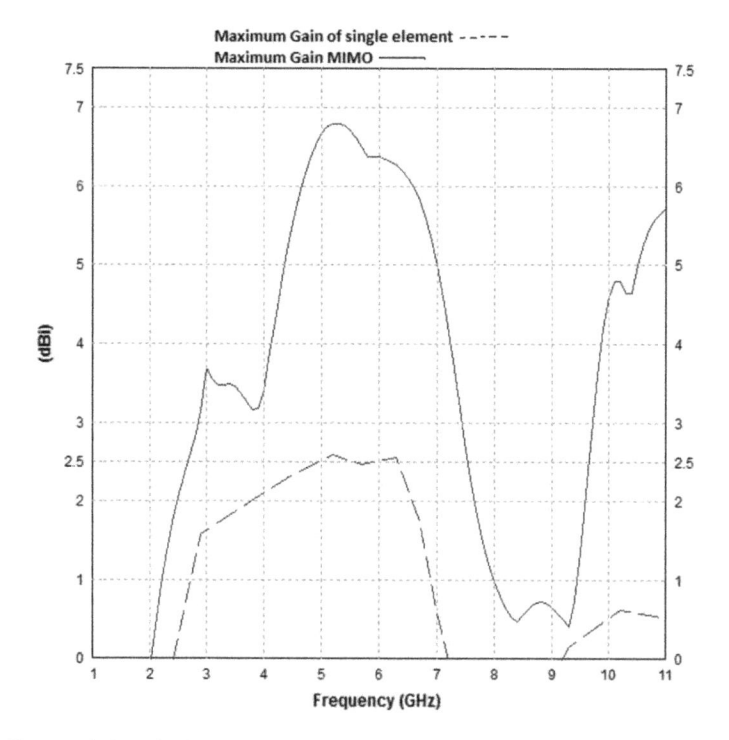

**Fig. 6** Characteristics of gain of MIMO antenna compare with single-element structure

## 4 Conclusion

A miniaturized spanner shape monopole MIMO antenna with low correlation coefficient is proposed for UWB application. The etched slot contributed to spanner shape leading to miniaturization of antenna. Isolation between two radiating elements is increased by keeping space of 0.25 $\lambda$. The ultra-wideband characteristics are achieved due to monopole structure. The proposed antenna has a compact dimension of 59.5 mm × 52 mm. There is a good agreement between simulated and measured results. The proposed antenna has more advantages in terms of gain compare to other structures reported in literature. The study of ECC of proposed antenna shows that the proposed antenna system is a good candidate for UWB MIMO/diversity applications.

**Acknowledgements** The authors would like to thank Sameer, IIT Bombay for providing testing facility for this work using vector network analyzer.

# References

1. Najam, A., Duroc, Y., Tedjni, S.: Ultra Wideband Current Status and Future Trends (chapter 10), pp. 209–236. Intech publishers
2. Federal Communications Commission (FCC): Revision of Part 15 of the Commission's Rules Regarding Ultra-Wideband Transmission Systems FCC. FCC Technical Report, ET-Docket FCC02-48, pp. 98–153, Feb 2002
3. Najam, A., Duroc, Y., Tedjni, S.: UWB-MIMO antenna with novel stub structure. Prog. Electromagnet. Res. C **19**, 245–257 (2011)
4. Dossche, S., Blanch, S., Romeu, J.; Optimum antenna matching to minimize signal correlation on a two-port antenna diversity system. IEEE Electron. Lett. 1164–1165 (2004)
5. Sievenpiper, D., Zhang, L., Broas, R.F.J., Alexopolous, N.G., Yablonovitch, E.: High impedance electromagnetic surfaces with a forbidden frequency band. IEEE Trans. Microw. Theor. Tech. 2059–2074 (1999)
6. Gao, P., He, S., Wei, X., Xu, Z., Wang, N., Zheng, Y.: Compact printed UWB diversity slot antenna with 5.5-GHz band-notched characteristics. IEEE Antennas Wirel. Propag. Lett. 376–379 (2014)
7. Yuan, D., Zhengwei, D., Gong, K.: A novel dual-band printed diversity antenna for mobile terminals. IEEE Trans. Antennas Propag. 2088–2096 (2007)
8. Ray, K.P., Thakur, S.S., Kakatkar, S.S.: Bandwidth enhancement techniques for printed rectangular monopole antenna. IETE J. Res. **60**(3), 249–256 (2014)
9. Menzel, W., AI-Tikriti, M., Lopez, M.B.E.: Common aperture, dual frequency printed antenna (900 MHz and 60 GHz). IEEE Electron. Lett. **37**(17), 1059–1060 (2001)
10. Chiu, C.Y., Shum, K.M., Chan, C.H., Luk, K.M.: Bandwidth enhancement technique for quarter-wave patch antennas. IEEE Antennas Wirel. Propag. Lett. **2**, 130–132 (2003)
11. Wong, K.L, Lin, Y.C, Tseng, T.C.: Thin internal GSM/DCS patch antenna for a portable mobile terminal. IEEE Trans. Antennas Propag. **54**(1), 238–241 (2006)
12. Ray, K.P., Thakur, S.S., Deshmukh, R.A.: Broadbanding a printed rectangular monopole antenna, In: Proceedings of IEEE Conference on Applied Electromagnetic, pp. 1–4, Kolkata-India, Dec (2009). https://doi.org/10.1109/aemc.2009.5430695
13. Ray, K.P., Thakur, S.S.: Ultra wide band vertex truncated printed pentagon monopole antenna. Microw. Opt. Technol. Lett. **56**, 2228–2234 (2014)
14. Ray, K.P.: Design aspect of printed monopole antennas for ultra-wide band application. Int. J. Antennas Propag. (Hindawi Publishing Corporation) **2008**, Article ID 713858
15. Mandal, T., Das, S.: Microstrip feed spanner shape monopole an tennas for ultra wide band applications. J. Microw. Opto-Electron. Electromagnet. Appl. 15–22 (2013)
16. Ahmed, O., Sebak, A.R.: Mutual coupling effect on ultrawideband linear antenna array performance. Int. J. Antennas Propag. (Hindawi Publishing Corporation) **2011**, Article ID 142581
17. Ren, J., Mi, D., Yin, Y.-Z.: Compact ultra wideband MIMO antenna with WLAN/UWB bands coverage. Prog. Electromagnet. Res. C **50**, 121–129 (2014)
18. Jian, R., Wei, H., Yingzeng, Y., Rong, F.: Compact printed MIMO antenna for UWB applications. IEEE Antennas Wirel. Propag. Lett. 1517–1520 (2014)
19. Moosazadeh, M., Kharkovsky, S.: Compact and small planar monopole antenna with symmetrical L-and U-shaped slots for WLAN/WiMAX applications. IEEE Antennas Wirel. Propag. Lett. 388–391 (2014)
20. Tripathi, S., Mohan, A., Yadav, S.: A compact octagonal fractal UWB MIMO antenna with WLAN band rejection. Microw. Opt. Technol. Lett. 1919–1925 (2015)
21. Qin, H., Liu, Y.F.: Compact UWB MIMO antenna with ACS-fed structure. Prog. Electromagnet. Res. 29–37 (2014)
22. Toktas, A., Akdagli, A.: Compact multiple-input multiple-output antenna with low correlation for ultra-wide-band applications. Microwaves Antennas Propag. IET **9**, 822–829 (2015)
23. Satam, V., Nema, S: Dual notched, high gain diversity antenna for wide band applications. Microw. Opt. Technol. Lett. **59**, 1222–1226 (2017)

# Analysis of Multi-resonant Rectangular Microstrip Antenna Embedded with Multiple Slots

Amit A. Deshmukh, A. P. C. Venkata and Aarti G. Ambekar

**Abstract** By introducing slot at an appropriate position on the rectangular patch, wider bandwidth has been realized. Multi-resonator design of rectangular microstrip antenna embedded with rectangular and half U-slot is reported that yields bandwidth of 280 MHz (28%). In this paper, a detailed resonant modal analysis is presented for the above multi-resonant, multiple slots cut rectangular microstrip antenna. The present study shows that combination of half U-slot and rectangular slot reduces the resonance frequencies of $TM_{01}$ and $TM_{11}$ modes with reference to $TM_{10}$ mode and also optimizes the impedance at them, which exhibits loop formations inside the VSWR = 2 circle, showing a wider bandwidth. Slots also modify current distributions at those modes, to yield broadside pattern. Further based upon optimized reported design, formulations for various patch parameters for multi-resonant slot cut antenna are presented. Antennas designed using them at different resonant mode frequencies yield similar wideband response.

**Keywords** Compact microstrip antenna · Broadband microstrip antenna Higher order resonant modes · Compact half U-slot · Rectangular slot

## 1 Introduction

Various configurations of compact microstrip antenna (MSA) can be obtained by making few changes in regular shapes like rectangular, circular and triangular by using shorting post or by cutting slots [1]. Without using shorting post or slot, compact MSA can be obtained using symmetry of these regular shape patches around feed point axis [1]. In case of such compact designs, the frequency of fundamental mode remains nearly the same in circular and triangular patch; however, it increases in rectangular MSA (RMSA) [1]. This increase in RMSA

---

A. A. Deshmukh (✉) · A. P. C. Venkata · A. G. Ambekar
Department of Electronics and Telecommunication Engineering,
SVKM's, D J Sanghvi College of Engineering, Mumbai, India
e-mail: amit.deshmukh@djsce.ac.in; amitdeshmukh76@gmail.com

frequency is attributed to dependency of resonance frequency on both the patch dimensions, i.e. length and width [1]. Since it was invented for the first time in 1995 [2], very frequently used method to enhance MSA's VSWR bandwidth (BW) is the use of slot. Over last two decades, many configurations have been reported using different variations of slots like U-slot, rectangular slot, V-shape slot and their modified variations [2–8]. In earlier reported work on slot cut MSAs, it was stated that slot introduces new resonant mode when its length is equated to half- or quarter-wave in length. When dimensions of patch are fixed then with respect to the boundary conditions as applied to patch geometry and further to the solution of wave equation applied to these conditions, number of modes in the structure is fixed [9]. The modifications in patch (in terms of slots) dimensions will only result in relative variations in individual modal frequencies. Therefore based on this theory, the detailed modal study of U-slot and two U-slots embedded RMSAs as reported in [2] and [3] was reported in [10]. It shows that slot reduces mode's resonant frequency, further optimizes input impedance variation at higher order orthogonal mode, to give wideband response [10]. Further symmetry in MSA embedded with slots have been used to realize compact configuration, e.g. compact and broadband RMSA with half U-slot cut was realized from RMSA embedded with U-slot [11, 12]. Further increase in BW of RMSA embedded with half U-slot is obtained by cutting rectangular slot on patch radiating edge [13]. As against half U-slot design, this multi-resonator design yields additional 14% BW with identical gain and pattern responses [13]. However in [13], in multi-resonant multiple slots cut design, which of the patch modes that yields wider BW in compact configuration were not mentioned specifically. Also, design formulation which will help in realize similar antenna for to have wideband response at different frequencies is also not provided.

In this paper, a study highlighting on resonant mode behaviour of RMSA embedded with half U-slot and also a rectangular slot is presented. First RMSA with only half U-slot was analysed which was followed by analysis with rectangular slot. Using this analysis it was observed that two slots reduce the resonance frequencies of $TM_{01}$ and $TM_{11}$ modes in the RMSA. Also, they optimize the impedance variations at them which results in wideband response. Further in the proposed work, with respect to optimized reported design of multiple slots cut antenna, formulations for various patch parameters in terms of $TM_{10}$ mode operating wavelength of un-slotted patch are presented. The antennas were designed using these equations at different $TM_{10}$ mode frequencies on thicker air substrate. The redesigned antenna shows wideband response showing loop formation inside VSWR = 2 circle. Thus, the proposed work in this paper gives detailed insight into the functioning of antenna in terms of patch resonating modes, and further, it gives design formulations. They will be useful for redesigning the MSA at different frequencies. The work presented in following sections was analysed using simulations by using IE3D, on infinite ground plane. Here choice of infinite ground was selected for reducing the computational time in the analysis. With finite ground plane, only impedances at various resonant modes will get affected but the modes involved in wideband response will not get changed.

## 2 Broadband Multi-resonant Multiple Slots Cut RMSA

The reported configuration of multiple slots cut multi-resonant antenna is given in Fig. 1a, b [13]. Reported antenna optimized using air substrate of thickness 23 mm, gives simulated and measured BW of around 280 MHz (28%) as given in Fig. 1c [13]. Wideband response is reported due to optimum mutual coupling between two slots. In this design, RMSA with half U-slot yields BW of 135 MHz (14.8%) whereas RMSA with rectangular slot gives 130 MHz (14.5%) of BW. Thus in their combination, reported multiple slots cut RMSA yields 14% additional BW. In [13], an explanation for wideband response by showing variation in modal frequencies using resonance curve plots is shown, but modal identification is missing. Also work in [13] does not give any formulation for modes or any design guidelines which can be helpful in redesigning of identical antenna at other frequencies. Thus, this is the research gap present for the work reported in [13]. The antenna reported

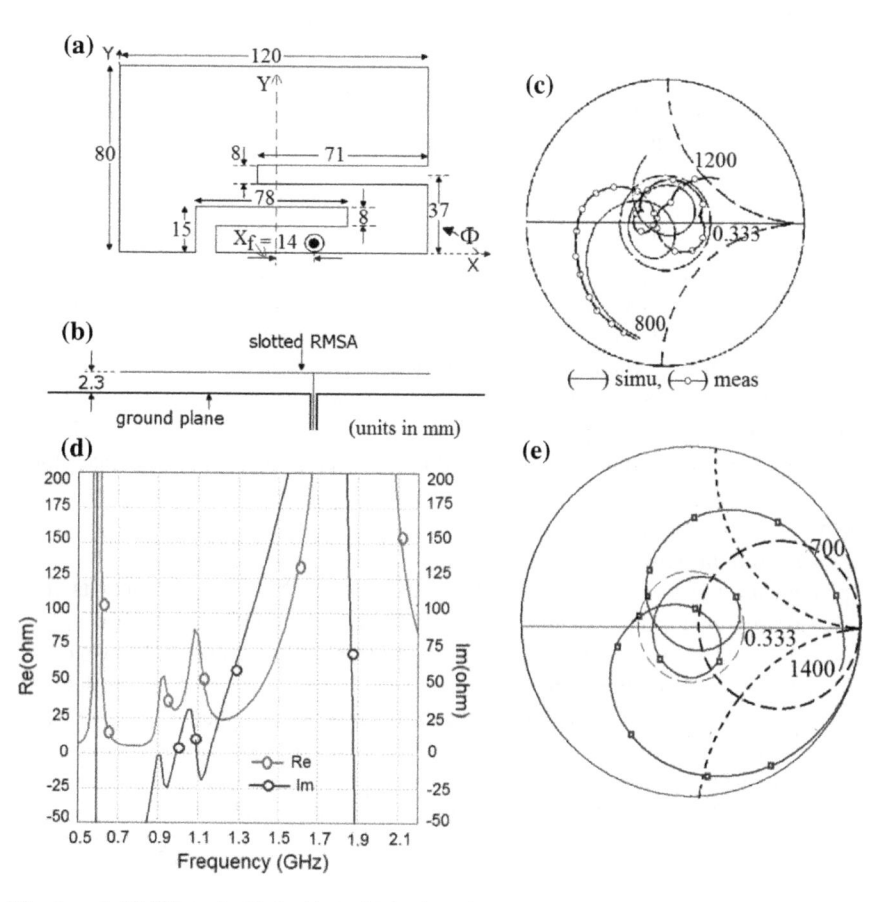

**Fig. 1** **a**, **b** RMSA embedded with multiple slots, its **c** measured Smith chart [13], **d** resonance curve and **e** Smith chart using simulations for RMSA embedded with multiple slots

in [13] is simulated, and its impedance curves (resonance and Smith chart) obtained using IE3D are as shown in Fig. 1d, e. In resonance curve, three peaks are present whereas Smith chart shows the presence of two loops inside VSWR = 2 circle. These loops are nearer in frequency to the three resonant peak points. Distribution of surface currents at three modes shows current variations along patch length over the slotted MSA. Hence, pattern in this design shows broadside radiation with lower cross-polarization content [13]. However, which of the modes that yield wider BW is not mentioned in [13]. To investigate this, its analysis is carried out as discussed below.

## 3 Analysis of Multi-resonant Multiple Slots Cut RMSA

Equivalent RMSA dimensions for the reported design are, length '$L$' = 120 mm and width '$W$' = 80 mm. This leads to the aspect ratio of ($L/W$) of 1.5. Using reported equation for modal frequencies in RMSA, resonance frequencies are calculated as $f_{TM10}$ = 1.010 GHz, $f_{TM01}$ = 1.310 GHz, $f_{TM11}$ = 1.651 GHz, $f_{TM20}$ = 2.030 GHz and $f_{TM21}$ = 2.410 GHz. The RMSA is coaxially fed with N-type connector with a diameter of inner wire as 3.2 mm that realizes 50 $\Omega$ impedance. The impedance plot for RMSA is shown in Fig. 2a. Here three peaks in the resonance curve are observed; those are at 0.998 GHz ($TM_{10}$), 1.278 GHz ($TM_{01}$) and 1.730 GHz ($TM_{11}$). Ratio for various observed modal frequencies is $f_{01}/f_{10}$ = 1.28 and $f_{11}/f_{01}$ = 1.34, which is smaller due to higher aspect ratio. Surface current plots for modal frequencies are given in Fig. 2b–d. As per modal indices numbers, current variations for half wavelength are present at $TM_{10}$ mode along length of the patch. Variations in half wavelength for currents along width of the patch are present at $TM_{01}$ mode. Whereas half wavelength currents variations along both patch dimensions are present for resonant $TM_{11}$ mode. Next half U-slot is introduced along non-radiating edge of RMSA as given in Fig. 3a. Impedance plots for varying dimensions of slot ('$L_V$' is vertical length and '$L_h$' being horizontal length) are given in Fig. 3b, c. For this length analysis, initially by keeping '$L_V$' = 15 mm, in steps of 10 mm, slot length '$L_h$' is increased. The length '$L_h$' is at perpendicular direction with reference to currents at $TM_{01}$ mode hence its frequency decreases. Further, it comes nearer to frequency of $TM_{10}$ mode. It is stated in literature that U-slot yields compensation in probe inductance [2]. However as observed here, with reference to coaxial position, with increase in '$L_h$', effective patch width decreases. This reduction leads to increase in capacitive impedance which is formed in between patch with slot and the backed ground. This capacitance compensates for inductance due to longer probe with air substrate of thickness 23 mm. A wideband response is obtained for optimum spacing between $TM_{01}$ and $TM_{10}$ modes that yield loop formation which is completely inside VSWR = 2 circle. Here wideband is obtained for dimensions of '$L_V$' = 15 mm and '$L_h$' = 78 mm. Simulated BW as observed for RMSA with U-slot is more than 150 MHz (>14%).

# Analysis of Multi-resonant Rectangular Microstrip ...

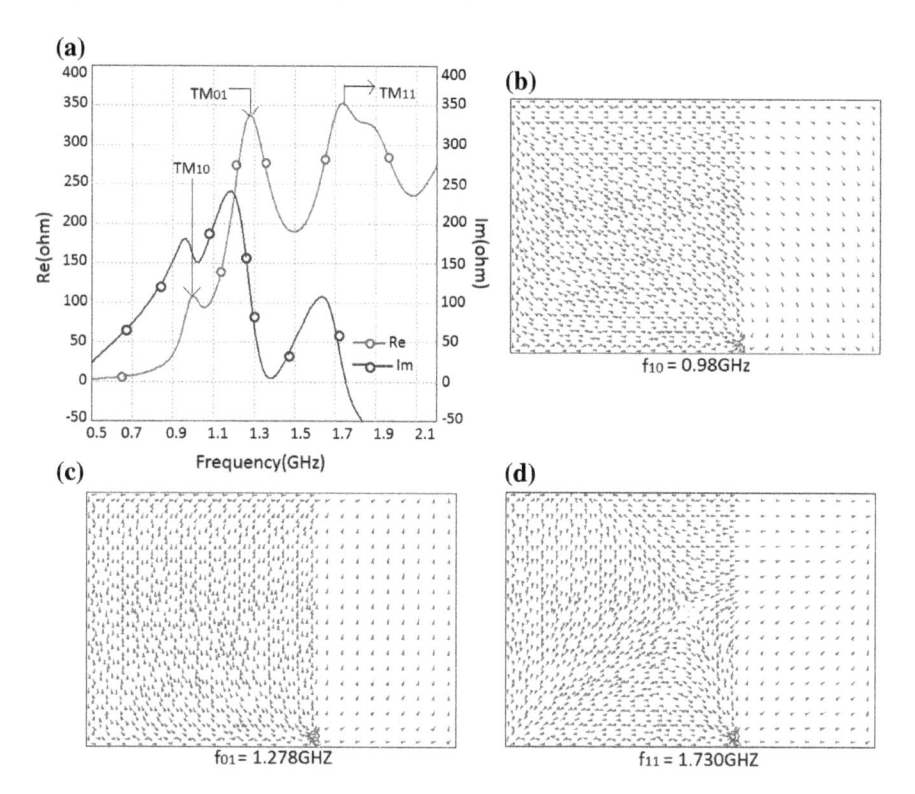

**Fig. 2** **a** Impedance plots and **b–d** current distributions at observed modes in RMSA

In further analysis a rectangular slot of length '$R_L$' at a position of '$R_H$' is cut, as shown in Fig. 4a. The impedance plots highlighting the variation in real and imaginary part of impedance against varying rectangular slot dimensions ('$R_H$' and '$R_L$') are provided in Fig. 4b, c. Here, initially by keeping '$R_H$' = 37 mm, slot length '$R_L$' is increased from 10 to 70 mm. Marginal effect is observed on $TM_{01}$ and $TM_{11}$ mode frequencies till length '$R_L$' is equal to 30 mm. As '$R_L$' increases beyond 30 mm, effective rectangular patch length that is seen by coaxial probe reduces further. Similarly, effective rectangular patch width that is seen by probe increases; hence, current flowing in the direction of width experiences a longer path length which reduces the corresponding modal frequencies. Hence, with increase in '$R_L$', $TM_{01}$ and $TM_{11}$ mode frequencies reduce further. The larger dimension of rectangular slot '$R_L$' is at perpendicular to surface currents at $TM_{01}$ mode and at parallel to the $TM_{10}$ mode. Rectangular slot further reduces the patch length and increases the width of the patch which is seen by coaxial probe. So current flowing in the direction of width experiences a longer path which reduces the corresponding $TM_{01}$ and $TM_{11}$ mode frequencies.

This also increases the capacitive impedance as formed between slot embedded RMSA and the ground plane. The spacing between $TM_{01}$ and $TM_{11}$ mode is

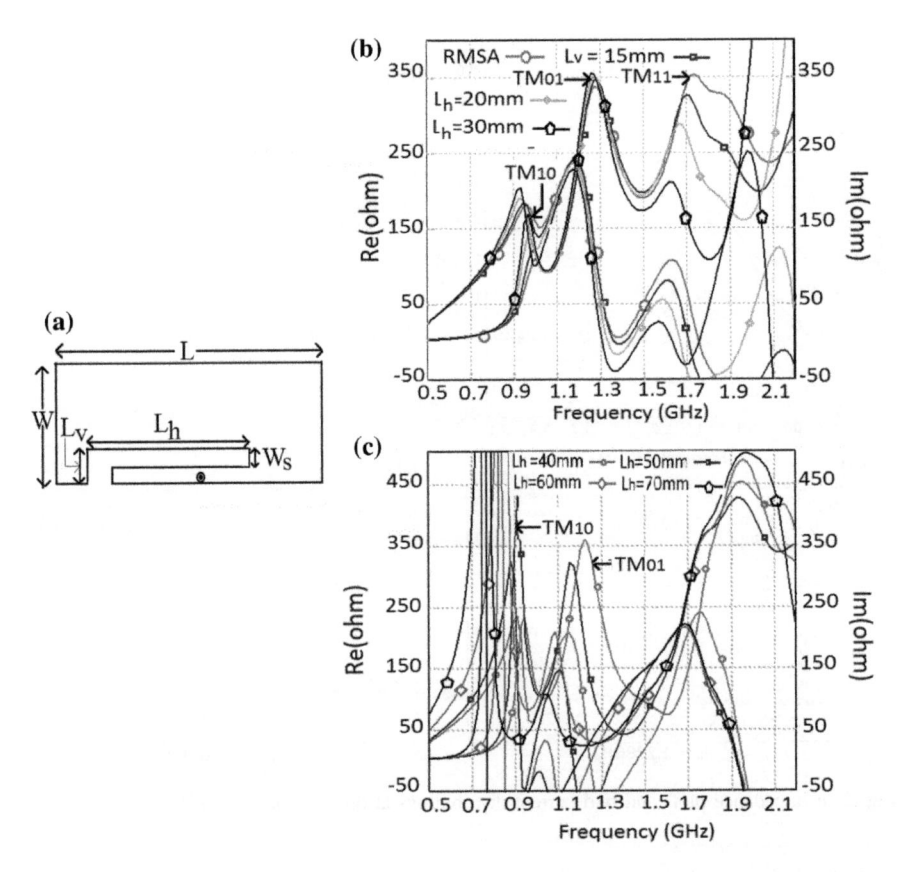

Fig. 3 a RMSA with half U-slot and its b, c impedance plots for varying U-slot dimensions

optimized for realizing broadband response so that second loop formed is fully inside VSWR = 2 circle. This is obtained at '$R_L$' = 71 mm and '$R_H$' = 37 mm. For optimized configuration two loops are observed inside the VSWR circle indicating wider BW. Further analysis is carried out by keeping '$R_L$' = 71 mm and varying value of '$R_H$' as 35, 37 and 39 mm. The impedance plots and Smith chart for this variation are explained in Fig. 5a, b. As observed, variations in '$R_H$' do not affect the frequencies of respective modes. It only slightly changes the impedances at them and this impedance variation is slightly more at $TM_{01}$ mode compared to $TM_{11}$ mode. Further changes in impedance are observed by variation in feed point for the optimized configuration. The feed point variations with impedance plots are provided in Fig. 6a–c. With change in feed point reduction in effective patch length seen by probe changes which affect the impedance. This variation in impedance is observed both at $TM_{10}$ and $TM_{11}$ modes. As the feed moves away from the open end inside half U-slot (from 'A' to 'C'), input impedance reduces. Thus, it can be inferred from above study that two slots reduces $TM_{01}$ and $TM_{11}$ mode frequencies with reference to $TM_{10}$ mode. Feed point location optimizes input impedances at

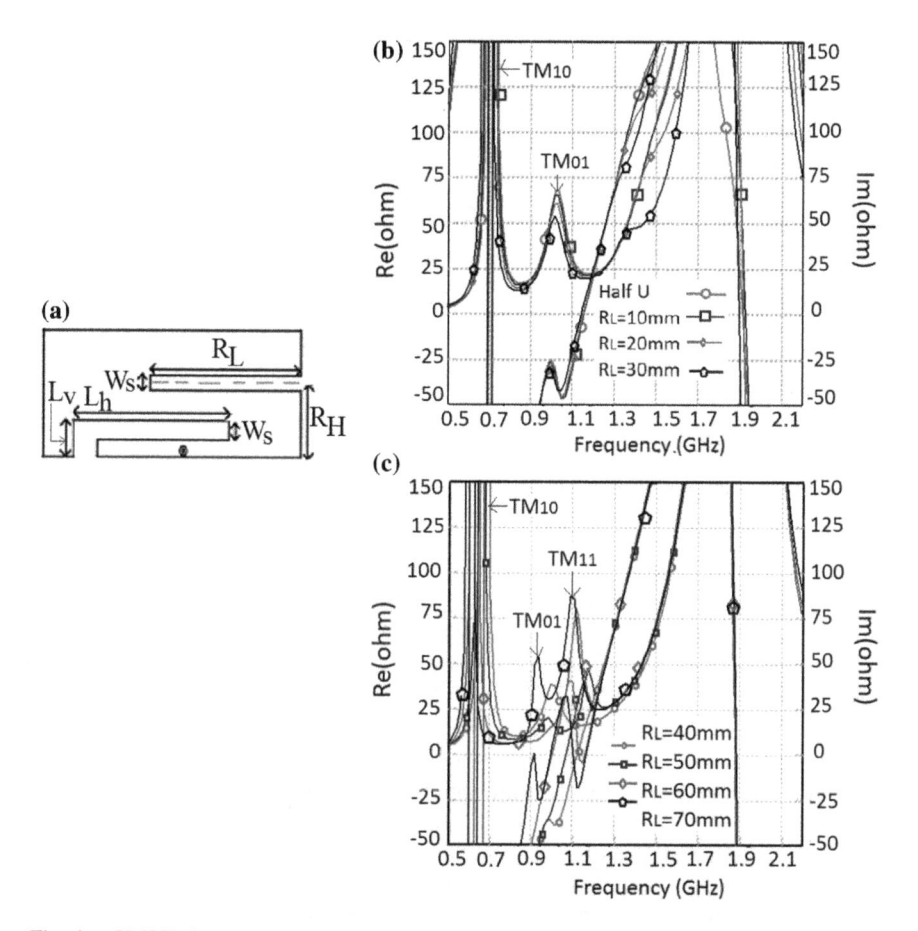

**Fig. 4** a Half U-slot cut RMSA loaded with rectangular slot and its **b**, **c** impedance plots against varying length of rectangular slot

respective modes. Also, the slots alter and realize the unidirectional current distributions at three resonant modes. These variations in frequency and impedance yield wideband response in multiple slots cut antenna with broadside pattern across the BW. Modified current distribution also ensures same directions of E and H planes, over the BW.

Further based upon reported optimum design, various design equations for multiple slots cut antenna are proposed. To realize the equations, first $TM_{10}$ mode resonance frequency of equivalent RMSA without slots is calculated. Further, all the patch parameters, i.e. substrate thickness, patch and slot dimensions and the coaxial feed point location, are expressed in terms of $TM_{10}$ mode operating wavelength as given in Table 1. To design similar antenna at different $TM_{10}$ mode frequencies, patch dimensions for substrate thickness of $0.075\lambda_0$, are calculated by using Eq. (1).

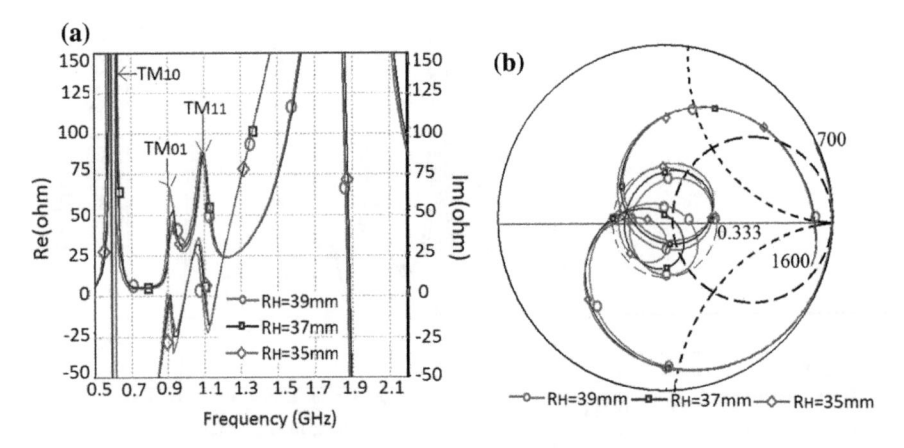

**Fig. 5 a, b** Resonance curve and Smith charts with varying rectangular slot position for RMSA embedded with half U-slot and a rectangular slot

$$L = 30/2f_r - 2(0.65h) \tag{1}$$

The patch width is selected such that aspect ratio (L/W) is 1.5. This parameter is important since it will decide spacing of $TM_{11}$ and $TM_{01}$ mode frequency with respect to $TM_{10}$ mode. Further using various equations of modified slots cut RMSA in terms of wavelength, parameter values are calculated at the desired $TM_{10}$ mode frequency and for 2 and 3 GHz they tabulated in Table 1. The antennas are designed using them and simulated input impedance plots for them are shown in Fig. 7a, b. At two $TM_{10}$ mode frequencies, wideband response is observed with formation of impedance loci loops inside VSWR = 2 circle. The simulated BWs at 2 and 3 GHz are 0.576 GHz (28%) and 0.945 GHz (30%), respectively. Thus, the proposed formulations can be used to realize similar antenna configurations at different resonance frequencies. Thus, novelty in proposed work lies in providing thorough understanding of wideband response in multiple slots cut antenna in terms of patch resonant modes. The design equations are also presented, which are useful in realizing similar configuration at different frequencies. In future scope of work, more detail analysis with respect to variations in patch dimensions (that changes intermodal frequencies) on the realized BW and more in-depth formulations of resonant length at $TM_{10}$, $TM_{01}$ and $TM_{11}$ modes as function of slot parameters will be presented. The similar designs will be experimented on suspended dielectrics instead of air as used in the present study.

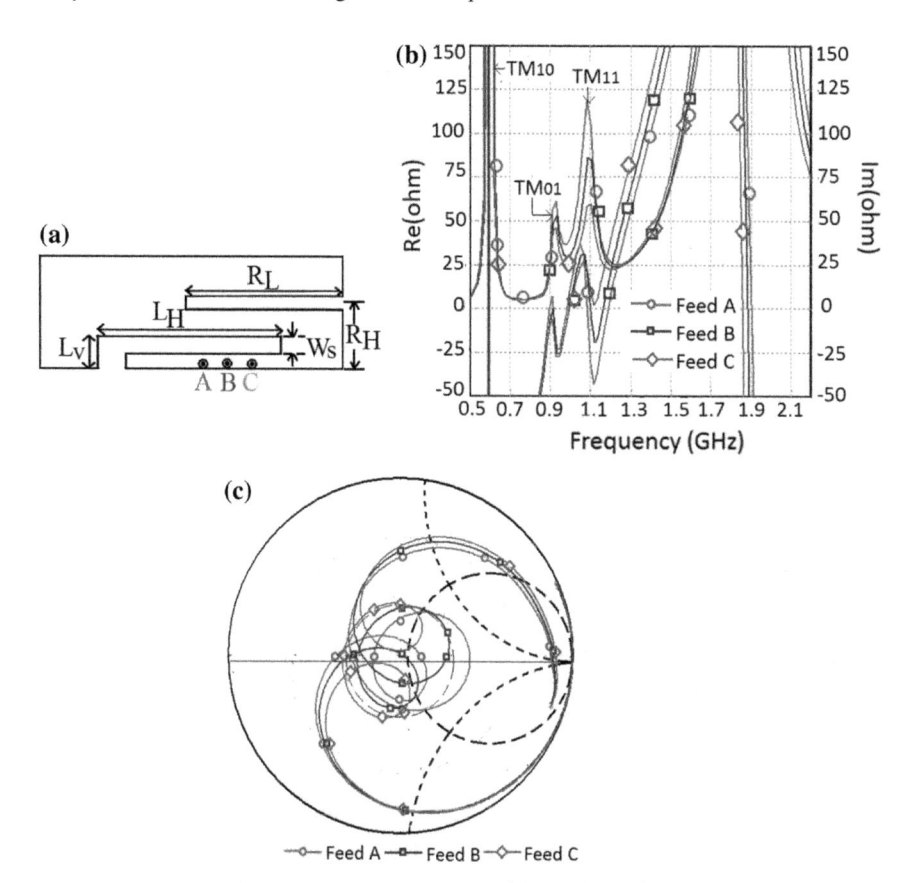

Analysis of Multi-resonant Rectangular Microstrip ...

**Fig. 6** a Half U-slot cut RMSA loaded with rectangular slot for variation in feed point location and its **b** impedance curve and **c** Smith charts

**Table 1** Various antenna parameters for multiple slots cut RMSA

| Parameter | Reported MSA (mm) | Equation as function of $\lambda_0$ | MSA at 2 GHz | MSA at 3 GHz |
|---|---|---|---|---|
| $L$ | 120 | 0.394 | 62 | 40.5 |
| $W$ | 80 | 0.263 | 41 | 27 |
| $L_h$ | 78 | 0.256 | 39 | 26 |
| $L_v$ | 15 | 0.049 | 7.5 | 5 |
| $R_L$ | 71 | 0.233 | 35 | 23 |
| $R_H$ | 37 | 0.121 | 18 | 12 |
| $H$ | 23 | 0.075 | 12 | 8 |
| $w_s$ | 8 | 0.026 | 4 | 3 |
| Feed point | $X_f = 14$, $Y_f = 1.6$ | $X_f = 0.0459$, $Y_f = 0.002$ | $X_f = 7$, $Y_f = 0.6$ | $X_f = 4$, $Y_f = 0.6$ |

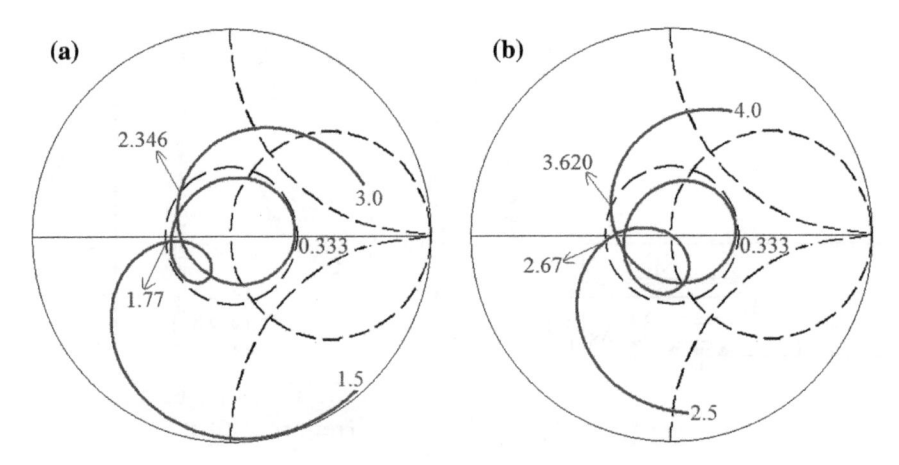

**Fig. 7** Smith charts for multiple slots cut RMSA for $TM_{10}$ mode at **a** 2 and **b** 3 GHz

## 4 Conclusions

A study that explains the wideband response in multiple slots cut rectangular patch which is embedded with half U and rectangular slot is presented. A wideband response is attributed to the reduction in the frequencies of higher order modes, $TM_{01}$ and $TM_{11}$. The frequency of $TM_{01}$ mode is reduced by introducing half U-slot, while frequency of $TM_{11}$ mode is reduced by introducing rectangular slot, along with half U-slot. Two slots also modify directions of surface current components at higher order resonant modes. It helps in obtaining radiation pattern in broadside direction. Further equations for various antenna parameters for multiple slots cut rectangular patch antenna are proposed in terms of $TM_{10}$ mode wavelength. The antennas designed using these equations at given $TM_{10}$ mode frequency shows similar wideband response. Thus, novelty of proposed works as against reported broadband slot cut MSAs is that it provides detail explanation in terms patch resonating mode, which is needed to realize similar antenna at different configurations, as well as formulations for different antenna parameters which are helpful in redesigning the antenna at given $TM_{10}$ mode frequency.

## References

1. Garg, R., Bhartia, P., Bahl, I., Ittipiboon, A.: Microstrip Antenna Design Handbook. Artech House, USA (2001)
2. Huynh, T., Lee, K.F.: Single-layer single-patch wideband microstrip antenna. Electron. Lett. **31**(16), 1310–1312 (1995)
3. Guo, Y.X., Luk, K.M., Lee, K.F., Chow, Y.L.: Double U-slot rectangular patch antenna. Electron. Lett. **34**(19), 1805–1806 (1998)

4. Islam, M.T., Shakib, M.N., Misran, N.: Broadband E-H shaped microstrip patch antenna for wireless systems. Prog. Electromagn. Res. **98**, 163–173 (2009)
5. Rafi, G., Shafai, L.: Broadband microstrip patch antenna with V slot. In: IEE Proceedings on Microwaves Antennas and Propagation, vol. 151, no. 5, pp. 435–440 (2004)
6. Deshmukh, A.A., Kumar, G.: Compact broadband E-shaped microstrip antennas. Electon. Lett. **41**(18), 989–990 (2005)
7. Ramadan, A., Kabalan, K.Y., El-Hajj, A., Khoury, S., Al-Husseini, M.: A reconfigurable U-koch microstrip antenna for wireless applications. Prog. Electromagn. Res. **93**, 355–367 (2009)
8. Ansari, J.A., Ram, R.B.: E-shaped patch symmetrically loaded with tunnel diodes for frequency agile/broadband operation. Prog. Electromagn. Res. B **1**, 29–42 (2008)
9. Balanis, C.A.: Antenna theory and design, 3rd edn. Wiley InterScience, USA (2005)
10. Deshmukh, A.A., Ray, K.P.: Analysis of broadband variations of U-slot cut rectangular microstrip antennas. IEEE Mag. Antennas Propag. **57**(2), 181–193 (2015)
11. Deshmukh, A.A., Kumar, G.: Compact broadband U-slot loaded rectangular microstrip antennas. Microw. Opt. Technol. Lett. **46**(6), 556–559 (2005)
12. Chair, R., Mak, C.L., Lee, K.F., Luk, K.M., Kishk, A.A.: Miniaturized wide-band half U-slot and half E-shaped patch antennas. IEEE Trans. Antennas Propag. **53**(8), 2645–2652 (2005)
13. Deshmukh, A.A., Ray, K.P.: Compact broadband slotted rectangular microstrip antenna. IEEE Antennas Wirel. Propag. Lett. **8**, 1410–1413 (2009)

# Broadband Rectangular Microstrip Antennas Embedded with Pairs of Rectangular Slots

Amit A. Deshmukh and Divya Singh

**Abstract** The wideband response in microstrip antenna is realized by cutting slot inside the patch. The proposed work here provides detailed study about the wideband behavior in one, two, and three pairs of slots (rectangular) cut rectangular microstrip antenna. The slots in rectangular patch tune resonance frequencies of higher order $TM_{02}$, $TM_{12}$, and $TM_{22}$ resonant modes with respect to fundamental $TM_{10}$ mode that yields wider bandwidth. The surface currents on the patch are modified in respective directions due to multiple slots that achieve broadside pattern over the VSWR bandwidth. Resonant mode explanation for wideband response was not available in the literature. For multiple slots cut patch antenna, the proposed work provides the same.

**Keywords** Broadband microstrip antenna · Rectangular microstrip antenna
Pairs of rectangular slots · Higher order mode

## 1 Introduction

Ever since reported in 1995, slot cut method has been frequently referred to obtain a wideband response in microstrip antenna (MSA) and different geometries of slots have been implemented [1–8]. Bandwidth (BW) enhancement is due to second resonant mode added by slot that is nearer to fundamental resonant mode. The slot cut designs are optimized for thicker substrates ($h \geq 0.06\lambda_0$) which are having lower dielectric constant. In most of the designs, air is used as the substrate. But for thickness more than $0.08\lambda_0$, BW is limited due to larger probe inductance. In these designs, proximity feeding technique is used which yields optimum BW on thicker substrates [9]. For substrate thickness in the range of $0.06$–$0.08\lambda_0$, BW in MSAs embedded with slot is increased by introducing additional slots [10]. Addition of

---

A. A. Deshmukh (✉) · D. Singh
Department of Electronics and Telecommunication Engineering,
SVKM's, D J Sanghvi College of Engineering, Mumbai, India
e-mail: amit.deshmukh@djsce.ac.in

the same introduces resonant modes nearer to modes of slot cut MSA that increases the BW. On thinner substrate ($0.036\lambda_0$), BW in rectangular MSA (RMSA) is increased by cutting multiple pairs of rectangular slots [11]. As per earlier literature, it was reported that slot introduces new mode when its length equals either half-wave or quarter-wave in length. However, that assumption did not explain the effects of antenna and slot parameters and feed point locations on realized BW. Also, no design guidelines/formulations for slot cut antennas were provided. A recent study on slot cut MSAs brings out that slot modifies patch's higher order orthogonal resonant mode and along with fundamental mode, yields wider BW [12]. For thicker substrate, designs formulations for slot cut MSAs are also provided [13]. When additional slots are cut inside slot cut MSA, it will modify next higher order mode frequency that enhances patch BW [12]. However, in those multiple slots cut MSA, patch aspect ratio was selected to be higher ($\sim 1.5$), which leads to close spacing of resonant mode frequencies [10, 12]. Since slot redistributes the frequency spacing between various patch modes, interspacing between modal frequencies is an important parameter in the design. Using smaller patch aspect ratio (<1.2) BW in RMSA has been increased by cutting pairs of slots [11]. It yields BW of more than 10% in 850 MHz frequency band [11]. However, in [11], resonant mode explanation about wideband behavior in multiple slots cut antenna is not provided. This is an important aspect as it will help to realize design guidelines for similar slots cut antennas at different frequencies.

In present work, a study to bring out resonant mode explanation about wideband behavior in RMSA that is embedded with single or double or three pairs of rectangular slots as presented in [11] is discussed. The study is carried out by analyzing impedance curve (real and imaginary) plots, modal current distributions with increasing slot length. The proposed work puts forward that in single pair of slots cut RMSA, $TM_{02}$ mode is tuned with respect to $TM_{10}$ mode, which gives 70 MHz ($\sim 8\%$) of BW. Further addition of second pairs of slot, RMSA's modified $TM_{12}$ mode frequency is tuned with reference to $TM_{10}$ and $TM_{02}$ modes which yields BW of around 90 MHz (>11%). Lastly, with the addition of third pair of slots, patch's modified $TM_{22}$ mode frequency is tuned with reference to modified $TM_{10}$, $TM_{02}$, and $TM_{12}$ modes, that gives BW of more than 120 MHz (>14%). With reference to modal current distributions at $TM_{10}$, next $TM_{02}$, $TM_{12}$, and $TM_{22}$ modes, exhibits orthogonal current variations. In all these pairs of rectangular slots cut RMSA, slots modify current directions at higher order resonating modes, to give pattern maxima in broadside direction with cross-polar levels less than 15 dB as compared to the co-polar levels. Thus, wideband response in multiple slots cut RMSA is the result of coupling between $TM_{10}$ mode with higher order $TM_{02}$, $TM_{12}$ and $TM_{22}$ modes. In reported literature analysis to explain wideband behavior in multiple slots cut MSA with smaller patch aspect ratio is not given. Thus, novelty here lies in presenting an in-depth explanation for wideband response in multiple slots cut antenna. In future scope, formulations at resonating length at modified modes will be developed as function of slot dimensions. Using them, design procedure to realize similar antenna design at different frequency ranges will be developed.

## 2 Analysis of Pairs of Rectangular Slots Cut RMSAs

The pair of rectangular slots cut designs of RMSA is shown in Fig. 1a–d. Here three layer suspended configuration is used [11]. The slot cut RMSA is fabricated on glass epoxy substrate ($\varepsilon_r = 4.3$, $h = 0.16$ cm, tan $\delta = 0.02$). The patch is fabricated on top glass epoxy layer whereas the ground plane is present on another side of bottom glass epoxy layer [11]. In coaxially fed RMSA, the presence of a loop in impedance locus indicates two resonant modes in the patch. In single pair of slots cut RMSA, a single loop is observed whereas in two pairs of slots cut RMSA, two loops are present [11]. This indicates the presence of two and three resonant modes, respectively, in slot cut antenna. In three pairs of slots cut design, four modes are present as three loops are present [11]. To find out which modes are responsible for wideband response, the modal analysis for three configurations is presented. Simulated resonance curve plot generated using IE3D software for equivalent RMSA shows the presence of $TM_{10}$ (0.812 GHz) and $TM_{02}$ (1.410 GHz) modes as shown in Fig. 1e. Peaks due to other higher order resonant modes are absent since impedance matching for same is not obtained for given feed position and also the simulation was carried out till 1.5 GHz.

Further pair of slots of length "$l_1$" is cut inside this RMSA and resonance curve plots and Smith chart for length variation are shown in Fig. 2a–d. The slot length is parallel to current directions at $TM_{10}$ mode hence its frequency remains unchanged.

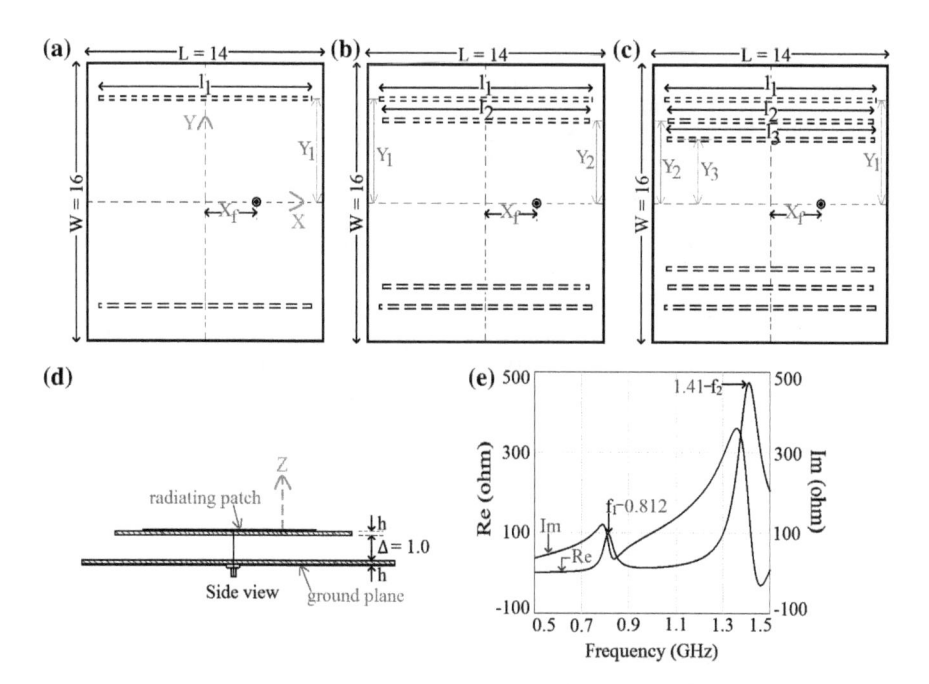

**Fig. 1** a–d Pairs of slots cut configurations of RMSA [11], and e resonance curve plot for RMSA

However, length "$l_1$" is orthogonal to surface currents at higher order $TM_{02}$, $TM_{12}$, and $TM_{22}$ modes; therefore, their frequencies decreases. Reduction observed for $TM_{02}$ mode frequency is larger. With increase in "$l_1$", its frequency comes closer to $TM_{10}$ mode frequency. The loop formed in Smith chart, because of optimum spacing in between $TM_{10}$ and $TM_{02}$ modes, lies inside VSWR = 2 circle, as given in Fig. 2d. At modified $TM_{02}$ mode, current distributions on patch for "$l_1$" = 7 cm is shown in Fig. 2e. Due to slot length, the contribution of currents has increased along the patch length. An optimum BW is obtained for "$l_1$" = 12.8 cm, as shown in Fig. 2f. The simulated BW is 68 MHz (8.8%) and the measured BW is

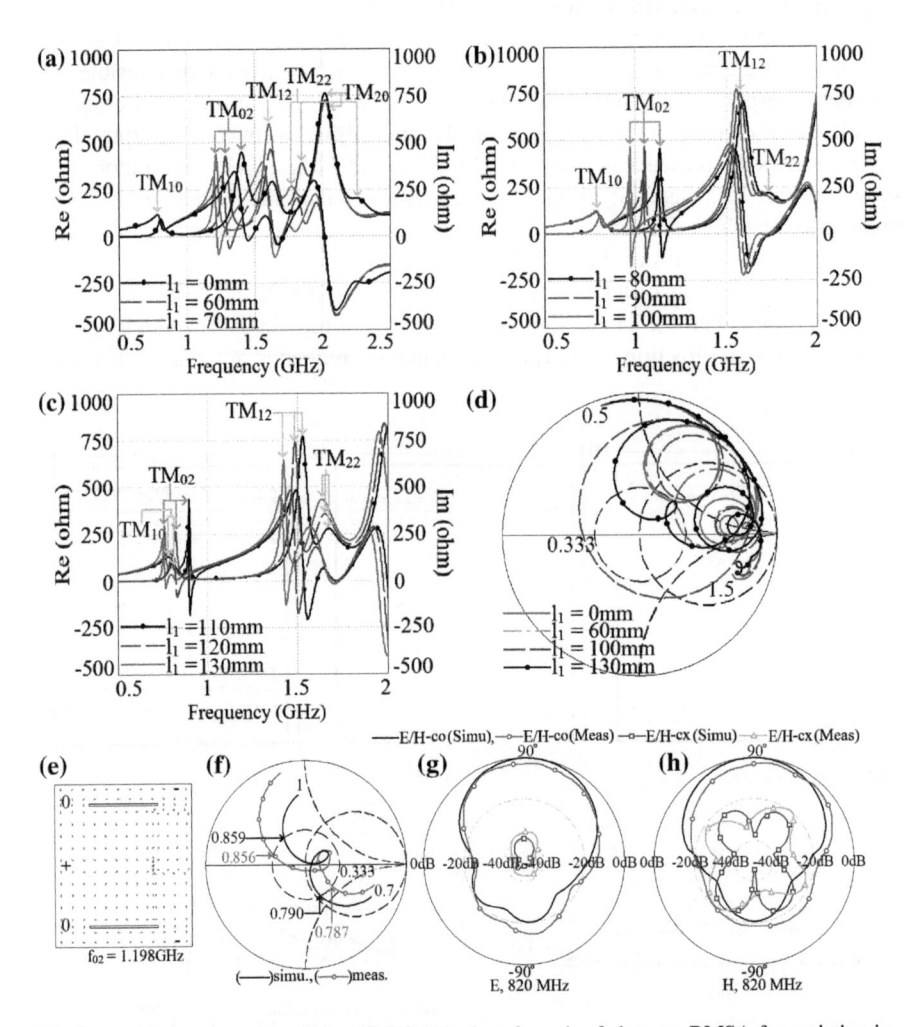

**Fig. 2** **a–c** Resonance curve plots and **d** Smith chart for pair of slots cut RMSA for variation in "$l_1$" and its **e** $TM_{02}$ mode current distribution on patch, **f** optimum Smith chart and **g**, **h** pattern at center frequency of VSWR BW in the optimum design

70 MHz (9%). The pattern at center frequency of BW is in the broadside direction with cross-polar levels less than 15 dB as compared to co-polar levels as shown in Fig. 2g, h. The pattern does not show any variation across the BW.

Further inside pair of slots cut RMSA, second pair of rectangular slots having length "$l_2$" is cut. For an optimum length of "$l_1$" = 12.8 cm, impedance curves against increasing "$l_2$" is given in Fig. 3a–d. Here $TM_{02}$ mode frequency nearly remains same since slot lengths "$l_1$" have modified its surface current distribution. An increase in "$l_2$" largely reduces $TM_{12}$ mode frequency and tunes it with respect to $TM_{10}$ and $TM_{02}$ modes, which optimizes the second loop because of $TM_{12}$ mode inside VSWR = 2 circle in impedance locus. Further second pairs of slots maximize modal current contribution at $TM_{02}$ and $TM_{12}$ modes along patch length as shown in Fig. 3e, f. The wideband response is obtained with an optimum spacing of $TM_{12}$ mode with $TM_{10}$ and $TM_{02}$ modes, which is obtained for "$l_1$" = 12.8 cm and "$l_2$" = 12.2 cm. Here two BWs, namely, simulated and measured, are 87 MHz (11%) and 93 MHz (12%), respectively, as shown in Fig. 3g. RMSA with two pairs of rectangular slots show an identical pattern and gain response to that exhibited by RMSA with single pair of rectangular slots. Thus, the second pair of slots tunes third higher order mode frequency to enhance patch BW. Similarly, analysis of three pairs of slots is carried out.

Resonance curve and Smith chart plots with varying "$l_3$" are provided in Fig. 4a–d. Length "$l_3$" mainly affects the frequency and impedance at fourth-order (for given feed point location) $TM_{22}$ mode and further, it yields the position of the loop because of same completely inside 0.333 circle of reflection coefficient. Modal surface current variations at three modified patch modes as given in Fig. 4e–g, are directed along the patch length.

For "$l_1$" = 12.8 cm, "$l_2$" = 12.2 cm and "$l_3$" = 12.4 cm, two BWs, namely, simulated and measured are, 142 MHz (16%) and 149 MHz (18%), respectively, as given in Fig. 5a. A fabricated prototype of three pairs of slots cut RMSA is shown in Fig. 5b. Due to nearly unidirectional surface currents at $TM_{10}$ and modified higher order modes, three pairs of slots cut RMSA exhibit an identical pattern and gain response to that observed in RMSA embedded with single pair of rectangular slots. Among three slot cut designs, RMSA embedded with three pairs of rectangular slots gives lower cross-polarization characteristics as three pairs of slots optimizes maximum currents along patch length. Thus, proposed work clearly brings out resonant mode explanation about wideband response in pairs of slots cut RMSA, which is not given in [11]. This explanation will be helpful in realizing formulations for resonant modes at fundamental and modified higher order modes in terms of slot and RMSA dimensions. This will help in realizing similar configurations on thinner suspended substrates.

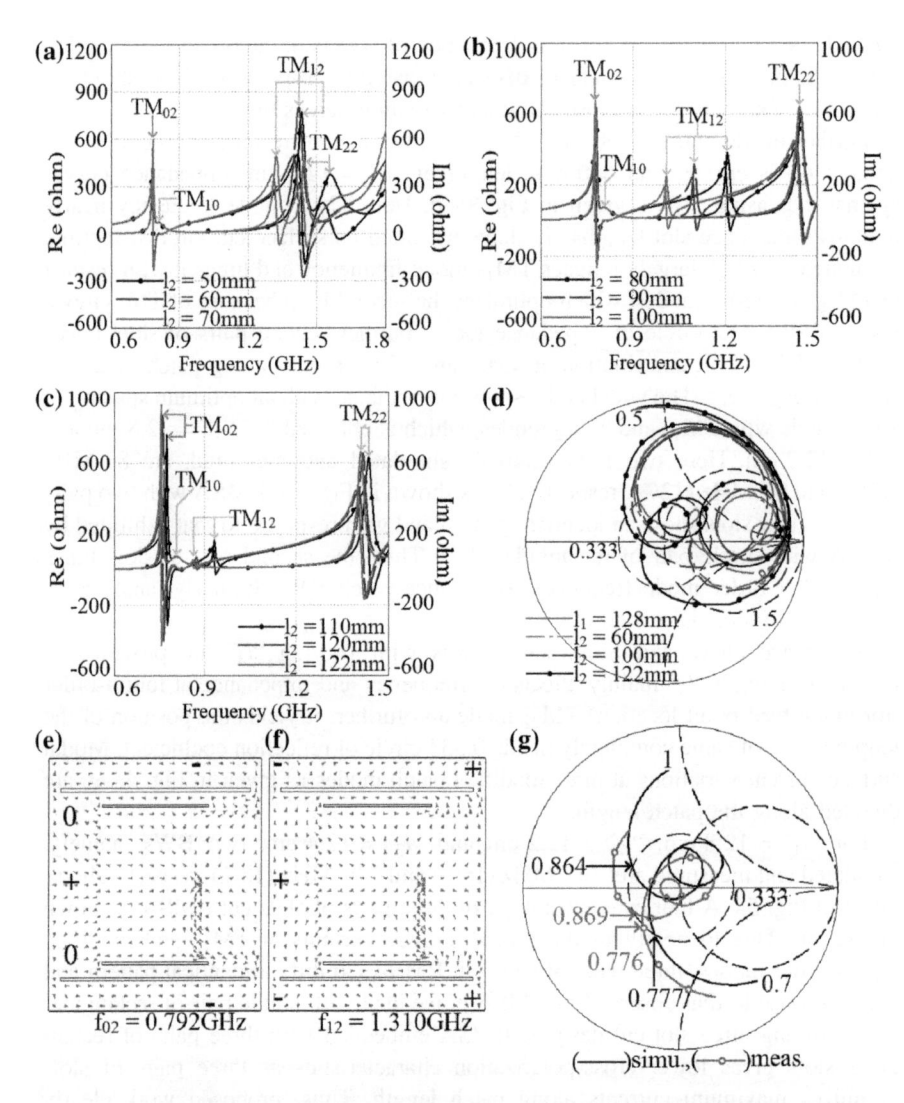

**Fig. 3** a–c Resonance curve plots and **d** Smith chart for two pairs of slots cut RMSA for variation in "$l_2$" and respective **e, f** modal surface current distributions at two modes and their **g** optimum Smith chart for "$l_1$" = 12.8 cm and "$l_2$" = 12.2 cm

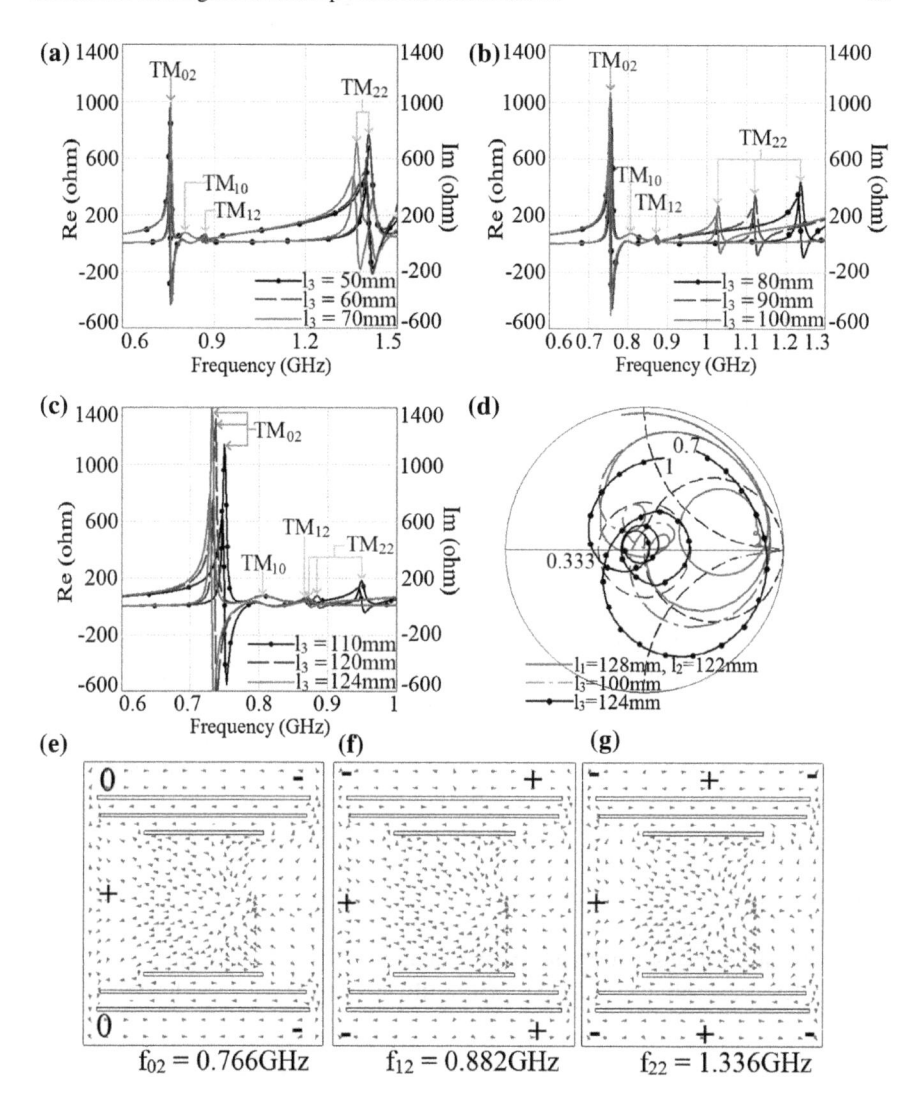

**Fig. 4** **a–c** Impedance curves and **d** input impedance loci using Smith charts for three pairs of slots cut RMSA for variation in "$l_3$" and its **e, f** current distributions at modified resonating modes for "$l_3$" = 7 cm

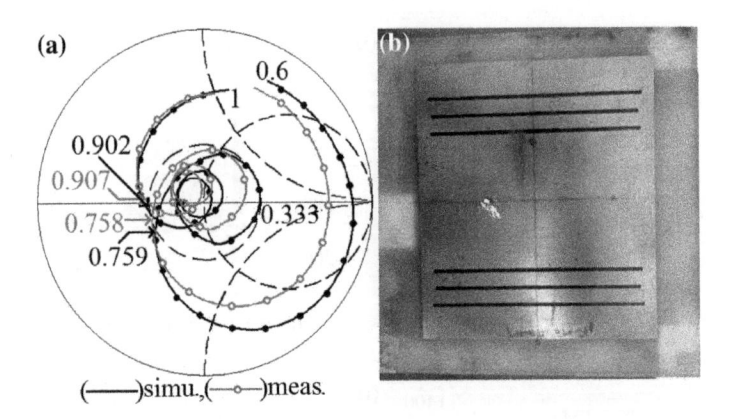

**Fig. 5** **a** Smith chart in optimum design and **b** fabricated prototype of three pairs of slots cut RMSA

## 3 Conclusions

A study to understand the wideband behavior in multi-resonant pairs of slots cut RMSA is presented. With respect to fundamental $TM_{10}$ mode, the first pair of slots tunes second-order orthogonal $TM_{02}$ mode resonance frequency whereas second and third pairs of slots tune $TM_{12}$ and $TM_{22}$ mode resonance frequency which yields wideband response. At higher order resonating modes, depending upon the surface current variations, radiation pattern shows conical to boresight pattern with higher cross-polarization levels. Pairs of slots modify the surface current distributions at higher order modes to give broadside radiation pattern across the VSWR BW. Due to optimum alignment of current components at modified three resonant modes along patch length, three pairs of slots cut design yields lower cross-polarization levels among all, with maximum BW. All the three slots cut design yields broadside gain of more than 8 dBi over most of VSWR BW.

## References

1. Huynh, T., Lee, K.F.: Single-layer single-patch wideband microstrip antenna. Electron. Lett. **31**(16), 1310–1312 (1995)
2. Islam, M.T., Shakib, M.N., Misran, N.: Multi-slotted microstrip patch antenna for wireless communication. Prog. Electromagn. Res. Lett. **10**, 11–18 (2009)
3. Yang, S.L.S., Kishk, A.A., Lee, K.F.: Frequency reconfigurable U-slot microstrip patch antenna. IEEE Antennas Wirel. Propag. Lett. **7**, 127–129 (2008)
4. Yang, F., Zhang, X., Rahmat-Samii, Y.: Wide-band E-shaped patch antennas for wireless communications. IEEE Trans. Antennas Propag. **49**(7), 1094–1100 (2001)
5. Bao, X.L., Ammann, M.J.: Small patch/slot antenna with 53% input impedance bandwidth. Electron. Lett. **43**(3), 146–147 (2007)

6. Ang, B.K., Chung, B.K.: A wideband E-shaped microstrip patch antenna for 5–6 GHz wireless communications. Prog. Electromagnet. Res. **75**, 397–407 (2007)
7. Khodaei, G.F., Nourinia, J., Ghobadi, C.: A practical miniaturized U-slot patch antenna with enhanced bandwidth. Prog. Electromagnet. Res. B. **3**, 47–62 (2008)
8. Wong, K.L.: Compact and Broadband Microstrip Antennas, 1st edn. Wiley, New York, USA (2002)
9. Deshmukh, Amit A., Ray, K.P.: Analysis of broadband $\Psi$-shaped microstrip antennas. IEEE Mag. Antennas Propag. **55**(2), 107–123 (2013)
10. Guo, Y.X., Luk, K.M., Lee, K.F., Chow, Y.L.: Double U-slot rectangular patch antenna. Electron. Lett. **34**(19), 1805–1806 (1998)
11. Deshmukh, Amit A., Kumar, G.: Broadband pairs of slots loaded rectangular microstrip antennas. Microw. Optic. Technol. Lett. **47**(3), 223–226 (2005)
12. Deshmukh, Amit A., Ray, K.P.: Analysis of broadband variations of U-slot cut rectangular microstrip antennas. IEEE Mag. Antennas Propag. **57**(2), 181–193 (2015)
13. Deshmukh, A.A., Ray, K.P.: Analysis and design of broadband U-slot cut rectangular microstrip antennas. In: Sadhana—Academy Proceedings in Engineering Science, Springer, New York

# CPW-Fed Printed Monopole with Plus Shaped Fractal Slots for Wider Bandwidth

Ameya A. Kadam and Amit A. Deshmukh

**Abstract** Broadband design of CPW-fed printed monopole with plus shaped fractal slots using offset feed is proposed. It yields impedance bandwidth of 66% in 6 GHz frequency band that covers most of the C-band. Further, detailed analysis of antenna structure and modes is carried out. The proposed antenna shows broadside radiation pattern over the bandwidth with a maximum gain of 7.5 dBi. Experiments have been performed on fabricate prototype, and the obtained results show a good agreement with the simulated results.

**Keywords** Coplanar waveguide · Microstrip · Monopole · Wireless

## 1 Introduction

Over last many decades, RF and antenna design engineers are exploring the various ways to enhance the narrow bandwidth (BW) of microstrip patch antennas like thicker substrates [1], introduction of slots [2], adding stubs [3], coplanar waveguide (CPW) feeding [4], and defected ground structure [5]. It is a well-known fact that CPW-fed microstrip antennas (MSAs) possess high impedance matching and wideband behavior. Recently, the use of slot antennas and CPW feeding technique in microstrip patch is rapidly increasing with the growth of telecommunication systems [6]. The CPW-fed printed slot antennas possess properties like low profile, wide impedance BW, omnidirectional radiation pattern, less dispersion, and high

---

A. A. Kadam · A. A. Deshmukh (✉)
Department of Electronics and Telecommunication Engineering,
SVKM's, D J Sanghvi College of Engineering, Mumbai, India
e-mail: amit.deshmukh@djsce.ac.in; amitdeshmukh76@gmail.com

A. A. Kadam
e-mail: ameya.kadam@djsce.ac.in

radiation efficiency with simple structures [7]. Therefore, CPW-fed planar slot antennas are considered a comparably promising design for wideband wireless applications. The recent advancements in wireless communication system are driving the engineering efforts to develop wideband and compact systems. Fractals, because of their space filling and self-similar properties, has motivated antenna design engineers to adopt this geometry as a sustainable option to meet the demand of multiband operations [8–13]. Fractal geometries for wideband operation are relatively less utilized so far.

In this paper, a new technique has been employed to increase the BW of the CPW-fed printed monopole antenna with plus shaped fractal slots geometry to facilitate impedance matching of antenna. The simulation and experimental results of the proposed antenna show the BW more than 66% at center frequency around 6 GHz thereby covering most of the C-band (4–8 GHz). The antenna was first simulated and optimized using IE3D software followed by measurements. The patch was fabricated on FR4 substrate of thickness "$h$" = 1.6 mm with dielectric constant "$\varepsilon_r$" = 4.4 and loss tangent tan $\delta$ = 0.02. The return loss and impedance loci were measured using $R$ and $S$ vector network analyzer (ZVH-8). The radiation and gain properties were measured using $R$ and $S$ RF source (SMB 100A) and spectrum analyzer (FSC 6) inside antenna lab with required minimum far field distance; between reference horn antenna and CPW-fed fractal antenna which is under test.

## 2 CPW-Fed Printed Monopole with Plus Shaped Fractal Slots

The geometry of the reported CPW-fed monopole antenna with plus shaped fractal slot geometry is shown in Fig. 1 [14]. The antenna is designed on FR4 substrate. A 50 $\Omega$ CPW-fed transmission line which consists of a single strip having a width $W_s$ of 3.6 mm is used to feed the antenna. The SMA connector having probe diameter of 1.2 mm with Teflon dielectric is used to feed the microstrip line which has a characteristic impedance of 50 $\Omega$. The length and width of the designed antenna are fixed at $L$ = 30.6 mm and $W$ = 30.6 mm, respectively.

Two finite ground planes with the same size are placed symmetrically at both sides of CPW line having dimensions $L_p$ = 15.6 mm and $W_p$ = 23.4 mm. The return loss and resonance curve of the reported structure are shown in Fig. 2b, c, respectively.

The VSWR $\leq$ 2, BW for the reported configuration was found to be 3.36 GHz at the center frequency of 6.0 GHz which gives a total bandwidth of 56%. Also, the current distribution at the resonant modes is shown in Figs. 3a, b and 4a–c.

The surface currents show half-wavelength variations along the patch and around inner plus shaped fractal slot dimensions which are due to fundamental and second-order modes. To understand the functioning of this antenna geometry and the effect of plus shaped fractal slots, an analysis of this antenna was carried out by

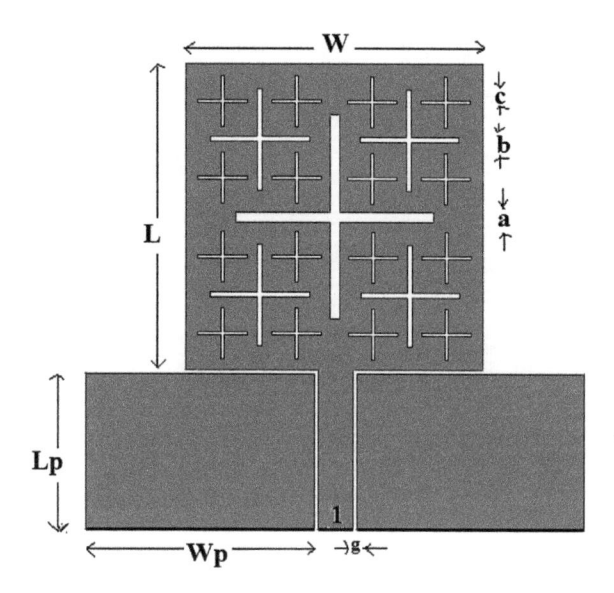

Fig. 1 CPW-fed monopole antenna with fractal slots (iteration 3) [14]

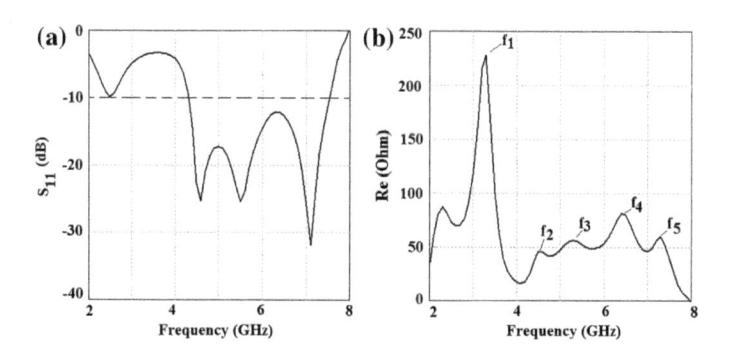

Fig. 2 Plots of a return loss and b resonance curve of the reported structure

reducing the iterative slots one by one as shown in Fig. 5a–c. The return loss and resonance curves of these modified fractal structures are shown in Fig. 6a, b, respectively.

As seen from Fig. 6a, the change in BW is not significant due to the addition of iterative slots. It is seen that from Fig. 6b that addition of fractal slots reduces the resonant frequencies of the higher order modes resulting in wider bandwidth. Further, a CPW-fed simple square MSA with offset without plus shaped fractal slots and with plus shaped fractal slots up to iteration 3 are investigated. The antenna structure is shown in Fig. 7a. The parametric study for variation in CPW-fed line position on the realized BW was carried out and the optimum result was obtained for strip position at 3 mm from the center. The effect of offset feed of iteration 3

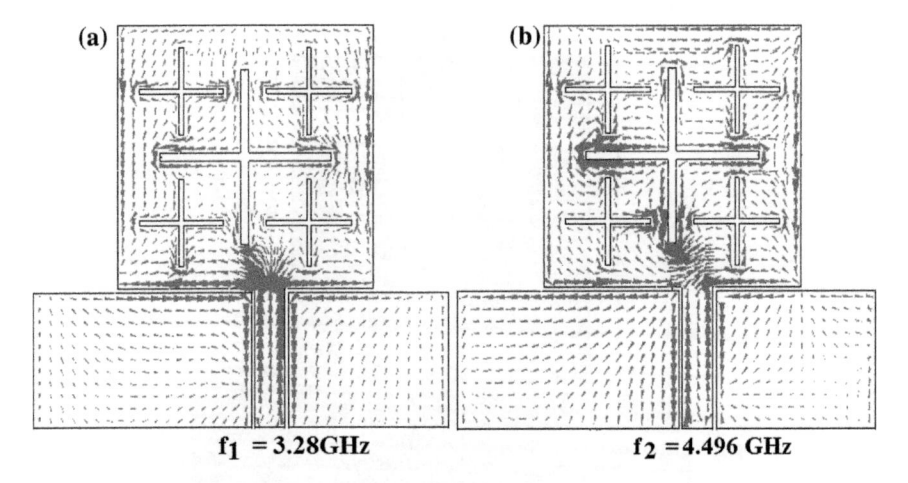

**Fig. 3** Current distribution at various resonant mode frequencies: **a** 3.28 GHz and **b** 4.496 GHz

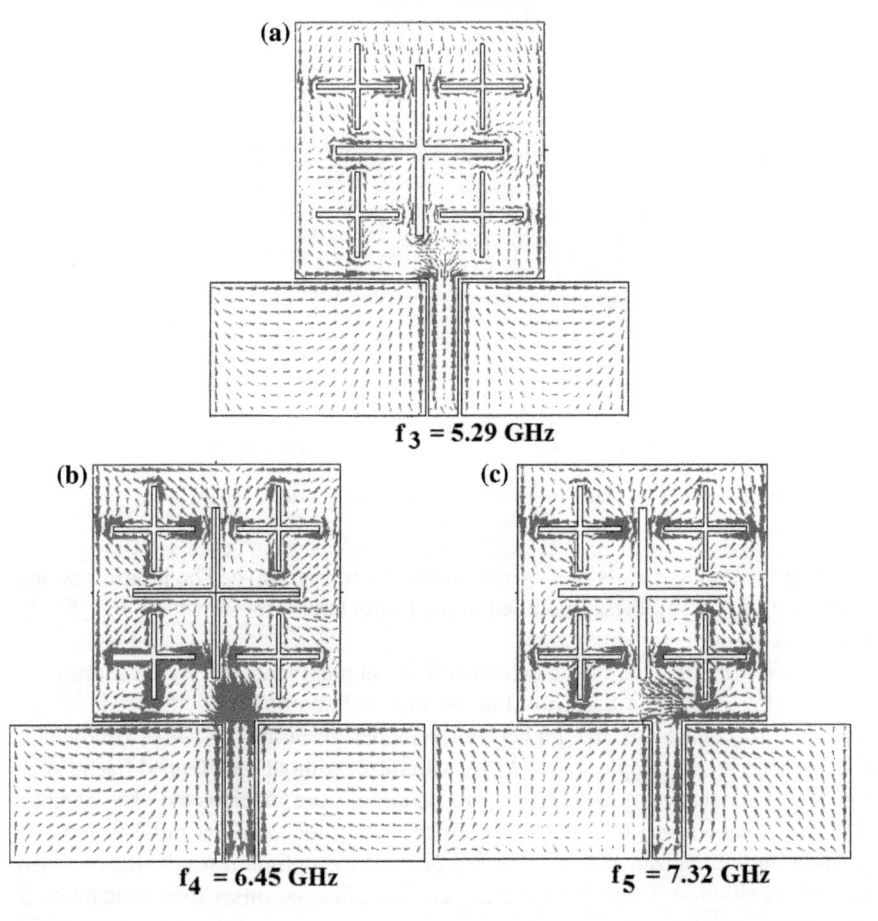

**Fig. 4** Current distribution at various resonant mode frequencies: **a** 5.29 GHz, **b** 6.45 GHz, and **c** 7.32 GHz

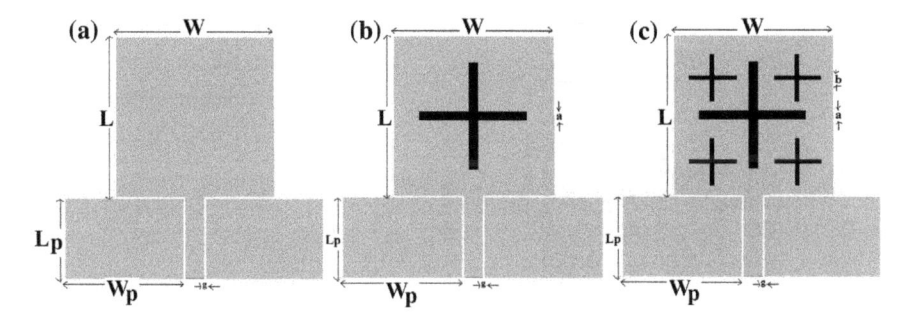

**Fig. 5** Geometrical construction of reported antenna for **a** zero iteration, **b** first iteration, and **c** second iteration

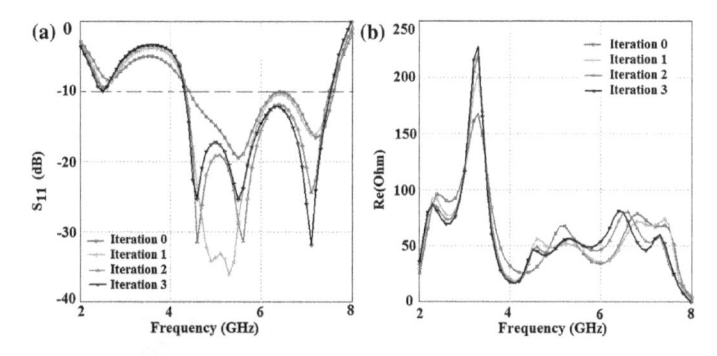

**Fig. 6** Plots of **a** return loss ($S_{11}$) and **b** resonance curves of the fractal iterative structures

geometry on resonance and return loss is shown in Fig. 7b, c. With an offset feed location at 3 mm the impedance is below 75 Ω for a wide range of frequencies indicating a good impedance match that results in more wide bandwidth. The effects of offset feed for higher iterations with plus shaped fractal slots are studied. The antenna with iteration 2 was fabricated on an FR4 substrate is shown in Fig. 7d and tested.

The measurement was carried out to validate the simulation. The simulated BW was found to be 3.52 GHz at center frequency of 6 GHz (60%) and measured BW is 3.6 GHz. The plot of measured return loss and input impedance loci are shown in Fig. 8a, b, respectively.

The measured copolar radiation pattern of the proposed antenna in y-z and x-z planes at lower resonant frequencies 4.5 and 5.5 GHz is shown in Fig. 9a, b, and that at high resonant frequencies 6.5 and 7.5 GHz is shown in Fig. 10a, b. The radiation pattern in E-plane is symmetrical to antenna axis. The cross-polar component of E-field and H-field are not mentioned in the radiation pattern of the reported structure. As seen from the radiation pattern level of cross-polar component is more as the orthogonal currents are present in the structure. At higher frequencies, the tilt in major lobe in radiation pattern is obtained due to the presence

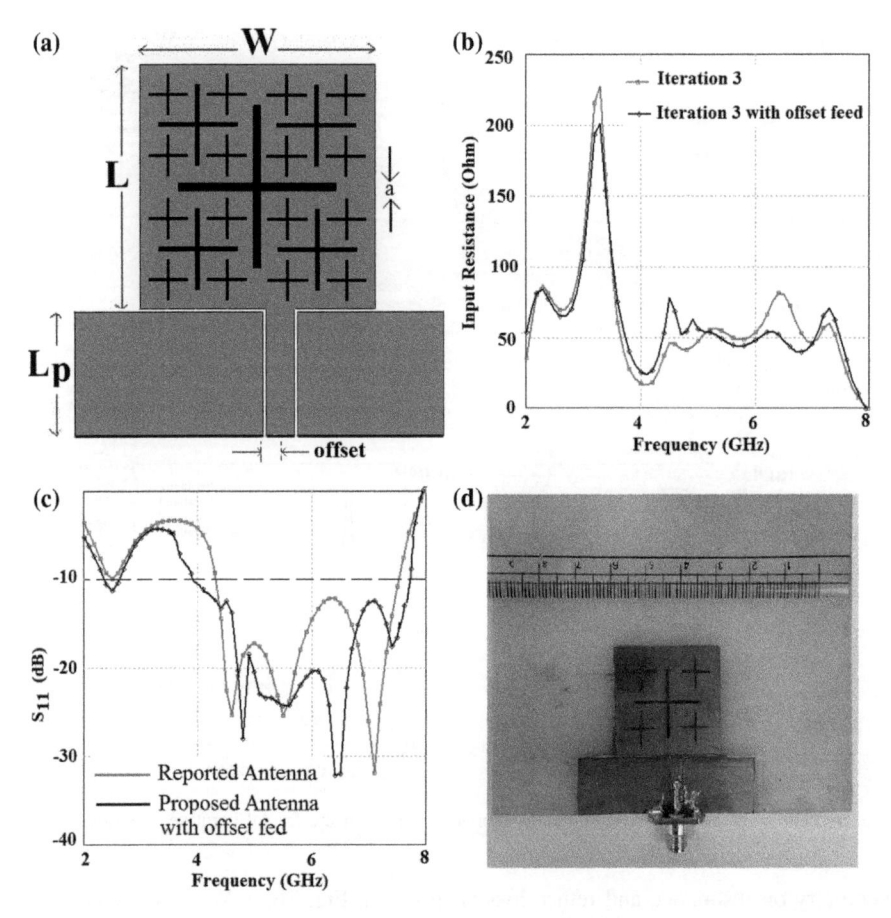

**Fig. 7** **a** Geometry of offset fed fractal antenna, **b** effect of offset feed on resonance plot, **c** effect of offset feed on return loss, and **d** fabricated prototype of offset CPW fed plus shaped fractal antenna (iteration 2)

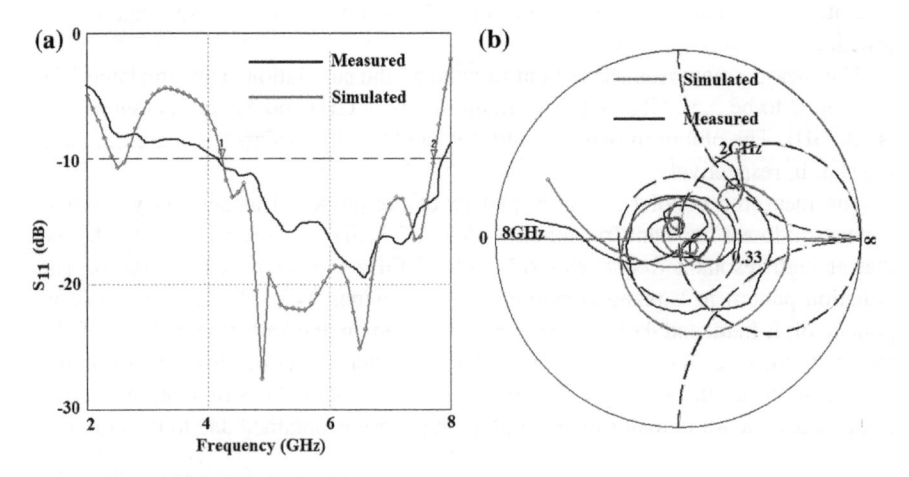

**Fig. 8** Measured and simulated plot of **a** return loss and **b** impedance loci of the proposed antenna

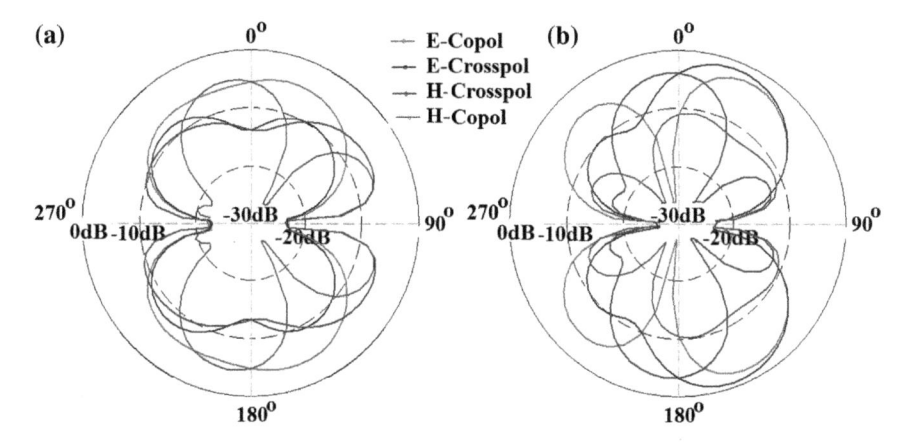

**Fig. 9** Radiation patterns at lower frequencies: **a** 4.5 GHz and **b** 5.5 GHz

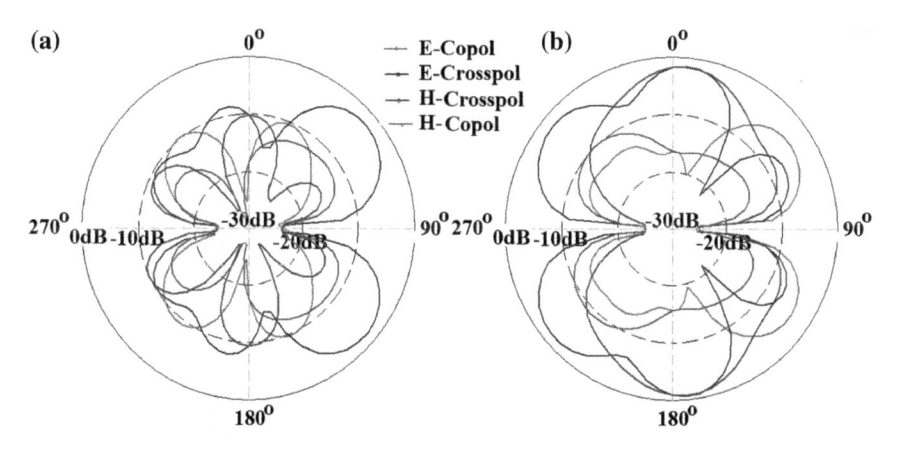

**Fig. 10** Radiation patterns at higher frequencies: **a** 6.5 GHz and **b** 7.5 GHz

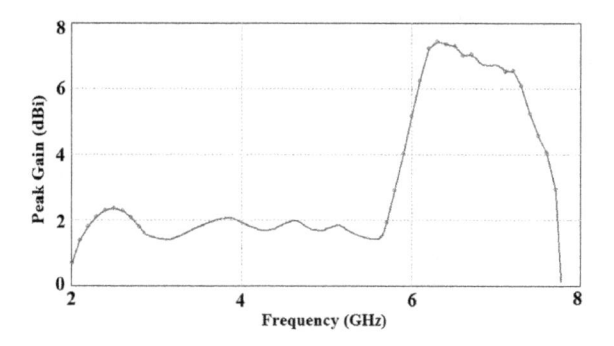

**Fig. 11** Simulated plot of peak gain of proposed antenna

of higher order modes. The simulated plot of gain for the proposed antenna is shown in Fig. 11. The proposed antenna shows broadside radiation pattern over the bandwidth with a maximum gain of 7.5 dBi.

## 3 Conclusions

A new wideband offset CPW-fed plus shaped fractal slot antenna is proposed. Using the offset feed location, an increased input impedance BW of 66% centered at around 6 GHz is obtained. The increased antenna bandwidth is due to the additional modes introduced in offset feed configuration. The antenna gives broadside radiation pattern over most of the BW.

## References

1. Jan, J., Su, J.: Bandwidth enhancement of a printed wide slot antenna with a rotated slot. IEEE Trans. Antennas Propag. **53**(6), 2111–2114 (2005)
2. Tseng, C.F., Huang, C.L.: Compact 2.5 GHz circularly polarized antenna using high permittivity substrates. In: Proceedings of the Asia–Pacific Microwave Conference, vol. 4, pp. 1–2. Suzhou, China, (Dec 2005)
3. Chaimool, S., Chung, K.L.: CPW fed mirrored-L monopole antenna with distinct triple bands for Wi-Fi and WiMAX applications. Electron. Lett. **45**(18), 928–929 (2009)
4. Deng, Z., Jiang, W., Gong, S., Xu, Y., Zhanng, Y.: A new method for broadening bandwidths of circular polarized microstrip antennas by using DGS & parasitic split ring resonator. Propag. Elect. Res. **136**, 739–751 (2013)
5. Jee, Y., Seo, Y.M.: Triple band CPW fed compact monopole antennas for GSM/PCS/DCS/WCDMA applications. Electron. Lett. **45**(9), 446–448 (2009)
6. Liu, W.C.: Design of a multiband CPW-fed monopole antenna using a particle swarm optimization approach. IEEE Trans. Antennas Propag. **53**(10), 3273–3279 (2005)
7. Werner, D.H., Ganguly, S.: An overview of fractal antenna engineering research. IEEE Antennas Propag. Mag. **45**(1), 38–57 (2003)
8. Vinoy, K.J., Abraham, J.K., Varadan, V.K.: On the relationship between fractal dimension and the performance of multi-resonant dipole antennas using Koch curves. IEEE Trans. Antennas Propag. **51**(9), 2296–2303 (2003)
9. Comisso, M.: Theoretical and numerical analysis of the resonant behaviour of the Minkowski fractal dipole antenna. IET Microw. Antennas Propag. **3**, 456–464 (2009)
10. Rani, S., Singh, A.P.: Modified Koch fractal antenna with asymmetrical ground plane for multi and UWB applications. Int. J. Appl. Electromagnet. Mech **42**, 259–267 (2013)
11. Rani, S., Singh, A.P.: On the design and optimization of new fractal antenna using PSO. Int. J. Electron. **100**(10), 1383–1397 (2012)
12. Orazi, H., Soleimani, H.: Miniaturisation of the triangular patch antenna by the novel dual-reverse-arrow fractal. IET Microw. Antennas Propag. **9**(7), 1–7 (2015)
13. Oraizi, H., Hedayati, S.: Miniaturization UWB monopole microstrip antenna design by the combination of Giusepe Peano and Sierpinski carpet fractal. IEEE Antennas Wirel. Propag. Lett. **10**, 67–70 (2011)
14. Kakkar, S., Rani, S.: Implementation of fractal geometry to enhance the bandwidth of CPW fed printed monopole antenna. IETE J. Res. **63**, 23–30 (2016)

# Multi-resonator Variations of 120° Sectoral Microstrip Antennas for Wider Bandwidth

**Amit A. Deshmukh, Pritish Kamble, Akshay Doshi and A. P. C. Venkata**

**Abstract** Sectoral Microstrip Antennas are realized by modifying the circular microstrip antennas. In this paper, proximity-fed design of 120° Sectoral microstrip antenna is discussed first. It yields bandwidth of nearly 650 MHz (~49%) with antenna gain of above 8 dBi in 1200 MHz frequency range. To enhance the bandwidth and gain of 120° Sectoral microstrip antenna, its various gap-coupled configurations coupled along the diagonal axes are proposed. Configurations with either single patch or two patches gap-coupled along diagonal axis are discussed with variations in the coupling edges. A maximum bandwidth of more than 1000 MHz (>70%) is realized for specific configurations of 120° patches, when they were coupled along both diagonal axes. In all gap-coupled variations, broadside MSA gain is nearly 9 dBi with pattern maximum in boresight direction is observed.

**Keywords** Broadband microstrip antenna · 120° sectoral microstrip antenna Gap-coupled configuration · Higher broadside gain

## 1 Introduction

Sectoral microstrip antenna (MSA) is a variation of circular MSA (CMSA) and for Sectoral angle in the range of 270°–340°, it gives additional resonating mode, whose frequency lies below the fundamental $TM_{11}$ mode frequency of circular patch [1, 2]. Bandwidth (BW) enhancement of Sectoral MSA (S-MSA) with angle in the range of 270°–340° has been obtained by using slot or tuning the Sectoral angle itself [1, 2]. But due to orthogonal resonating modal currents, radiation pattern exhibits higher cross polar content that leads to gain fluctuation in the boresight direction by 3–4 dBi [2]. However, as these antennas provide BW in the

---

A. A. Deshmukh (✉) · P. Kamble · A. Doshi · A. P. C. Venkata
Department of Electronics and Telecommunication Engineering,
SVKM's, D J Sanghvi College of Engineering, Mumbai, India
e-mail: amit.deshmukh@djsce.ac.in

range of 60–65%, even with higher cross-polar content, they may find usages in mobile communication systems, in which signal loss will be minimized due to larger cross-polarization levels. Over the past three decades, the simplest method to enhance the BW in MSA is the use of gap-coupled designs in which more than one resonating element has been used (parasitic patch) [3–5]. The BW in gap-coupled MSAs depends upon interspacing between frequencies of fed and gap-coupled elements, as well as the substrate parameters. Although gap-coupled technique increases antenna size, but it also provides gain enhancement. The 120° S-MSA is another variation of S-MSA and as compared with the CMSA it offers better results in terms of VSWR BW and gain characteristics [6].

In this paper, first the design of 120° S-MSA at fundamental mode frequency of 1200 MHz is discussed. It yields BW of 650 MHz (~49%) in 1200 MHz of frequency spectrum. The single 120° S-MSA yields peak gain of above 7.5 dBi. Further to enhance BW and the gain of 120° S-MSA, various gap-coupled variations of Sectoral patches are proposed. To reduce percentage increase in the patch area, parasitic elements are not gap-coupled along two coordinate axes, but they are gap-coupled along the diagonal axis, i.e., along $\Phi = 45°$ and 135° directions. Variations of gap-coupled designs using single or two patches coupled along varying patch edges are presented. Among all the proposed configurations, design using two 120° Sectoral patches coupled along both diagonal axis with their wider dimension toward fed 120° Sectoral patch, yields above 1000 MHz (>70%) of BW with peak antenna gain higher than 8.5 dBi. Although proposed gap-coupled variation requires larger patch size but it offers much wider BW as compared to that given by single circular element or other regular shape geometries as well as their slot cut variations. Various designs proposed in this paper are first studied using simulations using IE3D. Further experiments were carried out to validate simulations. Finite square ground plane having 30 cm side length is used in the experimentations. To feed the antennas, N-type connector in which to realize 50 $\Omega$ impedance, while using Teflon substrate in the coaxial cable, 0.32 cm of inner wire diameter is used. Further, in the experimentation, input impedance at an antenna input port and radiation pattern in the far-field distance, were measured using high-frequency $R$ and $S$ make instruments like "ZVH—8", "SMB 100A", and "FSC—6" inside an antenna lab.

## 2 Gap-Coupled Variations of 120° S-MSAs

Proximity fed design of 120° S-MSA on suspended glass epoxy substrate ($\varepsilon r = 4.3$, $h = 0.16$ cm, tan $\delta = 0.02$) is given in Fig. 1a, b. Use of air-suspended configuration improves antenna's BW and gain, whereas uses of suspended substrate improve the reliability of antenna parameters against substrate parameter variations. Proximity feeding technique is a variation of electromagnetic feeding method as employed in antennas [1]. Here instead of a resonating element, compact coupling strip is used to feed the energy to the radiating patch. Further, it is very simpler method to be implemented for use in the thicker substrate. For fundamental-mode

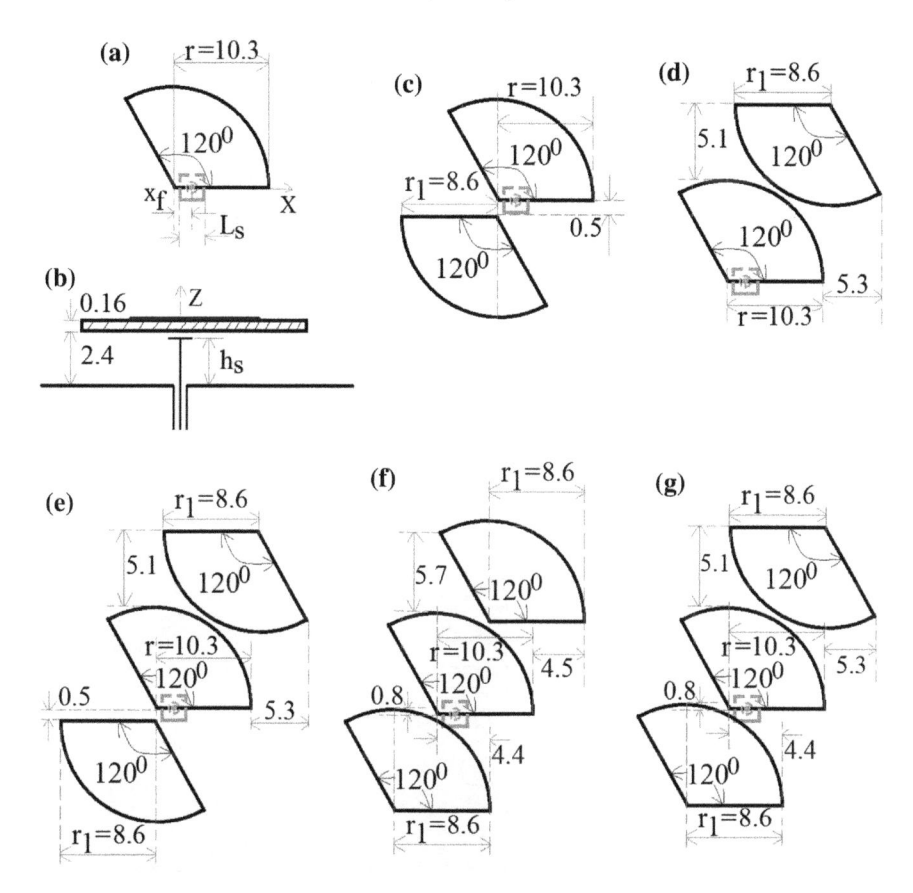

Fig. 1 a, b Proximity fed 120° S-MSA and c–g variations of proximity-fed gap-coupled 120° S-MSAs

frequency of 1200 MHz and the total substrate thickness of $0.1\lambda_0$, radius of 120° Sectoral patch is calculated using frequency equation for the same as reported in [6]. An optimized design of 120° S-MSA at 1200 MHz, yields two BWs namely simulated and measured, of 650 MHz (~49%). It exhibits pattern in broadside and shows a gain of above 7.5 dBi. Various gap-coupled configurations of 120° S-MSA were further studied for enhancement in BW and gain. The patches are coupled along the diagonal axis (to reduce overall increase in size) as given in Fig. 1c–g. In each of the gap-coupled variations radius of parasitic Sectoral patch is taken to be less than that of fed patch. For the configuration as shown in Fig. 1c, variations in modal frequencies on two Sectoral patches with parasitic patch radius variations are shown in Fig. 2a. With decrease in '$r_1$' fundamental-mode frequency on parasitic Sectoral MSA increases and maximum BW is obtained for '$r_1$' = 8.6 cm, when two-mode frequencies on fed and parasitic patches are optimally spaced. Simulated Smith chart for this variation is provided in Fig. 2b.

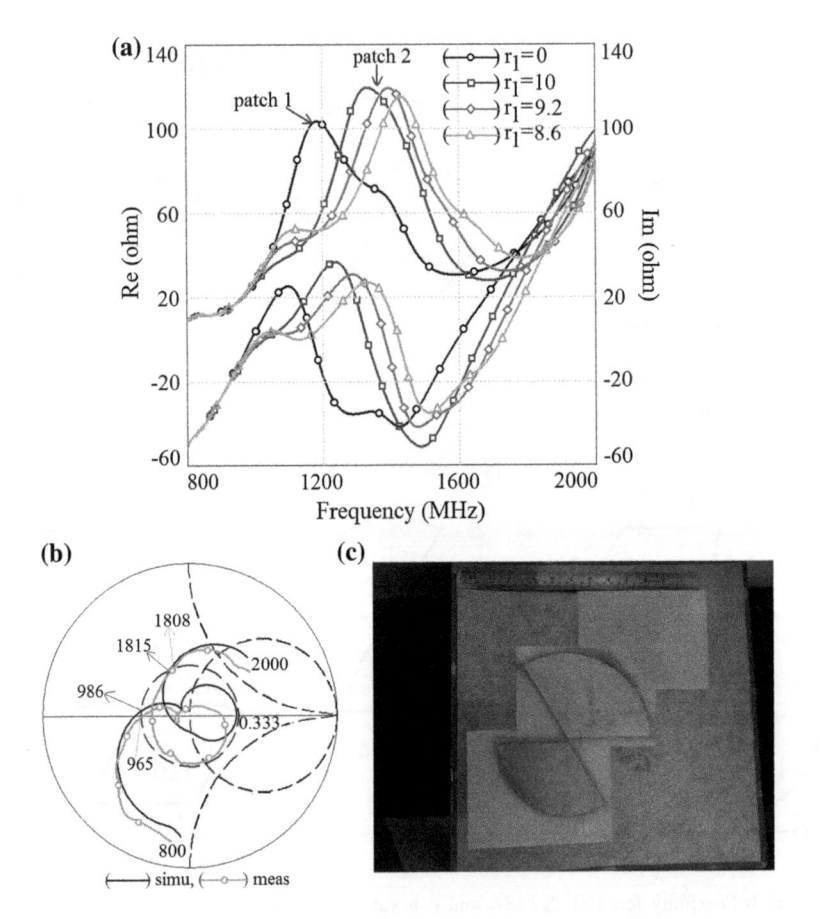

**Fig. 2** **a** Resonance curve plots for gap-coupled proximity-fed 120° S-MSA as shown in Fig. 1c, its **b** Smith chart and the **c** fabricated prototype

For this design, BW using simulation is 829 MHz (59.2%). An experiment is carried out by fabricating MSA on the suspended substrate. BW observed in experimentation is 843 MHz (61.2%). Antenna's fabricated prototype is given in Fig. 2c. For this gap-coupled design, pattern exhibits maxima in the boresight direction as shown in Fig. 3a, b. However, as the position of proximity strip is not symmetrical with respect to two patches, the radiation pattern shows maxima away from the boresight direction. Due to radial direction surface currents, E- and H-planes are aligned along $\Phi = 45°$ and $135°$, respectively. The simulated gain plot over BW for this gap-coupled design is shown in Fig. 3c. The peak antenna gain is more than 8 dBi. The reduction in gain toward higher frequencies of BW is attributed to the shifting of radiating pattern ways from the boresight direction, i.e., from along, $\Phi = 0°$ and $\theta = 0°$. Similarly gap-coupled variation of 120° S-MSA as shown in Fig. 1d is optimized for wider BW. In this variation, simulated and

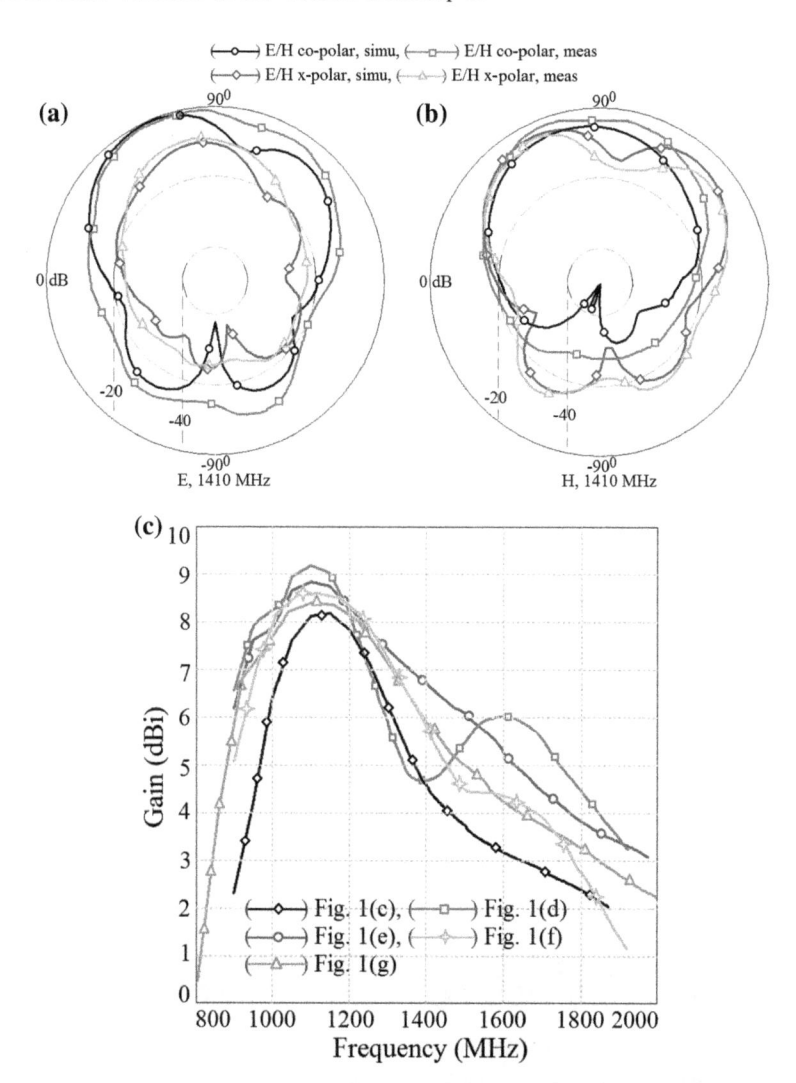

**Fig. 3 a, b** Pattern plots at BW's center frequency for MSA given in Fig. 1c and **c** plots for gain variations across BW using simulations in gap-coupled proximity-fed 120° S-MSAs

measured BWs are, 735 MHz (57.2%) and 755 MHz (59.6%), respectively as given in Fig. 4a. This configuration also exhibits boresight radiation pattern but it yields peak gain of more than 9 dBi.

Instead of using single-parasitic patch, BW enhancement is also studied by using two parasitic patches as shown in Fig. 1e–g. Here in three designs, radius of parasitic 120° S-MSAs is kept the same. This will yield nearly the same parasitic patch resonance frequency (difference in two frequencies may be present due to varying mutual coupling between fed Sectoral patch to parasitic Sectoral patches with

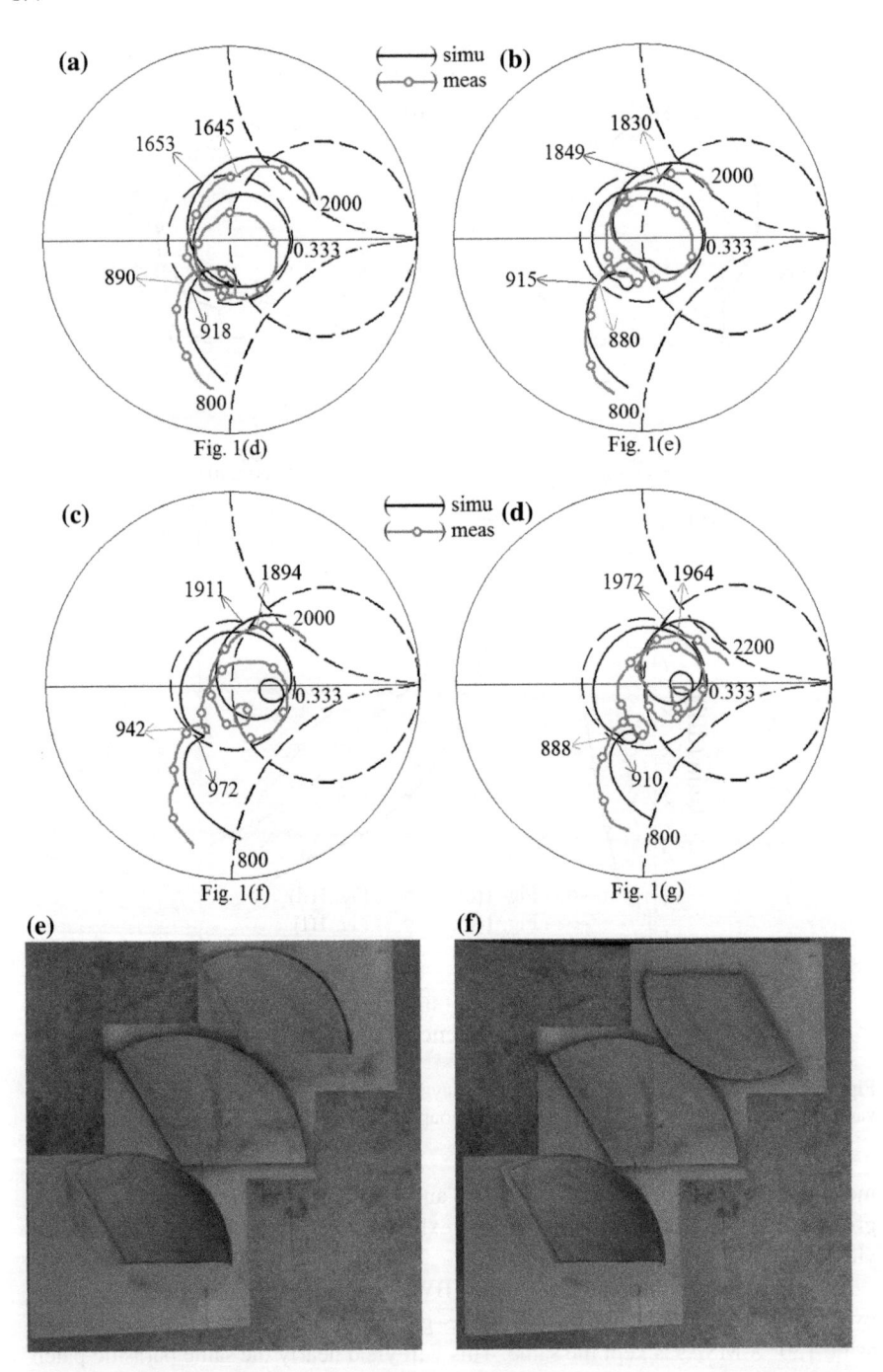

**Fig. 4 a–d** Optimum Smith charts highlighting the realized BW for gap-coupled designs as provided in Fig. 1d–g, and **e, f** respective fabricated prototypes

respect to changing field distributions around the patch periphery). The dimensions of the optimized MSAs are given in Fig. 1e–g.

For the antenna as given in Fig. 1e, BWs using simulation and measurement are 934 MHz (67.6%) and 950 MHz (70.11%), respectively, as provided in Fig. 4b. Similarly these two BWs for multi-resonator antenna as shown in Fig. 1f are 930 MHz (65.1%) and 952 MHz (67.13%), respectively, as shown in Fig. 4c. For the antenna as shown in Fig. 1g, a maximum BW is realized. Here the simulated BW is 1062 MHz (73.7%) as given in Fig. 4d.

Experiment has been carried out and measured BW is 1076 MHz (75.2%). Fabricated prototypes for some of the gap-coupled designs are presented in Fig. 4e, f. The pattern in these gap-coupled designs is in boresight directions. Due to patches present on both side of fed patch, pattern maxima variation away from boresight direction over BW is lower. All these designs yield peak gain of more than 8.5 dBi.

## 3 Conclusions

Broadband design of 120° S-MSA in 1200 MHz is discussed. The multi-resonator gap-coupling technique of BW and gain enhancement is further used. Various gap-coupled configurations of 120° Sectoral patches coupled along with diagonal axes are presented. In the configuration of 120° S-MSAs gap-coupled with their larger dimension toward the fed MSA, obtains maximum BW of larger than 1000 MHz (>70%). It yields boresight pattern showing peak gain of higher than 8.5 dBi. The VSWR BW is more than the BW obtained from regular shape MSAs and their gap-coupled and slot cut variations.

## References

1. Deshmukh, A.A., Jain, A.R., Ray, K.P.: Broadband 270° sectoral microstrip antenna. Microw. Opt. Technol. Lett. **56**(6), 1447–1449 (2014)
2. Deshmukh, A.A., Phatak Neelam, V.: Broadband sectoral microstrip antennas. IEEE Antennas Wirel. Propag. Lett. **14**, 727–730 (2015)
3. Lee, H.F., Chen, W.: Advances in Microstrip and Printed Antennas. Wiley, New York (1997)
4. Kumar, G., Ray, K.P.: Broadband Microstrip Antennas, 1st edn. Artech House, USA (2003)
5. Garg, R., Bhartia, P., Bahl, I., Ittipiboon, A.: Microstrip Antenna Design Handbook. Artech House, USA (2001)
6. Deshmukh, A.A., Kamble, P., Doshi, A., Issrani, D., Ray, K.P.: Proximity fed broadband 120° sectoral microstrip antenna. In: Proceedings of ICACC—2017, 22–24 Aug 2017, Kochi, India

# Partial Corner Edge-Shorted Rectangular Microstrip Antenna Embedded with U-slot for Dual-Band Response

Amit A. Deshmukh and Mohil Gala

**Abstract** A design of dual-band rectangular microstrip antenna embedded with conventional U-slot and loaded with a partial shorting edge is presented. The paper presents a detailed analysis, which brings out the resonant mode explanation for dual frequency response. The U-slot first reduces and tunes the spacing between second-order shorted $TM_{3/4,0}$ mode and fundamental shorted $TM_{1/4,0}$ mode. This optimum spacing between two frequencies yields dual frequency response. Due to unidirectional current variations across shorted and slotted patch, antenna exhibits broadside radiation pattern at dual frequencies.

**Keywords** Shorted rectangular microstrip antenna · Dual-band microstrip antenna U-slot · Higher order mode

## 1 Introduction

A design of U-slot cut microstrip antenna (MSA) was first invented in 1995 [1], which has the advantage of increased antenna bandwidth (BW) without any increments in the patch size [1]. Further U-slot and its modified variations have been extensively used in realizing wideband as well as dual-band response [2–7]. In all these broadband designs using U-slot, the position of slot is symmetrical with respect to the coaxial feed. This yields optimum input impedance for to realize impedance matching within VSWR = 2 circle, for maximum range of frequencies yielding wider BW [2–6]. In dual-band antennas employing U-slot, the slot is either cut symmetrical with respect to feed or it is cut at an offset point with respect to feed [7]. Using half wavelength approximation for U-slot length, resonant length formulation for the resonant mode due to the slot is reported [8]. However, the prominent research gap which was observed for U-slot cut antenna designs was an explanation for wider patch BW in terms resonating modes and design procedure to realize similar designs at different frequencies and on varying substrate thickness. Recently for U-slot cut designs, a detailed study highlighting upon the resonant mode explanation for the wideband behavior, as well as resonant length formulation for rectangular MSA (RMSA) embedded with symmetrical U-slot is reported [9, 10]. That study shows that higher order resonant modes are altered in their frequency and impedance, due to U-slot, which yields wider BW. The formulations as reported in [10] were proposed only for the thicker substrate designs ($>0.07\lambda_0$). In shorted patches, wideband and dual-band response has also been realized by cutting U-slot [11–13]. These compact MSAs have been obtained using shorting technique and a larger reduction in frequency is obtained when shorting is employed towards the corner of the patch with smaller shorting ratio [14]. The dual-band response for corner shorted rectangular MSA (RMSA) is obtained by embedded U-slot inside the patch [13]. It yields two frequencies in 2400 and 5200 MHz frequency spectrum. As given in [13], 2400 MHz is governed by shorting patch mode where the summation of shorted patch dimensions equals quarter wave in length. It is reported that 5200 MHz frequency is governed by U-slot, wherein the inner length of slot equals quarter wave in length [13]. However using slot formulation, prediction of calculated length is not provided as well as effects of variation in slot frequency with an aspect ratio of U-slot is not given. Thus a research gap in form of modal explanation about dual-band functioning for reported design is noticed.

In this paper, in order to bring out resonant mode explanation for reported variation of shorted U-slot cut RMSA, detailed analysis for equivalent corner partial edge shorted U-slot cut RMSA is presented here. In the proposed study as against single shorted RMSA with U-slot, here partial corner edge shorted RMSA with U-slot is considered. The resonant mode identification for shorted RMSA is reported in [15]. As per those modal descriptions, in the discussed configuration here, U-slot is found to be reducing second-order $TM_{3/4,0}$ mode frequency and due to its optimum spacing with reference to the $TM_{1/4,0}$ mode, dual frequency response was observed. Further feed position realizes input impedance matching at altered dual resonant modes which yield optimum BW at the dual frequencies. Thus work presented here explains functioning of partial corner-shorted rectangular patch embedded with U-slot, in terms of patch resonating modes. This was the research gap in available work on corner shorted dual spectrum antenna using U-slot. As explanations presented here gives insight into antenna working, based on these new findings, resonant length formulation for dual resonant modes as a function of patch and slot dimensions can be formulated. They can be helpful in designing of the identical antenna in the different frequency spectrum. The present study was carried out using the infinite ground plane. The selection of finite ground plane size can affect mainly the impedance at modified resonant modes than the frequency, hence the proposed study here is also applicable for the finite ground plane. Since in simulations, taking infinite ground plane reduces computation time, hence the same was considered.

## 2 Analysis of Dual-Band Shorted U-Slot Cut RMSA

Dual-band planar inverted F-antenna (PIFA) (corner shorted U-slot cut RMSA) for wireless applications as reported in [13] is shown in Fig. 1a. The dual band is realized for the frequency band at 2400 and 5200 MHz as given in Fig. 1b. To bring out a resonant modal explanation for dual-band antenna, its analysis is discussed in this work. As against single shorted design as given in Fig. 1a, partial corner edge-shorted RMSA embedded with U-slot as given in Fig. 1c is selected in the present study. Here as against single shorting, due to the selection of partial corner edge shorting, resonant modes present in the patch does not change. Here fundamental quarter wavelength-mode frequency will only increase due to a reduction in effective quarter wavelength on the partial shorted patch [14].

The MSA as shown in Fig. 1c is simulated and its smith chart and impedance plots exhibiting variation in real and imaginary part of impedance over frequencies are given in Fig. 1d, e. Here patch and slot dimensions are $L = 15$ mm, $W = 12.5$ mm, $L_2 = 10.85$ mm, $W_2 = 4.2$ mm, $L_3 = 1.25$ mm and $G_1 = G_2 = 1.4$ mm. The antenna is fed using a coaxial feed of diameter 1.2 mm. The resonance curve shows two resonant peaks at 2.096 and 4.6 GHz. Dual band is observed in impedance locus with frequencies inside VSWR = 2 circle in Smith chart, that give BW from 2.54 to 2.805 GHz for the first band and BW from 5.271 to 5.366 GHz for the second band in the simulations. In this design, largely currents at dual resonant modes are directed along shorted patch length as shown in Fig. 1f, g. But towards second frequency they are seen to be more circulating around U-slot length. To find more details about two resonant modes, analysis of slot cut-shorted antenna is discussed.

To further get a clear insight initially the shorted RMSA without U-slot is studied. Simulated resonance curve plot for shorted RMSA is shown in Fig. 2c. The current distribution at observed peaks in resonance curve plot is given in Fig. 2a, b. At first frequency peak, a single quarter wavelength variation for currents is noted along shorted patch length (from the diagonal corner and towards shorting post). This variation is the result of dominant $TM_{1/4,0}$ mode [15]. At next resonant mode, current shows three-quarter wavelength variation along the shortened length and hence it is a shorted $TM_{3/4,0}$ mode. At $TM_{1/4,0}$ and $TM_{3/4,0}$ modes, due to horizontal currents, E-plane is along $\Phi = 0°$. Further U-slot with dimensions '$L_h$' and '$L_v$' is introduced inside partially corner edge shorted RMSA, as shown in Fig. 1c. First vertical length '$L_v$' is introduced and subsequently the horizontal length '$L_h$'. For an increase in '$L_v$', both mode frequencies remain constant. This is because slot length is not present nearer to maximum current locations at two resonating modes. Further horizontal slot '$L_h$' is increased in length. Reduction in $TM_{1/4,0}$ mode frequency with increase in '$L_h$' is lesser as slot length is parallel to its modal surface currents. At $TM_{3/4,0}$ mode, initially when surface currents are being parallel to '$L_h$', it exhibits lesser reduction in frequency. But further increase in '$L_h$' the $TM_{3/4,0}$ mode frequency reduces and it shows three-quarter wavelength variation from inside the slot cut rectangular patch. Modal surface current and relevant field distributions against increasing slot dimensions are given in Fig. 3a–d. As observed

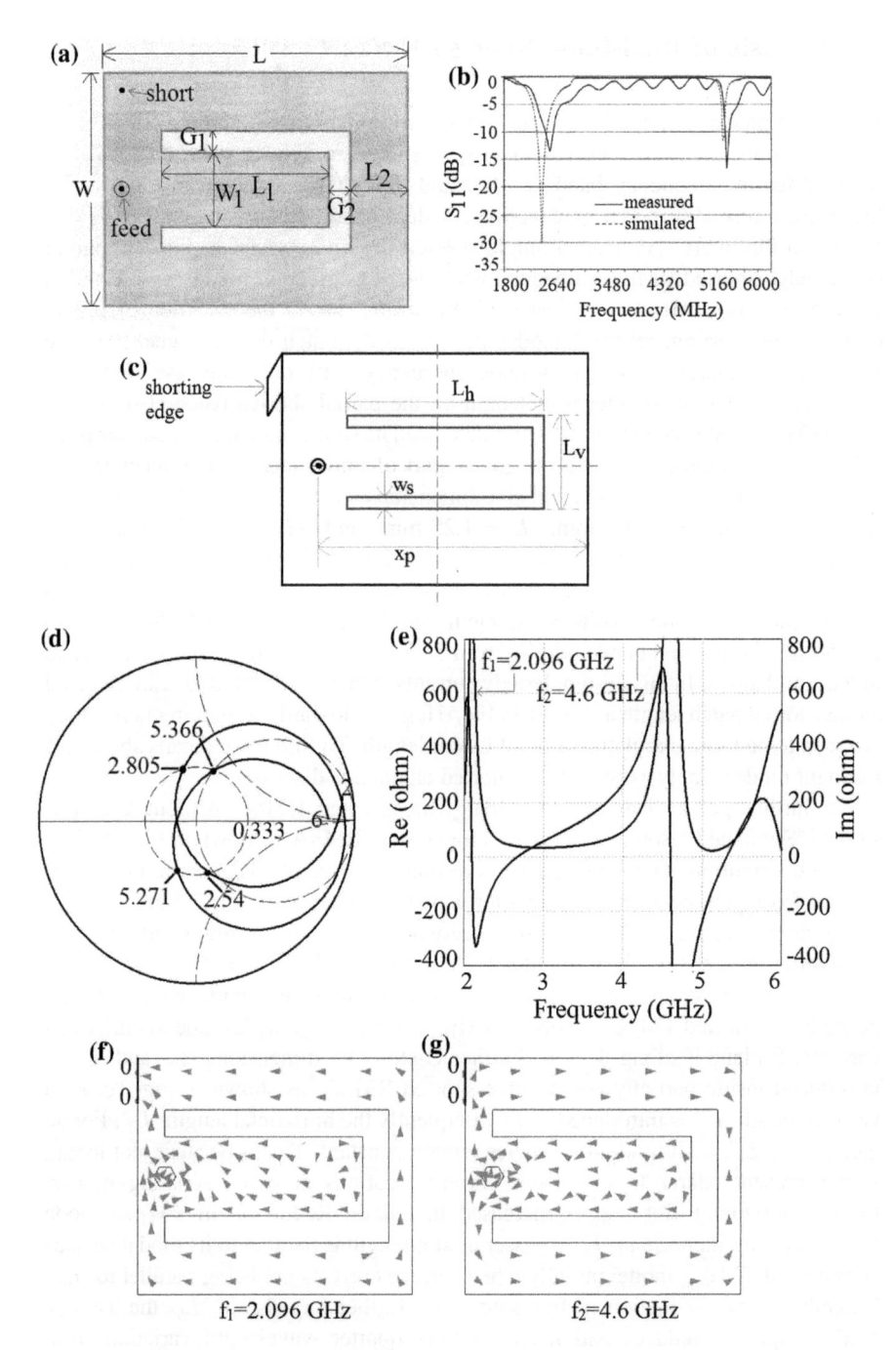

**Fig. 1** a Single-shorted U-slot cut RMSA and its **b** return loss plots [13], **c** partial corner edge-shorted U-slot cut RMSA, its **d** Smith chart and **e** resonance curve plots and its **f, g** surface current distribution at observed resonant modes

from current/field distributions, U-slot position aligns maximum of currents along the horizontal direction. Thus it can be inferred from this study that U-slot tunes the $TM_{3/4,0}$-mode frequency with reference to $TM_{1/4,0}$ mode, and results in dual-band response. The E-plane in radiation pattern at both the modes remains along $\Phi = 0°$ due to maximum horizontal current components. Thus a detailed analysis of the PIFA shorted U-slot cut RMSA as reported in [13] is presented here. This kind of modal explanation was not given in [13] and thus this is the novelty in this paper.

Based on these new findings, resonant length formulations for U-slot cut shorted antenna at modified $TM_{1/4,0}$ and $TM_{3/4,0}$ modes will be developed in the future scope of the present study. Using them some design guidelines will be realized that will give dual-band response at any desired frequency range. The similar study also can be further extended to shorted RMSA with varying shorting ratio (i.e. ratio of shorting edge to the patch width) or for a single shorting post loaded (corner or edge center shorted) RMSA.

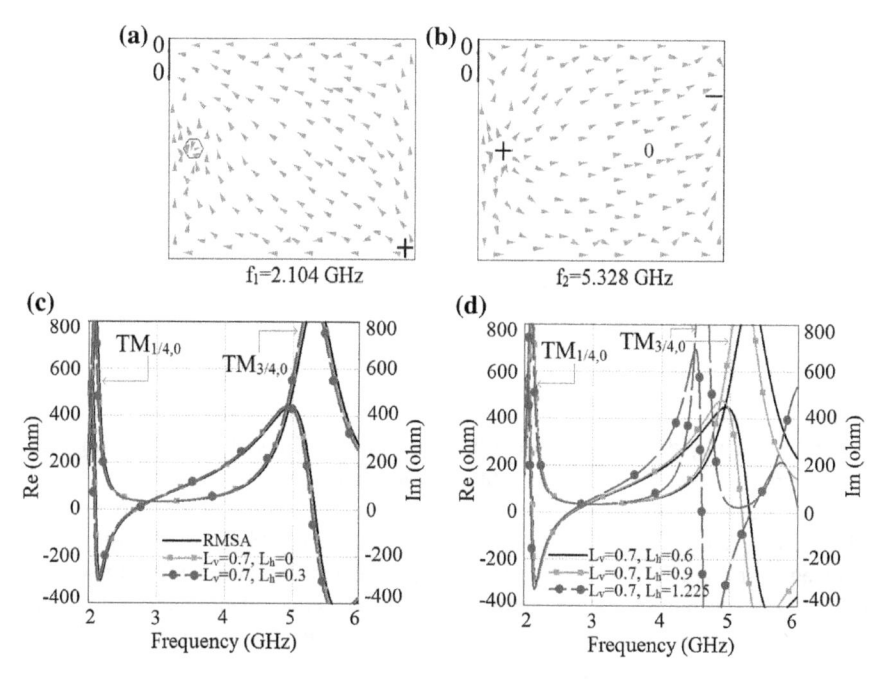

**Fig. 2** **a, b** Surface current distributions at observed resonant peaks and **c, d** impedance plots for partial corner edge shorted RMSA embedded with U-slot cut

**(a)** f₁=2.1 GHz, $L_h$=0.6 cm

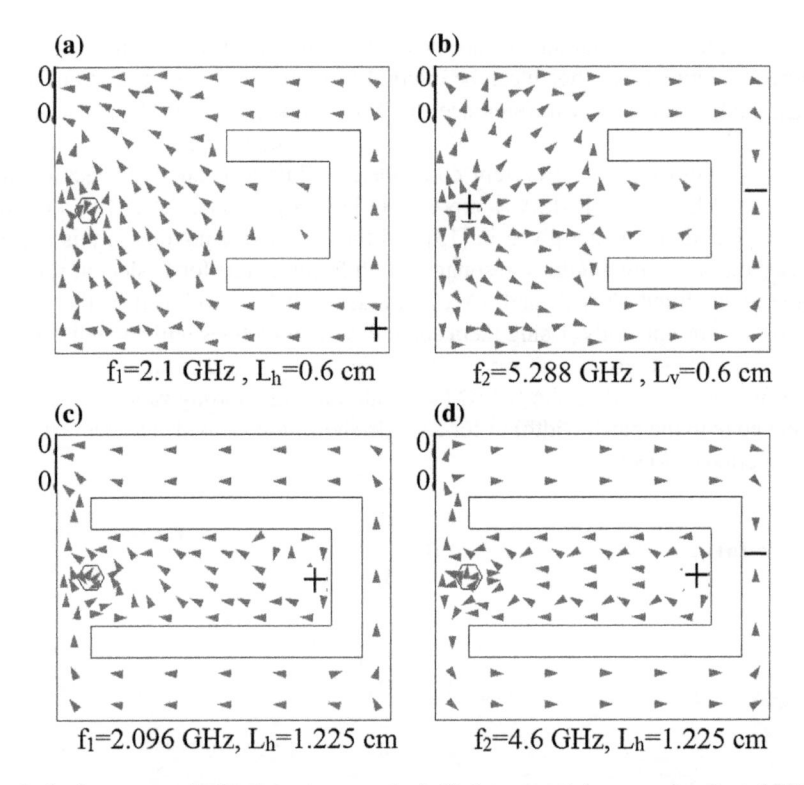

**(b)** f₂=5.288 GHz, $L_v$=0.6 cm

**(c)** f₁=2.096 GHz, $L_h$=1.225 cm

**(d)** f₂=4.6 GHz, $L_h$=1.225 cm

**Fig. 3** Surface current distributions at two modes in U-slot cut partial corner edge shorted RMSA

## 3 Conclusions

The dual frequency design using a combination of partial corner edge shorting and U-slot in RMSA is discussed. The detailed study is presented that gives an insight into the functioning of shorted patch dual frequency design. The U-slot realize tuning of frequency and input impedance at $TM_{1/4,0}$ and $TM_{3/4,0}$ modes of partial corner edge shorted RMSA that yields dual frequency response. Since slot yields unidirectional current variation across shorted slot embedded patch, broadside pattern is noticed at two frequencies. This modal analysis was not presented in reported work for corner shorted dual-band RMSA with U-slot, which is explained here. This is the novelty in presented work.

## References

1. Huynh, T., Lee, K.F.: Single layer single patch wideband microstrip antenna. Electron. Lett. **31**(16), 1310–1312 (1995)

2. Lee, H.F., Chen, W.: Advances in Microstrip and Printed Antennas. Wiley, New York (1997)
3. Kumar, G., Ray, K.P.: Broadband Microstrip Antennas, 1st edn. Artech House, USA (2003)
4. Wong, K.L.: Compact and Broadband Microstrip Antennas. Wiley, New York (2002)
5. Luk, K.M., Lee, K.F., Tam, W.L.: Circular U-slot patch with dielectric superstrate. Electron. Lett. **33**(12), 1001–1002 (1997)
6. Wong, K.L., Hsu, W.H.: Broadband triangular microstrip antenna with U-shaped slot. Electron. Lett. **33**(25), 2085–2087 (1997)
7. Lee, K.F., Luk, K.M., Mak, K.M., Yang, S.L.S.: On the use of U-slots in the design of dual and triple band patch antennas. IEEE Antennas Propag. Mag. **53**(3), 60–74 (2011)
8. Ghalibafan, J., Attari, A.R.: A new dual-band microstrip antenna with U-shaped slot. Prog. Electromagn. Res. C, **12**, 215–223 (2010)
9. Deshmukh, A.A., Ray, K.P.: Analysis of broadband variations of U-slot cut rectangular microstrip antennas. IEEE Mag. Antennas Propag. **57**(2), 181–193 (2015)
10. Deshmukh, A.A., Ray, K.P.: Analysis and design of broadband U-slot cut rectangular microstrip antennas. Accepted for publication in Sadhana—Academy Proceedings in Engineering Science. Springer, Berlin (2017)
11. Deshmukh, A.A., Kumar, G.: Compact broadband U-slot loaded rectangular microstrip antennas. Microw. Opt. Technol. Lett. **46**(6), 556–559 (2005)
12. Virga, K.L., Rahmat-Samii, Y.: Low profile enhanced-bandwidth PIFA antennas for wireless communications packaging. IEEE Trans. Microw. Theory Tech. 45, 1879–1888 (1997)
13. Salonen, P., Keskilammi, M., Kivikoski, M.: Single-feed dual-band planar inverted—antenna with U-shaped slot. IEEE Trans. Antennas Propag. **48**(8), 1262–1264 (2000)
14. Garg, R., Bhartia, P., Bahl, I., Ittipiboon, A.: Microstrip Antenna Design Handbook. Artech House, USA (2001)
15. Deshmukh, A.A., Ray, K.P.: Analysis of shorted plate compact and broadband microstrip antenna. IEEE Mag. Antennas Propag. **55**(6), 100–113 (2013)

# Novel π-Shape Microstrip Antenna Design for Multi-band Response

Amit A. Deshmukh, Archana Nishad, Gauri Gosavi, Priyanka Narayanan, Siddharth Nayak and Aarti G. Ambekar

**Abstract** The novel design of π-shape microstrip antenna for multi-band dual-polarized response is presented. The proposed work explains the modal distributions at various resonant modes in π-shape patch. By analyzing current distributions at its modes, the formulation for resonant length is presented. The frequencies calculated using them for a given patch dimension yields closer agreement with simulated data showing error less than 5%. Further by suitably selecting feed location, an optimum design of π-shape patch for triple frequency response is presented. It yields 1–1.5% of impedance bandwidth at every frequency. The pattern over the three frequencies exhibits maxima in boresight direction that also shows reduced cross-polar radiation component.

**Keywords** π-shape microstrip antenna · Triple-frequency microstrip antenna
Dual polarization · Higher order resonant mode · Resonant length formulation

## 1 Introduction

Frequently selected shapes of the radiating elements in microstrip antenna (MSA) are square/rectangular, circular, triangular (more commonly equilateral triangle) [1, 2]. The reason behind the selection of these regular shapes is in the ease of fabrication with simplicity in their mathematical modeling [1–3]. For e.g., for MSA on the thinner substrate and while carrying out multi-port network modeling of the same, Green's function that is used to model fields below the patch, is available for regular shape geometries which makes modeling of MSA less complex [1–3]. Multi-band antennas with single and dual polarizations are needed when signal trans-reception is

---

A. A. Deshmukh (✉) · A. Nishad · G. Gosavi · P. Narayanan · S. Nayak · A. G. Ambekar
Department of Electronics and Telecommunication Engineering, SVKM's,
D J Sanghvi College of Engineering, Mumbai, India
e-mail: amit.deshmukh@djsce.ac.in; amitdeshmukh76@gmail.com

A. G. Ambekar
e-mail: aarti.ambekar@djsce.ac.in

required at closely spaced frequencies [4]. By using slots embedded in the patch or by placing stubs on the patch edges, multiple frequency responses in regular shape MSA has been realized [5–10]. Without using these modifications dual or triple frequency response with similar radiation characteristics cannot be realized in regular shape geometries. Therefore different modified shapes are needed that can give multiple frequencies without the slot or stub. Another parameter that comes in the design of multiple frequency antennas is the proximity of each frequency with reference to each other. In regular shape geometries patch modal frequencies follow fix frequency separation (decided by sinusoidal progression in a rectangular patch or Bessel function roots as in the circular patch). Recently a design of π-shape MSA using shorting post for tunable dual frequency response was reported [11]. In [11], an explanation for resonant modes of π-shape patch as well as for its shorted variation is not provided. The results reported in [11] were just supported by the parametric study but an insight into antenna functioning is completely absent.

In this paper, π-shape MSA as reported in [11] is discussed first. Further design of π-shape MSA without using shorting post is investigated for resonance in 1000 MHz frequency spectrum. In the present study, FR4 substrate with parameters, $\varepsilon_r = 4.3$, tan $\delta = 0.02$, $h = 0.16$ cm, is chosen. In π-shape MSA, due to circulation of modal surface currents around modified patch length and stubs that are placed towards the corners of modified shape patch, triple frequency response was observed. A detailed study to explain the effects of various patch parameters that forms π-shape patch, using impedance/resonance curve plots and modal surface current distributions is presented. It is observed that at various patch modes, with variation in feed positions with reference to the selected location of feed, the impedance across fundamental and higher order modes either increases or decreases. Therefore for optimum response proper positioning of feed location is important. By selecting the same, an optimum triple frequency response with 1–1.5% of VSWR bandwidth (BW) at each frequency, is obtained. Across first three resonant modes, surface current shows orthogonal variations in the maximum amplitudes of surface currents. Due to this, radiation pattern exhibits dual-polarized response with bore-sight radiation pattern (with maximum along the normal to the plane of MSA) at triple frequencies. Measurements were carried out for the optimized design that shows closer agreement with the simulated data. By investigating the variation in modal surface currents around patch length, formulation for resonant mode length, as function of π-shape patch dimension is presented. The frequencies which are evaluated using these equations at three modes agree closely with results obtained using simulations. The MSA presented here is first studied using simulations by using the method of moments based IE3D which was followed by the measurements. Using high-end instruments like, "ZVH—8", "SMB 100A", and "FSC—6", inside Antenna lab experimentation was completed. In simulations and measurements, finite ground plane size is selected. In fabrication, a ground plane size is taken to be more than six times the substrate thickness in all directions with respect to patch boundaries. The SMA panel type connector is used to feed MSA, in which inner wire diameter of 0.12 cm is used, which for the dielectric constant of 2.1 (Teflon substrate) yields 50 Ω impedance.

Novel π-Shape Microstrip Antenna Design … 187

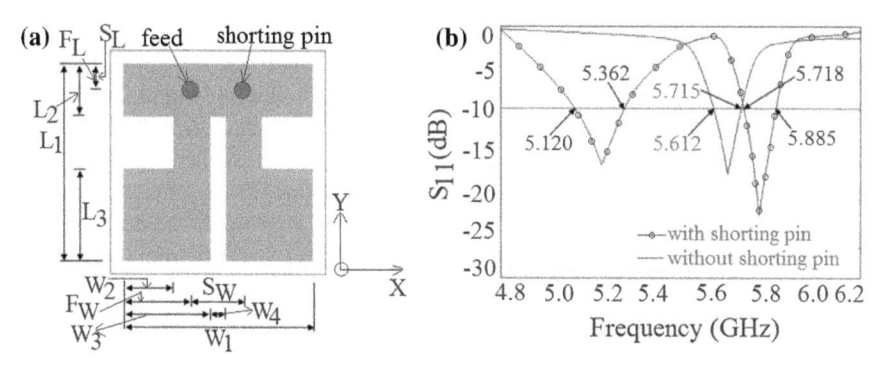

Fig. 1 a π-shape MSA and its b return loss ($S_{11}$) plots [11]

## 2 Shorted π-Shape MSA

Reported design of shorted π-shape MSA is explained in Fig. 1a [11]. Here MSA is realized using Taconic substrate ($h = 0.157$ cm, $\varepsilon_r = 2.17$). The response of MSA for the dimension as given in [11] yields dual-band response in 5.5 GHz frequency spectrum as given in Fig. 1b. The π-shape MSA without shorting yields single band frequency response as shown in Fig. 1b. Firstly in [11] which resonant modes of π-shape MSA that give this closely spaced dual frequency response is not explained in detail. Also whenever new radiating shape is presented, the discussion about its fundamental and higher order modes should have been presented, which reflects the novelty of proposed structure against conventional regular shape geometries. However in [11], no such discussion is outlined that gives an insight into modal field variations. Therefore in present work here, π-shape MSA is studied.

## 3 π-Shape MSA

For to resonate in 1000 MHz frequency band, a design of π-shape MSA is given in Fig. 2a. To investigate into the resonant mode behavior of this novel structure, detail study with parametric variations for different patch parameters is investigated. The resonance/impedance plots for π-shape MSA that provides the variations in real and imaginary part of input impedance against frequency are shown in Fig. 2b–d. Also surface current distributions at first three observed resonant modes are given in Fig. 2e–g. In Fig. 2b, excitation of different resonant modes in patch is explained. The impedance at first peak increases for feed at location "D", whereas at other feed locations, impedance reduces. It is observed that for feed variation from "C" to "A", impedance reduces to zero. Impedance is higher at feed locations of "A" and "D" at second peak but it reduces for feed at locations "B" and "C". At third peak, impedance seems to be smaller for feed at positions "A", "B" and "D", whereas it is around

30 Ω for feed at point "C". Thus as seen from the impedance variation and current distribution along patch length, they exhibits variation in half wavelength at the first mode. Along the same length, variation of two and three half wavelengths, at second and third peak is observed, in currents. At first mode, maximum magnitude of currents are directed along horizontal direction (i.e., along x-axis), hence E-plane of radiation is along the same axis. Due to wavelength variation across patch, maximum of surface currents is directed along Y-axis that makes E-plane directed along the same. At third peak, again effective current maximum (considering three half wavelength variations) is directed along horizontal direction hence E-plane is along X-axis. Thus as seen from these field/current distributions, E-plane of radiation across three resonant mode varies. Therefore, π-shape patch will give dual-polarized multi-band response. Further with the variation in stub length "Lst", three frequencies reduces. This will give rise to multi-band dual-polarized response with frequency tuning.

Due to multiples of half wavelength variations along π-shape MSA realizing impedance matching at three frequencies is important. The impedance scaling at three frequencies can be realized by controlling slot parameter "x". This length helps into realizing matching of impedance across three frequencies for inside VSWR = 2 circle. For the MSA dimensions shown in Fig. 2a and for "x" = 0.6 cm, "Lst" = 1.9 cm, an optimum triple-frequency response is realized and Smith charts for the same is provided in Fig. 3a. Frequencies and respective BW using simulations are, 620, 935, 1432 MHz and 10, 15, 19 MHz. The fabrication of MSA was done and measurement for same was carried out using ZVH—8. The corresponding measured values are, 605, 920, 1420 MHz and 11, 17, 21 MHz. The picture of the fabricated antenna is given in Fig. 3b. Radiation pattern across three frequencies is measured inside Antenna Lab. In pattern measurement, necessary minimum far field distance between antenna under test and reference Horn antenna is kept [4]. Against simulated pattern, measured pattern plots are given in Figs. 3c, d and 4a–d.

Across three frequencies, the pattern maximum is in the boresight directions. At first and third frequencies, E-plane has directed along Φ = 0°. It is directed along Φ = 90°, at the second frequency. Since the antenna is fabricated on FR4 substrate its gain at three frequencies is lower. Improvement in gain can be obtained using suspended configuration. Further by studying surface current distributions at three resonant modes, formulations for modal resonant lengths are proposed. The formulations for lengths are based upon the respective effective current lengths at three frequencies as given in Fig. 5a–c. The resonant length equations at three modes are realized by using Eqs. (1)–(3).

At $TM_{10}$ mode,

$$l_{r10} = 2\left(l_{st} + \frac{l_2}{2}\right) + 2\left\{w_1 - \left(\frac{w_4 + w_1 - w_2}{2}\right)\right\} + l_3 + 2\left(\frac{l_2}{2}\right) \quad (1)$$

Novel π-Shape Microstrip Antenna Design ...

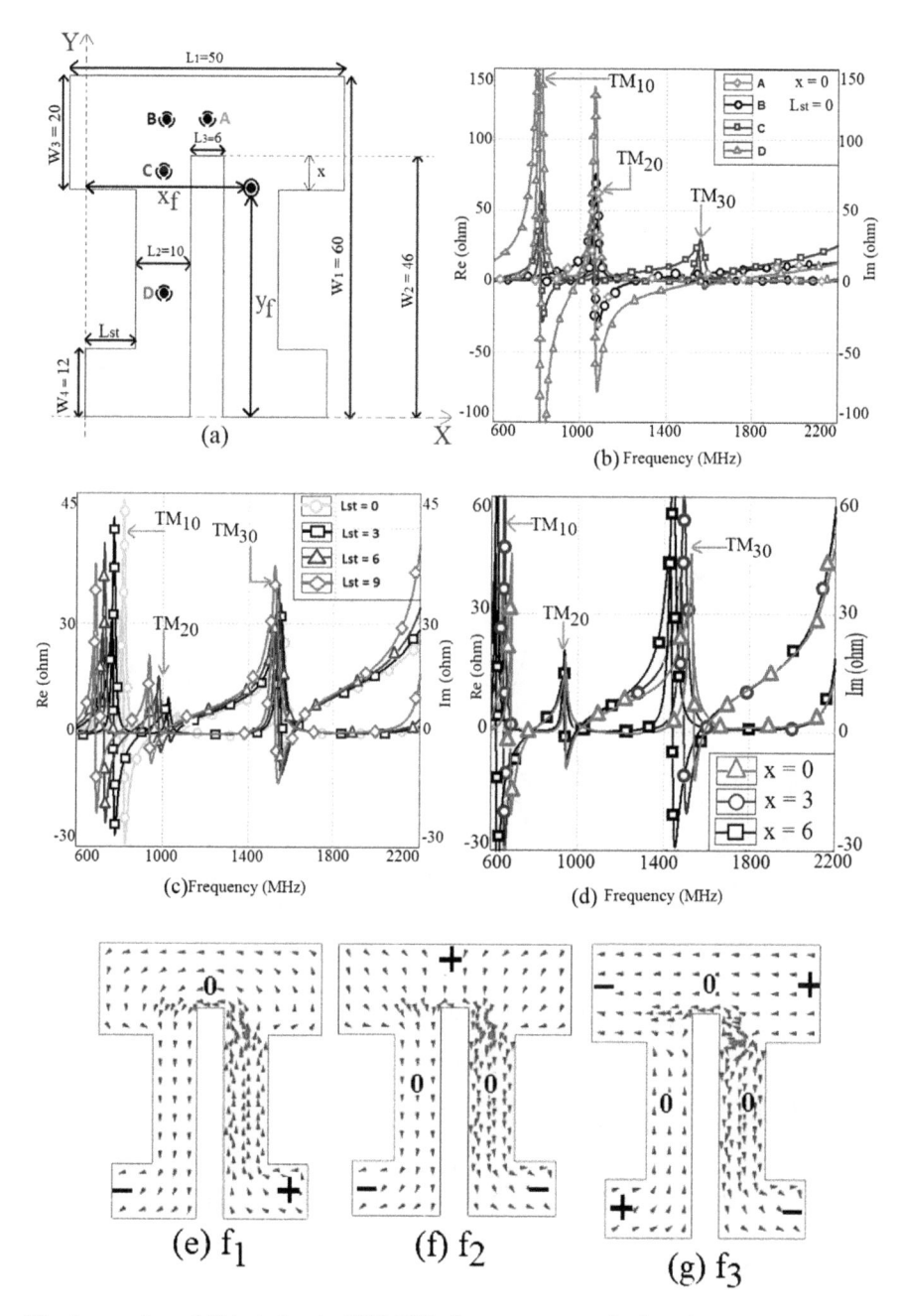

**Fig. 2** **a** π-shape MSA design in 1000 MHz frequency range, its impedance curve plots for variation in **b** feed location, **c** stub length and **d** slot length "$x$", **e–g** surface current distributions at three modes for π-shape MSA

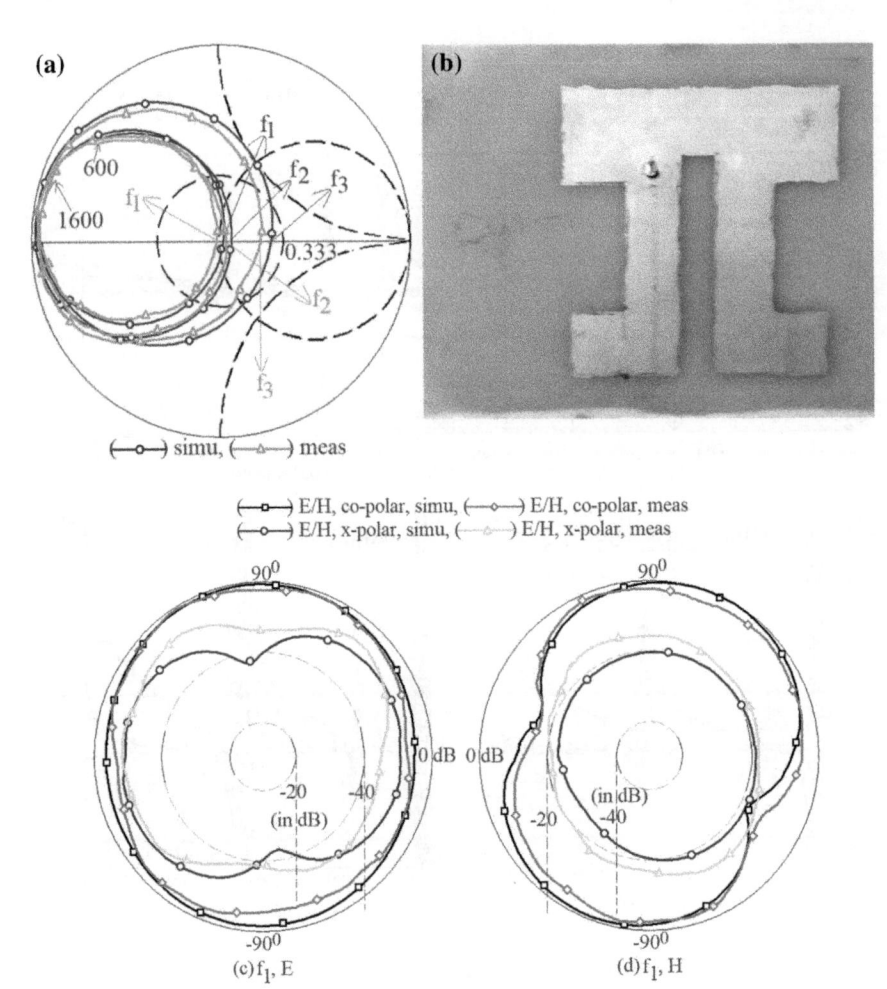

**Fig. 3** a Input impedance plots, **b** fabricated MSA, and **c, d** radiation pattern at first frequency in triple band dual-polarized π-shape MSA

At $TM_{20}$ mode,

$$l_{r20} = \left(l_{st} + \frac{l_2}{2}\right) + \left(w_1 - \frac{w_4}{2}\right) + \left\{\frac{l_1}{2} - \left(\frac{l_3 + l_2}{2}\right)\right\} \quad (2)$$

At $TM_{30}$ mode,

$$l_{r30} = l_1 + 2\left\{w_1 - \left(\frac{w_1 - w_2}{2} + \frac{w_4}{2}\right)\right\} + 2\left(l_{st} + \frac{l_2}{2}\right) \quad (3)$$

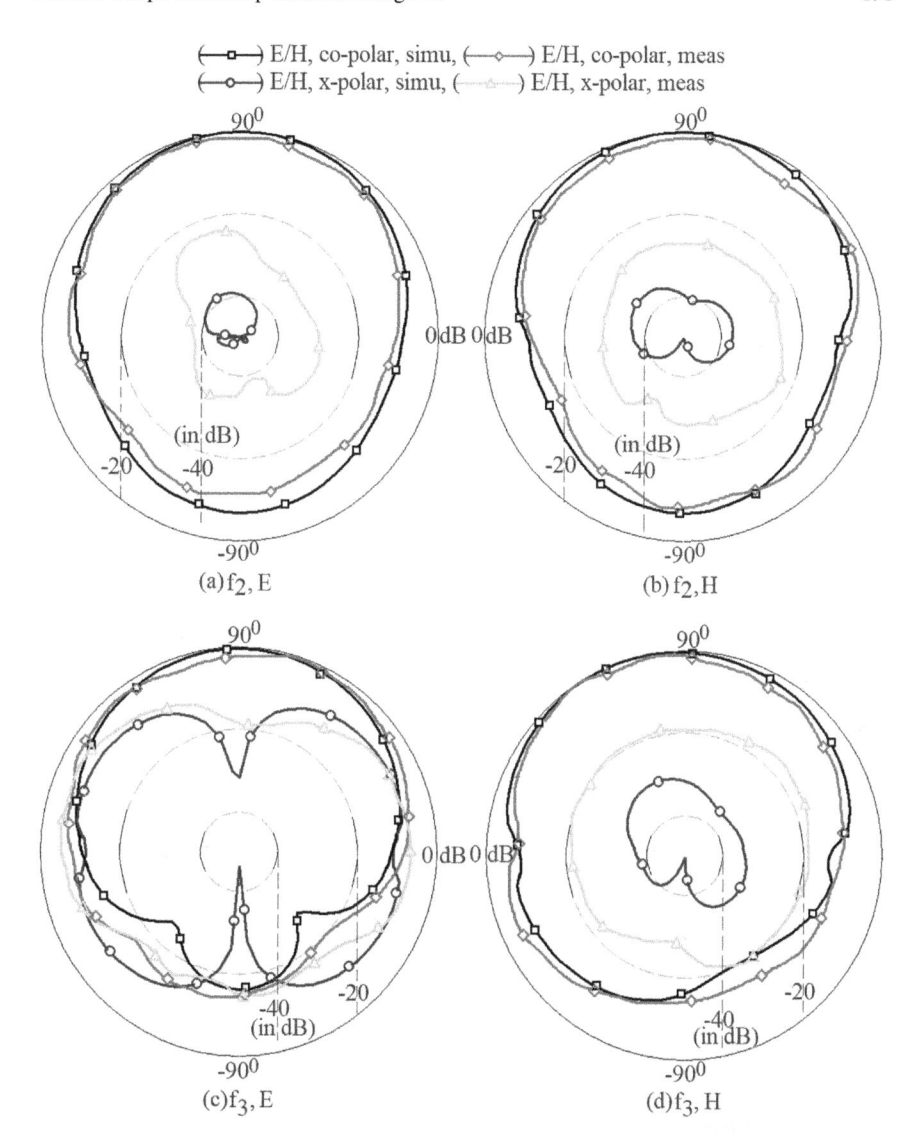

**Fig. 4** a–d Pattern plots at second and third frequency in triple-band dual-polarized π-shape MSA

At TM$_{10}$ mode, length as calculated by using Eq. (1) is equated to half wavelength, whereas length as obtained from Eq. (2), at TM$_{20}$ mode, it is equated to wavelength variation. Further, at TM$_{30}$ mode, resonant length is made equal to three times half the wavelength. At three modes, the effective value of dielectric constant is calculated using Eq. (4). This takes into account effective patch width as seen by surface currents along π-shape length. The frequency is calculated by using Eq. (5), where $m = 1$ for TM$_{10}$, $m = 2$ for TM$_{20}$ and $m = 3$ for TM$_{30}$ mode. The % error

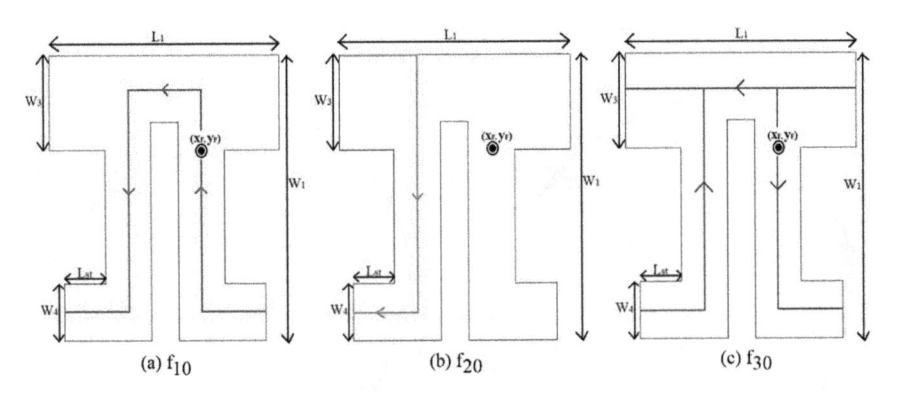

**Fig. 5** a–c Effective resonant lengths at three resonant modes for π-shape MSA

between simulated and calculated values is obtained by using Eq. (6). At first mode i.e. $TM_{10}$, calculated frequency is found to be 598 MHz as against simulated frequency of 591 MHz (% error 1.14%). At $TM_{20}$ mode, simulated and calculated frequencies are, 880 and 928 MHz, showing % error of 1.3%. Finally, at $TM_{30}$ mode, simulated and calculated frequencies are, 1391 and 1425 MHz that shows % error of 1.32%. Thus, proposed formulations for resonant length yields closer agreement of respective frequency with the simulated value.

$$\varepsilon_{re} = \frac{\varepsilon_r + 1}{2} + \frac{\varepsilon_r - 1}{2\sqrt{1 + \frac{12h}{w}}} \quad (4)$$

$$f_{m,0} = \frac{m * c}{2 * l_{m,0} * \sqrt{\varepsilon_{re}}} \quad m = 1, 2, 3 \quad (5)$$

$$\%\text{error} = \left[\frac{f_{\text{calculated}} - f_{\text{simulated}}}{f_{\text{simulated}}}\right] * 100 \quad (6)$$

## 4 Conclusions

Multi-band dual-polarized design of novel π-shape MSA is proposed. In this design, multi-band response is obtained without using the additional slot that is embedded in patch or without using additional stub that is placed on the patch edges. The narrow slit and additional patch length (that acts like a tuning stub) on π-shape patch realizes tuning of frequencies and impedance at them that yields optimum VSWR BW of around 1.5%, at respective frequencies. At each of the frequency, antenna shows broadside pattern that shows dual polarization across three modal frequencies. Further for three modes in π-shape patch, their formulation at respective resonant length is presented. Frequencies calculated using them at different modal frequencies agree closely with the simulated data. As the antenna is investigated on the lossy FR4 substrate, the antenna gain is poor. The same can be increased by using its suspended variation that will enhance the antenna BW.

# References

1. Carver, K.R., Mink, J.W.: Microstrip antenna technology. IEEE Trans. Antennas Propag. **29**, 2–24 (1981)
2. James, J.R., et al.: Some recent development in microstrip antenna design. IEEE Trans. Antennas Propag. **29**, 124–128 (1981)
3. James, J.R., Hall, P.S.: Handbook of Microstrip Antennas, vol. 1. Peter Peregrinus Ltd, London (1989)
4. Balanis, C.A.: Antenna Theory: Analysis and Design, 2nd edn. Wiley, New York (1997)
5. Garg, R., Bhartia, P., Bahl, I., Ittipiboon, A.: Microstrip Antenna Design Handbook. Artech House, USA (2001)
6. Kumar, G., Ray, K.P.: Broadband Microstrip Antennas, 1st edn. Artech House, USA (2003)
7. Wong, K.L.: Compact and Broadband Microstrip Antennas, 1st edn. Wiley, New York (2002)
8. Dahele, J.S., Lee, K.F.: Theory and experiment on microstrip antennas with air gap. In: IEE Proceedings on Microwaves, Antennas & Propagation, vol. 132, pp. 455–460 (1985)
9. Chen, J.S., Wong, K.L.: A Single-layer dual-frequency rectangular microstrip patch using a single probe feed. Microw. Opt. Technol. Lett. **11**(2), 83–84 (1996)
10. Daniel, A.E., Kumar, G.: Tunable dual and triple frequency stub loaded rectangular microstrip patch antenna. In: IEEE AP-S International Symposium Digest, pp. 2140–2143 (1995)
11. Choi, S.H., Kwak, D., Lee, H.C., Kwak, K.S.: Design of a dual-band π-shaped microstrip patch antenna with a shorting pin for 5.2/5.8 GHz WLAN systems. Microw. Opt. Technol. Lett. **52**(4), 825–827 (2010)

# The Design of Wideband E-shape Microstrip Antennas on Varying Substrate Thickness

Amit A. Deshmukh and Divya Singh

**Abstract** Slot cut patch antennas like C-shape, double C-shape, E-shape and U-slot cut designs are widely reported. Recently comparison between these designs is reported. However, that study does not provide any modal comparison for a wideband response. Also, identical patch dimensions were not considered in the comparison. In proposed work, for same patch dimensions, comparison for C-shape, double C-shape, E-shape and U-slot cut RMSA in terms of their resonant mode functioning is presented. Based on the explanation for resonating modes present in E-shape patch, resonant length formulations for modified patch modes in terms of slot and patch dimensions is proposed. The frequencies obtained using them agrees closely with simulated results. Further using proposed formulation design procedure to realize similar E-shape patch antenna at different frequencies and on varying substrate thickness is presented. This procedure yields wideband response for different slot and substrate parameters.

**Keywords** Broadband microstrip antenna · Compact microstrip antenna
C-shape microstrip antenna · Double C-shape microstrip antenna
U-slot · Pairs of rectangular slots · Higher order mode

## 1 Introduction

By embedding slot/slit in the patch, compact microstrip antenna has been realized [1, 2]. By suitably selecting slot position, wideband response in MSA has been obtained [3–8]. The role of slot in compact MSA is to alter the surface current distribution at fundamental mode, yielding reduction in frequency. Whereas in broadband slot cut MSA, it introduces additional resonant mode that enhances the bandwidth (BW). To provide resonant mode explanation, a detailed study for

---

A. A. Deshmukh (✉) · D. Singh
Department of Electronics and Telecommunication Engineering,
SVKM's, D J Sanghvi College of Engineering, Mumbai, India
e-mail: amitdeshmukh76@gmail.com

© Springer Nature Singapore Pte Ltd. 2018
H. Vasudevan et al. (eds.), *Proceedings of International Conference on Wireless Communication*, Lecture Notes on Data Engineering and Communications Technologies 19, https://doi.org/10.1007/978-981-10-8339-6_22

U-slot-cut rectangular MSA (RMSA) is reported in [9, 10]. It is shown that wideband response is the result of coupling between patch's fundamental mode to that with reduced frequency of orthogonal higher order mode. A comparative study for compact C and double C-shape MSAs along with wideband E-shape and U-slot cut RMSA is reported in [11]. However, based on the resonant modes, comparison for them in terms of radiation characteristics, polarization, VSWR BW, is not provided. Also, the reported comparative study is not presented for identical patch dimensions in every case.

In this paper, first, detailed study that gives comparison between compact C-shape MSA variations and broadband E-shape and U-slot cut RMSA variations for resonance frequency of 1000 MHz is presented. In C and double C-shape MSAs observed frequency is due to reduced $TM_{01}$ mode, whereas widened BW in E and U-slot introduced antenna is because of coupling between $TM_{10}$ and $TM_{02}$ modes. Wideband response with dual polarization has been realized in C-shape MSA due to coupling between $TM_{10}$ and $TM_{01}$ modes. Wideband E-shape MSA is variation of double C-shape MSA but due to symmetrical feed position, E-shape design gives wider BW with a single polarization. Further detailed study for wideband E-shape MSA that provides effects of slot and substrate thickness dimensions on VSWR BW, is presented. By analyzing the variation in modal frequencies against slot and substrate parameters, resonant length formulations for E-shape MSA is provided. Calculated frequencies using the same agree closely with simulated results. Using proposed formulations, design procedure that yields similar antennas at a different frequency on varying substrate thickness and slot parameters is presented which gives wideband response in each case. For the given patch dimensions, C and double C-shape MSAs offers size reduction by 60 and 70%, respectively, with lower BW, in comparison with square MSA. For high BW applications, U-slot and E-shape designs are preferred. To select the definite type of antenna for given application, comparative details giving pros and cons of these antennas would be quite helpful. Hence the novelty in proposed paper lies in bringing out that detailed study. Proposed equations can be used to design E-shape MSA at different frequencies on given substrate thickness.

## 2 Comparison of Slot Cut Compact and Wideband RMSAs

Comparison of C and double C-shape MSAs, E and U-slot cut RMSAs is presented for equivalent RMSA dimension of length '$L$' = 130 and width '$W$' = 150 mm on air substrate. First C-shape MSA as shown in Fig. 1a, b is studied and its impedance plot for $h$ = 10 mm, coaxial feed at $x_f$ = 60, $y_f$ = 10 and for varying slot length of '$l_s$', with width '$w_s$' = 8 mm and y = 45 mm, using IE3D is shown in Fig. 1c. For a given feed location, resonant peaks due to $TM_{10}$, $TM_{01}$ and $TM_{11}$ modes are observed in equivalent RMSA. With the increase in '$l_s$', $TM_{01}$ and $TM_{11}$

mode frequencies decrease since they have current contributions orthogonal to the slot. Here, $TM_{10}$ mode frequency remains unchanged. Current distributions at these modes for $l_s$ = 40 mm, is shown in Fig. 1d–f. At reduced $TM_{01}$ mode, currents are along patch width. In modified $TM_{11}$ mode, they show variations in length and width. The contribution of currents along length increases with slot length at modified $TM_{11}$ mode. For slot length of 60 mm, $TM_{01}$ mode frequency is lower than $TM_{10}$ mode frequency. Further frequency of modified $TM_{11}$ mode is nearer to $TM_{10}$ mode frequency. This close proximity of frequencies yields loop position inside VSWR = 2 circle in Smith chart as shown in Fig. 1g yielding wideband response. Further, along with modified $TM_{01}$ mode, C-shape patch yields dual frequency wideband response with dual polarization. At $TM_{01}$ mode (i.e. $f_1$), BW is 12 MHz ($\geq 1.6\%$) and due to $f_2$ ($TM_{10}$) and $f_3$ ($TM_{11}$), broadband response with single polarization is achieved, yielding the BW of 121 MHz ($\geq 10\%$). Here average antenna gain over BW is above 8 dBi.

Further double C-shape MSA is studied as shown in Fig. 2a, b. Here input impedance curves are shown in for $h$ = 10 mm, $x_f$ = 50, $y_f$ = 40 and for varying slot length of '$l_s$', and width '$w_s$' = 8 mm, and $y$ = 45 mm. Here feed position is selected which is similar to the position as given in [11] in double C-shape MSA. For equivalent RMSA, here resonant peaks due to $TM_{10}$ and $TM_{01}$ modes are noticed over a frequency range of 800–1200 MHz. With increasing slot length, $TM_{01}$ modal frequency reduces but the frequency of $TM_{10}$ mode remains

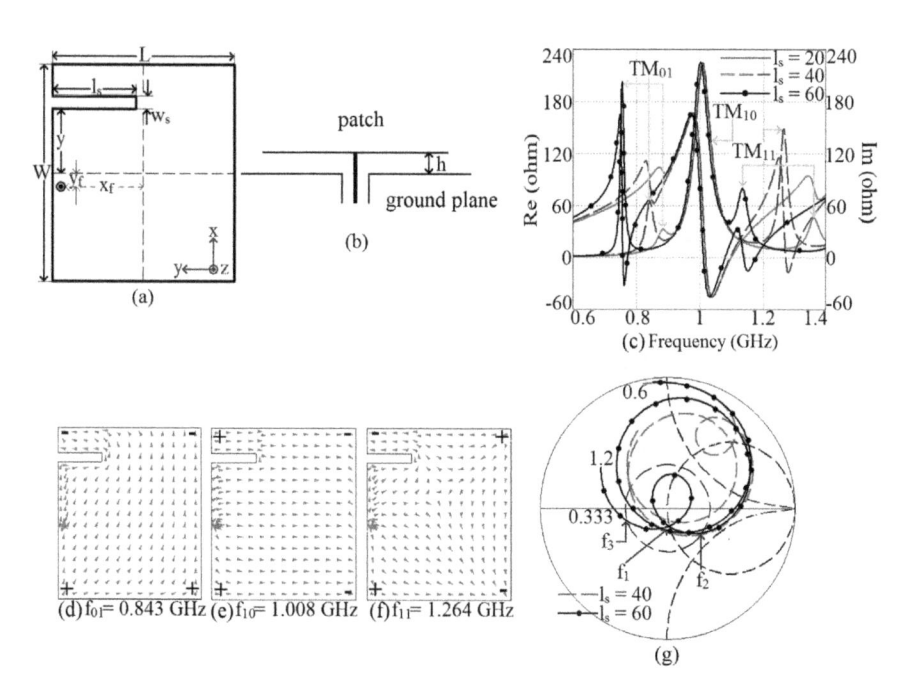

**Fig. 1** **a, b** C-shape MSA its **c** impedance plots with varying '$l_s$', **d–f** current distributions at three resonant modes and **g** smith charts for '$l_s$' = 60 mm

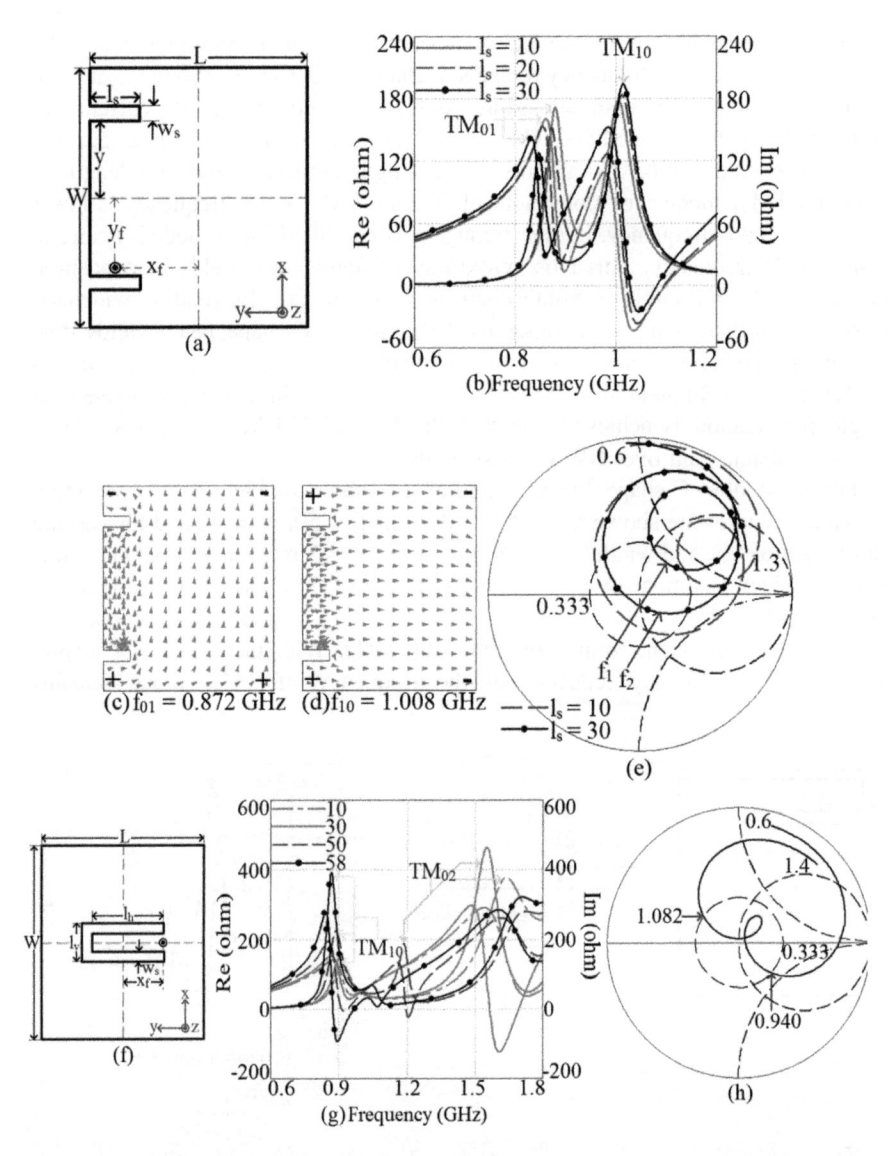

**Fig. 2 a** Double C–shape MSA, its **b** resonance curve plots, **c, d** surface current distributions at two modes, **e** optimum smith charts, **f** RMSA with U-slot, its **g** impedance plots and **h** optimum smith chart

unchanged. The Smith charts for increase in '$l_s$' are shown in Fig. 2c. For slot length '$l_s$' = 30 mm, dual frequency response showing the formation of impedance loci inside the VSWR = 2 circle at two modes, is observed. MSA gives BW of 11 MHz ($\geq 1.2\%$) at $TM_{01}$ mode i.e. $f_1$, and at $TM_{10}$ mode i.e. $f_2$, it gives BW of 52 MHz ($\geq 4.8\%$). As currents are orthogonal to two resonant modes, radiation pattern shows dual polarized response with broadside pattern. Here peak antenna gain is higher than 8 dBi across two frequencies. In comparison, C-shape MSA yields dual polarized dual and wideband response whereas double C-shape MSA can give dual polarized dual or multi-band response. Pattern and gain responses are similar in two cases. Since the MSA yields compact size (reduction for $TM_{01}$ mode frequency), antenna BW at two modes is smaller. The further similar study is carried out for U-slot cut RMSA as shown in Fig. 2f. As reported in [9, 10], reductions in $TM_{02}$ mode frequency with reference to $TM_{10}$ mode is observed as shown in Fig. 2g which yields wider BW as shown in Fig. 2h. Here BW of 142 MHz ($\geq 14\%$) is obtained. The pattern across BW is boresight with lower cross-polar radiation component and an average gain of above 7 dBi. As compared with C and double C-shape MSAs, this design yields better BW and pattern characteristics.

The analysis for U-slot has also been carried out for varying air substrate thickness with changing the aspect ratio for U-slot, which shows similar variations in modal frequencies. Also, formulations for modified resonant modes were developed for changing U-slot aspect ratio on varying substrate thickness. Those equations are similar to the formulations as given in [10]. Lastly, similar study is carried out for E-shape MSA as shown in Fig. 3a. Here detailed study is carried out for varying substrate thickness and slot parameters. For $h$ = 20 mm and $y$ = 30 mm, resonance curve plots indicate reduction in $TM_{02}$ mode frequency with reference to $TM_{10}$ mode as given in Fig. 3b. Also they yield re-orientation for current components at $TM_{02}$ mode to yield unidirectional current variation to realize wider BW of 166 MHz ($\geq 18\%$) as given in Fig. 3c. Similarly wideband E-shape MSA is optimized for wideband characteristics for different $h$ and two slot positions 'y', and impedance curves and Smith charts for these variations are shown in Fig. 3d–g. Results for all variations are summarized in Table 1.

With the increase in 'y' impedance at $TM_{02}$ mode as well as the VSWR BW increases. On average value, frequency ratio between two modal frequencies is around 0.7 when slots are closely spaced (lower 'y') whereas for when the spacing between slots is wider, then frequency ratio is around 0.8. Further after the studying surface current distributions at modified modes in E-shape patch, the formulation for resonating length at two frequencies is proposed. At $TM_{10}$ mode, slot width perturbs the surface current components, whereas at $TM_{02}$ mode slot length increases modal currents lengths. This variation in lengths is formulated using Eqs. (1)–(6).

$$\Delta l = \left( \frac{0.7 * h}{\sqrt{\varepsilon_{re}}} \right) \tag{1}$$

$$L_e = L + (2 * \Delta l) + \frac{w_s}{2} \tag{2}$$

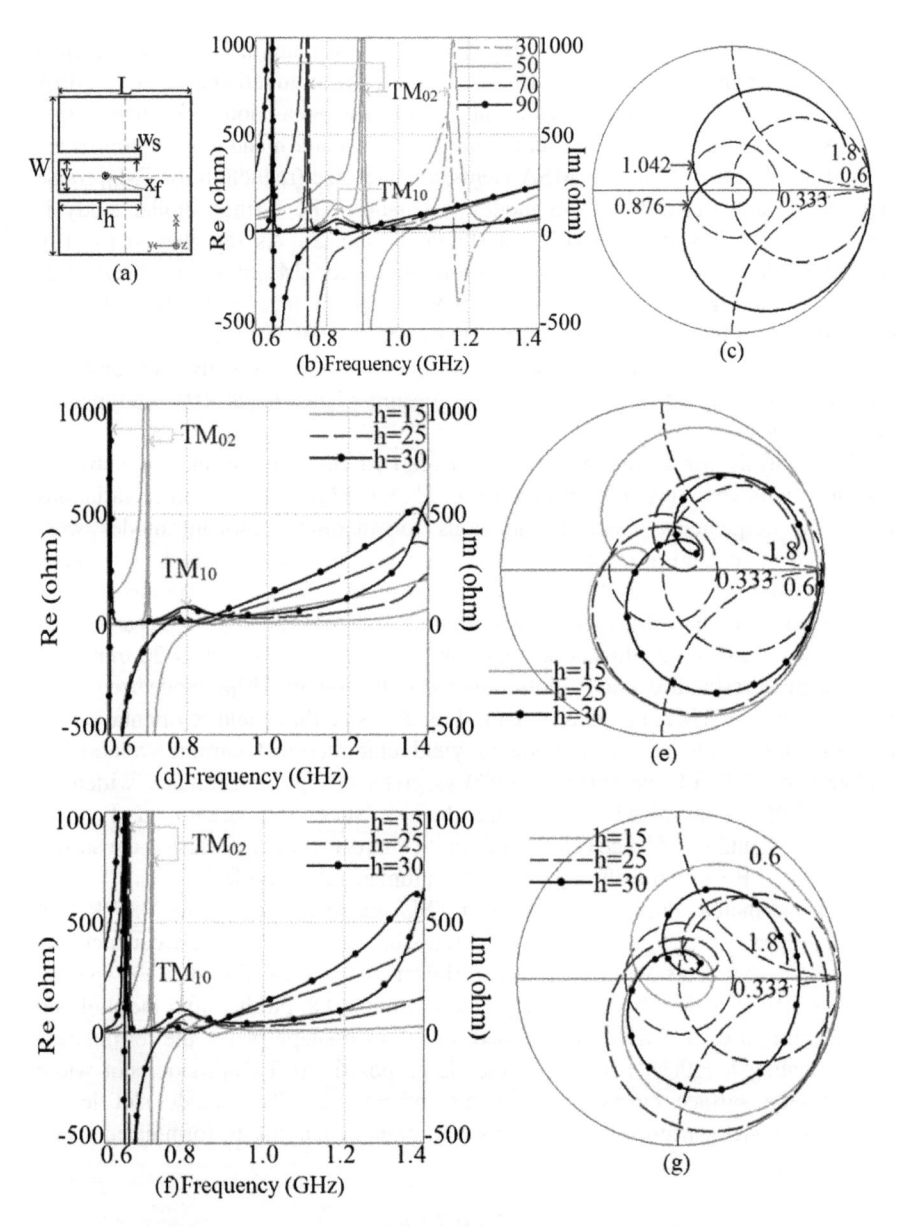

**Fig. 3** **a** E-shape MSA, its **b** resonance and **c** impedance plots for $h$ = 20 mm and $y$ = 30 mm, **d** resonance curve and **e** smith chart for E-shape MSA for variation in '$h$' and $y$ = 30 mm, **f** resonance curve and **e** smith chart for E-shape MSA for variation in '$h$' and $y$ = 50 mm

**Table 1** Optimized E-shaped MSAs for varying substrate thickness and slot position

| H (mm) | y (mm) | $l_h$ (mm) | $x_f$ (mm) | BW (GHz) | $f_{10}$ (GHz) | $f_{02}$ (GHz) | $f_{02}/f_{10}$ ratio |
|---|---|---|---|---|---|---|---|
| 15 | 30 | 80 | −20 | 67 | 0.96 | 0.76 | 0.79 |
| 20 | 30 | 90 | −30 | 166 | 0.92 | 0.69 | 0.75 |
| 25 | 30 | 100 | −35 | 170 | 0.90 | 0.62 | 0.69 |
| 30 | 30 | 100 | −35 | 175 | 0.88 | 0.6 | 0.68 |
| 15 | 50 | 73 | −30 | 197 | 0.96 | 0.80 | 0.83 |
| 20 | 50 | 80 | −35 | 184 | 0.92 | 0.74 | 0.80 |
| 25 | 50 | 85 | −35 | 186 | 0.90 | 0.70 | 0.78 |
| 30 | 50 | 86 | −45 | 194 | 0.88 | 0.68 | 0.78 |

$$f_{10} = \frac{c}{(2*\sqrt{\varepsilon_{re}}*L_e)} \quad (3)$$

$$W_e = W + (2*\Delta l) + \left(\left(\frac{1.2*l_h}{L}\right)*(4*l_h)*\sin\left(\frac{2*\pi*y}{W}\right)\right) \quad (4)$$

$$f_{02} = \frac{c}{(\sqrt{\varepsilon_{re}}*W_e)} \quad (5)$$

$$\text{Error} = \left(\frac{f_{cal} - f_{sim}}{f_{sim}}\right) \quad (6)$$

For effective dielectric constant ($\varepsilon_{re}$) of unity using above equations, calculated frequencies are obtained and against simulated frequencies they are plotted in Figs. 4 and 5. In those figures plots, simulated two frequencies are also shown that indicates wideband response being realized for 0.7 ratio in narrower slot and 0.8, in wider slots. For different slot dimensions and substrate thickness, two frequencies are closer that gives error less than 5%. Using the above formulations, E-shape antenna is designed at a different frequency and substrate thickness as discussed below.

The design steps for E-shape MSA are as given below.

1. The MSA is designed for two different substrate thickness, $h = 0.07\lambda_0$ and $0.09\lambda_0$ at which fringing field extension length is selected to be $\Delta l = 0.7$ h.
2. Using following equation, RMSA length 'L' is calculated. Patch width 'W' is taken to be 1.15L. Slot parameters i.e. width ($w_s$) and slot position (y) are selected to be, $0.026\lambda_0$ and $0.1\lambda_0$ (narrow slots), $0.166\lambda_0$ (wider slots), respectively.

$$L = \frac{c}{(2f_r)} - (2\Delta l) \quad (7)$$

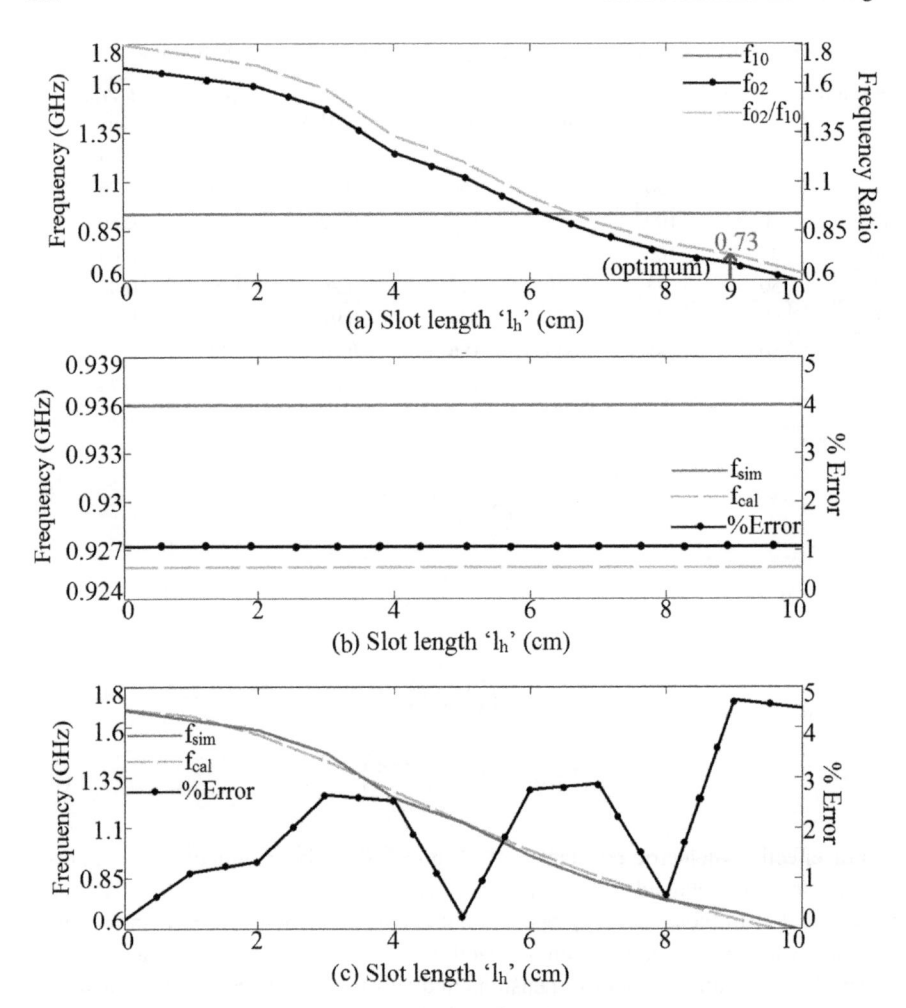

**Fig. 4 a** Simulated frequency and ratio plots, simulated and calculated frequencies with respective error plots at **b** $TM_{10}$ and **c** $TM_{02}$ mode for E-shape MSA with $h = 20$ mm and $y = 30$ mm

3. Using proposed formulations, plots for $TM_{02}$ and $TM_{10}$ mode frequencies are generated against varying slot length $l_h$ at desired $TM_{10}$ mode frequency. Form frequency plot, slot length is noted for which frequency ratio is 0.65–0.75 in narrow slots and 0.78–0.82 in wider slots. The antenna is simulated for these dimensions. These re-designing steps can be followed to design similar E-shaped MSA configuration at any other desired frequency.

Antenna dimensions for $TM_{10}$ of 4 GHz are, $h = 6$ or 7 mm ($0.07\lambda_0$ or $0.07\lambda_0$), $L = 30$ mm, $W = 35$ mm, $y = 7.5$ mm (narrow slot design) 12.5 mm (wider slot

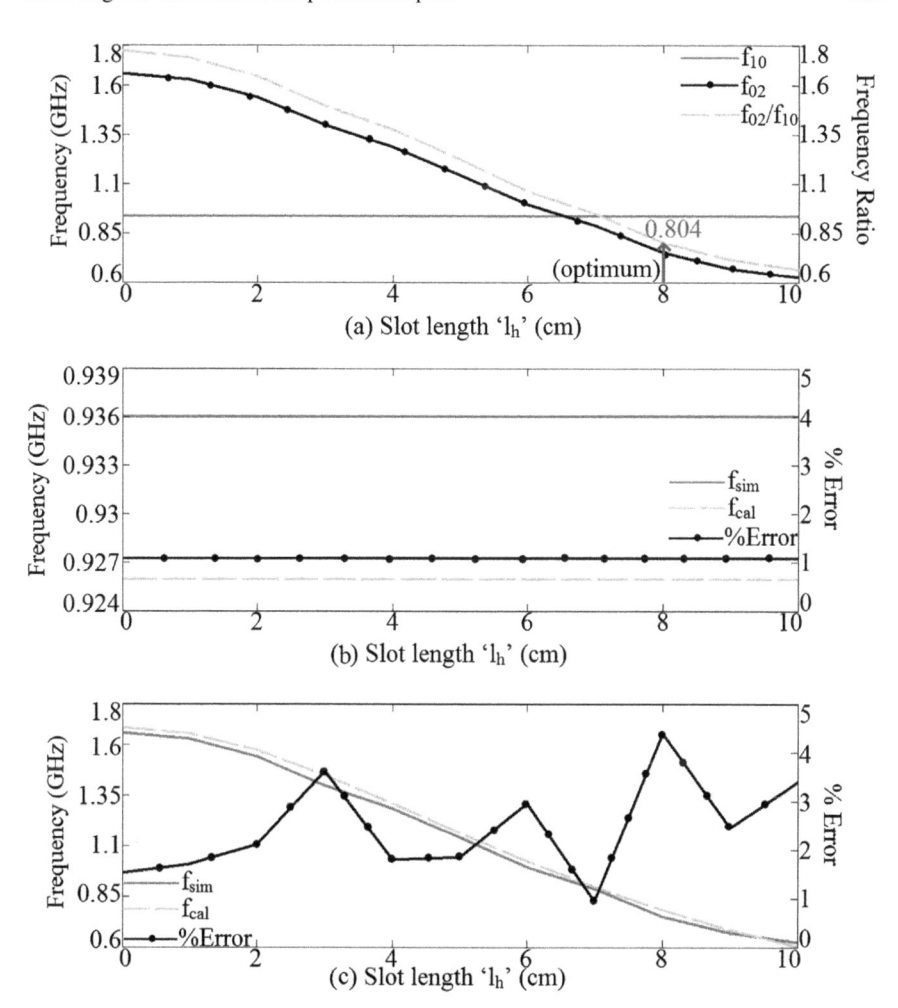

**Fig. 5 a** Simulated frequency and their ratio plots, simulated and calculated frequencies with respective error plots at **b** $TM_{10}$ and **c** $TM_{02}$ mode for E-shape MSA with $h = 20$ mm and $y = 50$ mm

design) and $w_s = 2$ mm. Simulated plots for re-designed E-shape patch is shown in Figs. 6 and 7. In all the cases wider BW (more than 0.7 GHz) showing loop formation inside VSWR = 2 circle, in Smith chart is observed. Thus proposed formulation and the design procedure can be selected for to obtain wideband response from E-shape patch on varying thicker substrates and for different slot parameters.

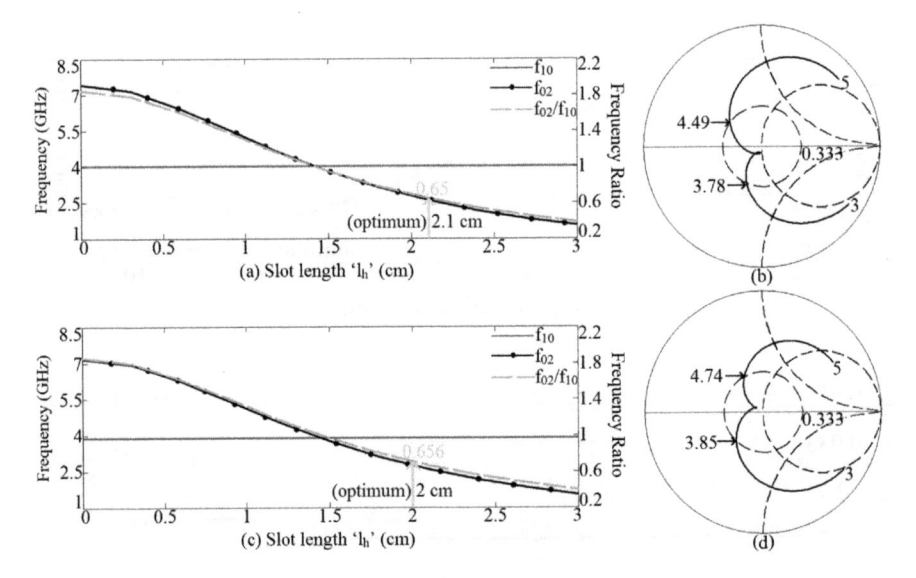

**Fig. 6** **a** Calculated frequency ratio plot using formulation, its **b** optimum input impedance plot for $h = 0.07\lambda_0$, **c** calculated frequency ratio plot using formulation, and its **d** results showing optimum input impedance plot for $h = 0.09\lambda_0$, for narrower slot ($y = 0.1\lambda_0$)

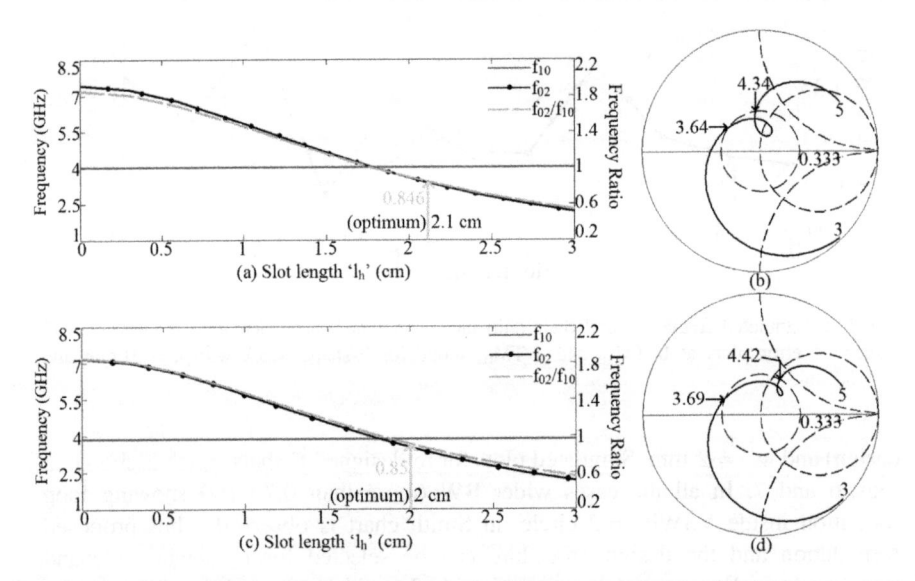

**Fig. 7** **a** Calculated frequency ratio plot using formulation, its **b** optimum input impedance plot for $h = 0.07\lambda_0$, **c** calculated frequency ratio plot using formulation, and its **d** results showing optimum input impedance plot for $h = 0.09\lambda_0$, for narrower slot ($y = 0.166\lambda_0$)

## 3 Conclusions

Proposed study in this paper presents a comparison between widely reported compact and wideband MSAs like C-shape, double C-shape, E-shape and U-slot cut RMSAs, for identical patch dimensions. Dual polarized dual and wideband response in C-shape variations is due to optimum spacing's of modified $TM_{10}$, $TM_{01}$ and $TM_{11}$ modes. Optimum spacing between $TM_{10}$ and $TM_{02}$ modes in U-slot and E-shape patches yields wider BW. Resonant length formulations for varying substrate thickness as well as slot positions in E-shape antenna are presented. The calculated frequency using them exhibits closer match with simulated results. The design procedure to obtain wider BW in E-shape MSA using formulations is explained. It yields wide band response that shows the formation of loop inside VSWR = 2 circle.

## References

1. Kumar, G., Ray, K.P.: Broadband Microstrip Antennas, 1st edn. Artech House, USA (2003)
2. Wong, K.L.: Compact and Broadband Microstrip Antennas, 1st edn. Wiley, New York (2002)
3. Deshmukh, A.A., Kumar, G.: Broadband pairs of slots loaded rectangular microstrip antennas. Microw. Opt. Technol. Lett. **47**(3), 223–226 (2005)
4. Islam, M.T., Shakib, M.N., Misran, N.: Multi-slotted microstrip patch antenna for wireless communication. Prog. Electromag. Res. Lett. **10**, 11–18 (2009)
5. Deshmukh, A.A., Ray, K.P.: Analysis of broadband Ψ-shaped microstrip antennas. IEEE Mag. Antennas Propag. **55**(2), 107–123 (2013)
6. Bao, X.L., Ammann, M.J.: Small patch/slot antenna with 53% input impedance bandwidth. Electron. Lett. **43**(3), 146–147 (2007)
7. Khodaei, G.F., Nourinia, J., Ghobadi, C.: A practical miniaturized U-slot patch antenna with enhanced bandwidth. Prog. Electromagn. Res. B, **3**, 47–62 (2008)
8. Ang, B.K., Chung, B.K.: A wideband E-shaped microstrip patch antenna for 5–6 GHz wireless communications. Prog. Electromagn. Res. **75**, 397–407 (2007)
9. Deshmukh, A.A., Ray, K.P.: Analysis of broadband variations of U-slot cut rectangular microstrip antennas. IEEE Mag. Antennas Propag. **57**(2), 181–193 (2015)
10. Deshmukh, A.A., Ray, K.P.: Analysis and design of broadband U-slot cut rectangular microstrip antennas. In: Sadhana—Academy Proceedings in Engineering Science, vol. 42, no. 10, pp. 1671–1684. Springer, Berlin (2017)
11. Bhardwaj, S., Samii, Y.R.: A comparative study of C-shaped, E-shaped, and U-slotted patch antenna. Microw. Opt. Technol. Lett. **54**(7), 1746–1756 (2012)

# Modified Circular Shape Microstrip Antenna for Circularly Polarized Response

Amit A. Deshmukh, Anuja Odhekar, Akshay Doshi and Pritish Kamble

**Abstract** The novel design of circularly polarized square microstrip antenna embedded with a pair of circular and rectangular slots and loaded with rectangular stubs on the corner edges of the square patch is proposed. The systematic study that presents degeneration of orthogonal modes supported by a parametric study on various patch parameters is presented. The modifications in patch yield two orthogonal modes along two diagonal axes of square patch and presence of slots and stubs yield optimum spacing between them that yields circularly polarized response. The VSWR and axial ratio bandwidth of 404 MHz (>40%) and 32 MHz (3.7%) in 900 MHz frequency band is obtained. The proposed antenna yields broadside pattern with gain greater than 7 dBi.

**Keywords** Circularly polarized microstrip antenna · Pair of circular slots
Pair of rectangular slots · Open-circuit stub · Resonant mode degeneration

## 1 Introduction

With advancements in mobile communication systems due to numerous advantages like low profile design, which can be easily integrated with microwave integrated circuits, microstrip antenna (MSA) finds many applications [1, 2]. The MSA in regular shape offers better radiation characteristics in terms of cross polarization levels [1, 2]. However, apart from size constraints, main limitations in regular shapes are lower cross-polar levels, that their usage leads to a larger single loss in multipath propagation environment due to uncertainty in directions of incoming wave polarization. In such applications antennas with higher cross polarization content are preferred. To completely minimize the signal loss, circularly polarized (CP) antennas are preferred since they receive signals from any directions with

---

A. A. Deshmukh (✉) · A. Odhekar · A. Doshi · P. Kamble
Department of Electronics and Telecommunication Engineering,
SVKM's, D J Sanghvi College of Engineering, Mumbai, India
e-mail: amit.deshmukh@djsce.ac.in; amitdeshmukh76@gmail.com

equal strength [3]. While using conventional antennas like dipole and its variations, Horn Antenna, etc., two antennas are needed that ensure three conditions (time and space orthogonality with equal amplitude) for CP response [3]. A distinct advantage of MSA is that it yields CP response without using additional patch [4]. Using single radiating patch, CP response in MSA is obtained by using slit or slot cut design that is fed by diagonal coaxial feed [4–6]. Wideband CP response with higher axial ratio bandwidth (BW) is obtained by using suspended configurations (thicker substrate designs) [7–10]. By using U-slots or rectangular slots, which has been widely reported to realize wideband response with a single polarization, CP response has also been obtained [11, 12]. In these slots cut designs, to realize equal contributions of orthogonal surface current components on the modified patch, unequal length slots have been used. By trimming the patch corners in a slot cut square MSA (SMSA), CP response has been obtained [13]. Although many papers have been reported using single feed modified geometries designs, but detail modal explanations about CP response is not outlined. Also in the majority of the reported work formulations for modified patch modes and then further design procedure to realize similar antenna at a different frequency is not mentioned.

In this paper, the novel design of stub-loaded SMSA using offset circular slots and a diagonally cut pair of rectangular slots to obtained CP response in 900 MHz frequency range is proposed. The 900 MHz frequency band is chosen so as to target application of mobile communication systems in the same frequency range. Here proximity feeding technique is selected since it is the simplest method for implementation on ticker substrate. For SMSA, detailed study explaining effects of circular slots and rectangular slots in stub-loaded SMSA on the degeneration of two modes is presented. The feed point location further optimizes impedance at two degenerated modes that yield CP response. On suspended glass epoxy substrate having parameters, $h = 0.16$ cm, $\tan \delta = 0.02$, $\varepsilon r = 4.3$, proposed antenna yields VSWR BW of more than 400 MHz (>40%) with AR BW of 32 MHz (3.7%). The antenna yields broadside pattern that gives co-polar gain greater than 6 dBi across most of the frequency spectrum. Thus with above pattern and BW characteristics, proposed design can be selected in applications of mobile communication in the frequency spectrum of 900 MHz. Here against reported variations of corner-trimmed patches, stub-loaded slot cut variations are discussed. Different to corner-trimmed design, here frequency of mode that is reduced by circular slots is tuned further to yields CP response. In the future scope of present work, a similar study will be carried out for rectangular MSA with an aspect ratio close to that of SMSA as well as resonant length formulations will be developed that will be useful for re-designing similar stub- and slot-loaded MSA on the thicker substrate. The antenna proposed in this paper is first studied using simulations using IE3D that is followed by the measurements. The input impedance, radiation pattern and gain measurements were carried out using instruments like ZVH—8, SMB 100A and FSC—6, inside an Antenna lab.

## 2 Stub-Loaded SMSA Embedded with Circular and Rectangular Slots

The air suspended design of SMSA is given in Fig. 1a, b. Here length ($L$) of the square patch is decided such that its $TM_{10}$ ($TM_{01}$) mode frequency is nearer 1000 MHz. This frequency was selected since after embedding slots, patch frequency will reduce and then modified SMSA will resonate at around 900 MHz of frequency. SMSA that is etched on glass epoxy substrate of thickness "$h$", is suspended above the ground plane with air thickness of "$ha$". The proximity strip is placed at height "$hs$" below patch. For $ha = 2.8$ cm, $hs = 2.6$ cm, proximity-fed SMSA is simulated for different feed positions as given in Fig. 1a and their plots showing resonance behavior of antenna is explained in Fig. 1c. Due to square geometry, $TM_{10}$ or $TM_{01}$ mode frequency nearer to 1000 MHz is observed. Inside SMSA, circular slot is cut and resonance plots for $L_s = 2.0$ cm, $x_s = y_s = 2.5$ cm, are provided in Fig. 1d, e. It is observed that with an increase in slot radius from 1.4 to 2.2 cm, modal frequency reduces but two modes are not getting separated.

Further, for slot radius of 2.2 cm, the position of slots is shifted more towards patch center and impedance curves for the same are explained in Fig. 2a. In this variation also degeneration of two modes was not observed. Further for slot position of $x_s = 2.0$ cm, slot radius is further increased from 2.2 to 2.6 cm. For this radius variation, resonance curves are shown in Fig. 2b. With the increase in radius, the frequency of resonance peak reduces and for radius of 2.6 cm, additional mode nearer to 950 MHz of frequency is noticed as shown by the arrow in Fig. 2b. The surface current distributions at two observed modes for $r_s = 2.6$ cm, is provided in Fig. 2c, d.

At first resonant mode currents are varying along diagonal length i.e. along $\Phi = 45°$ direction. At second mode they are showing variation along the orthogonal diagonal axis, i.e., along $\Phi = 135°$. Since the orthogonal mode is weakly excited, the prominent peak due to same is absent. To separate these two modes, an additional pair of slots was introduced as given in Fig. 1d. For this slot length variation, resonance curves are given in Fig. 3a. As observed from those plots, a prominent peak that explains proper excitation of the orthogonal mode is still not observed in the curve. This is due to a lesser reduction in the resonance frequency of mode that is varying along a diagonal axis, $\Phi = 45°$. The lesser reduction in frequency is present as rectangular slots are not present nearer to the maximum current location of the mode, along $\Phi = 45°$. Therefore to reduce this frequency further, additional stubs are employed at the corners of the square patch as explained in Fig. 1d. Due to stubs, variation in real and imaginary part of impedance against frequency using resonance curve plots is provided in Fig. 3b. With the introduction of stubs, frequency along $\Phi = 45°$, further reduces and two modes are separated that shows

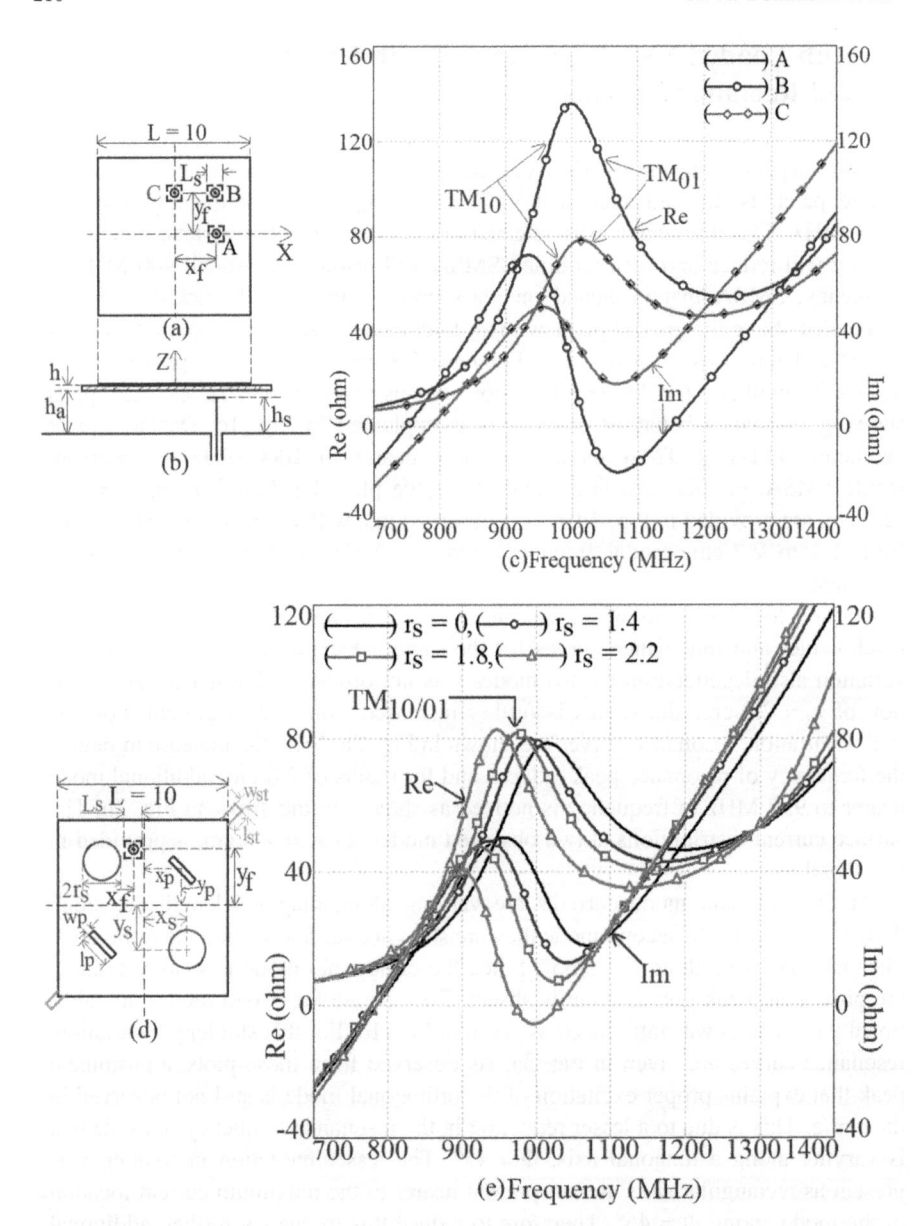

**Fig. 1** **a, b** Proximity-fed SMSA, its **c** impedance curve against feed locations, **d** stub-loaded and slot cut SMSA and its **e** impedance curve against slot radius "$r_s$"

# Modified Circular Shape Microstrip Antenna …

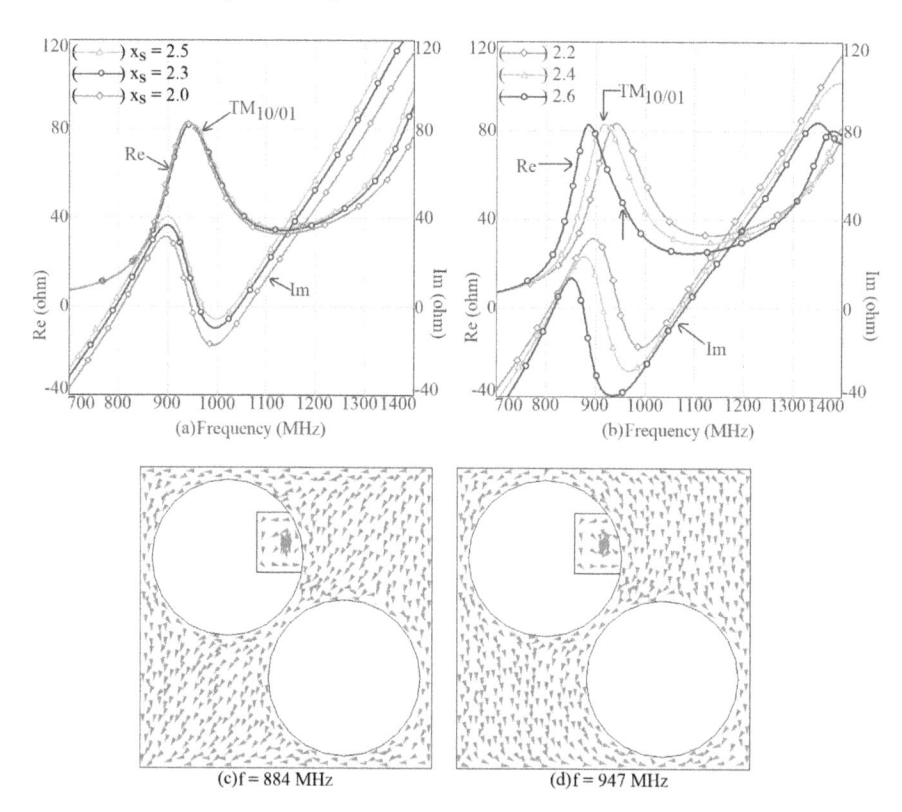

**Fig. 2** Resonance curve explaining variations in modal frequencies against circular **a** slot position and **b** slot radius, **c**, **d** surface current plots at two observed modes for proximity-fed SMSA embedded with circular slots

prominent peaks due to them. However, the second loop formed due to the orthogonal mode in the impedance locus is smaller that will not give optimum CP response as explained in Fig. 3c. For optimizing the loop size due to orthogonal modes effects of variation in proximity feed position are studied and Smith chart and impedance curves for same are shown in Fig. 3c, d.

As observe from Fig. 3c, d, the impedance at two modes increases that yields larger loop size to give an optimum response. For optimum design, Smith chart, radiation pattern at the center frequency, gain and AR variation, polarization plot and picture of the fabricated antenna is given in Figs. 4a–d and 5a, b.

The simulated BW is 404 MHz (41.6%). The antenna was fabricated. Using foam spacers which were kept towards antenna substrate periphery, it was suspended above ground plane. Experimentation was carried out using ZVH—8.

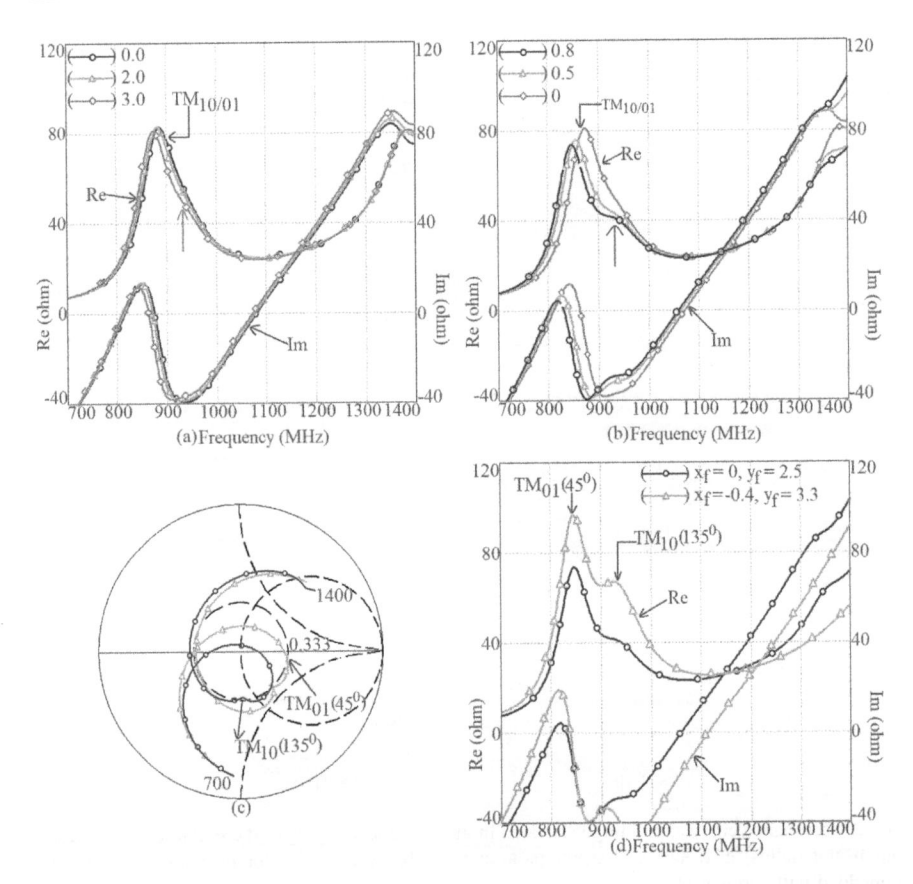

**Fig. 3** Impedance curves for variation in **a** slot length "lp" and **b** stub length "lst" and **c** Smith chart and **d** impedance curves for variation in feed parameters for proximity-fed pair of circular and rectangular slots cut stub-loaded SMSA

An experimental BW is 417 MHz (43.7%). The antenna's simulated AR BW is 32 MHz centered around 875 MHz. It offers a gain of above 6 dBi over complete VSWR BW. The antenna pattern exhibits a maximum in boresight direction that shows the difference between co and cross-polar levels of less than 3 dB, which indicates the presence of CP. The E and H-planes are aligned along $\Phi = 45°$ and 135°, respectively. Also, MSA offers left-hand CP response. Here AR BW is less than 5%. In further study, similar slot and stub-loaded designs will be studied for nearly square MSA (rectangular MSA with an aspect ratio less than 1.15). Formulations at two resonant modes will be realized which will provide a guideline to design identical antenna at the specific resonant frequency. Variations of similar configurations with square slots instead of a circular slot, stubs placed either along x- or y-axis will be studied for further enhancement in AR BW.

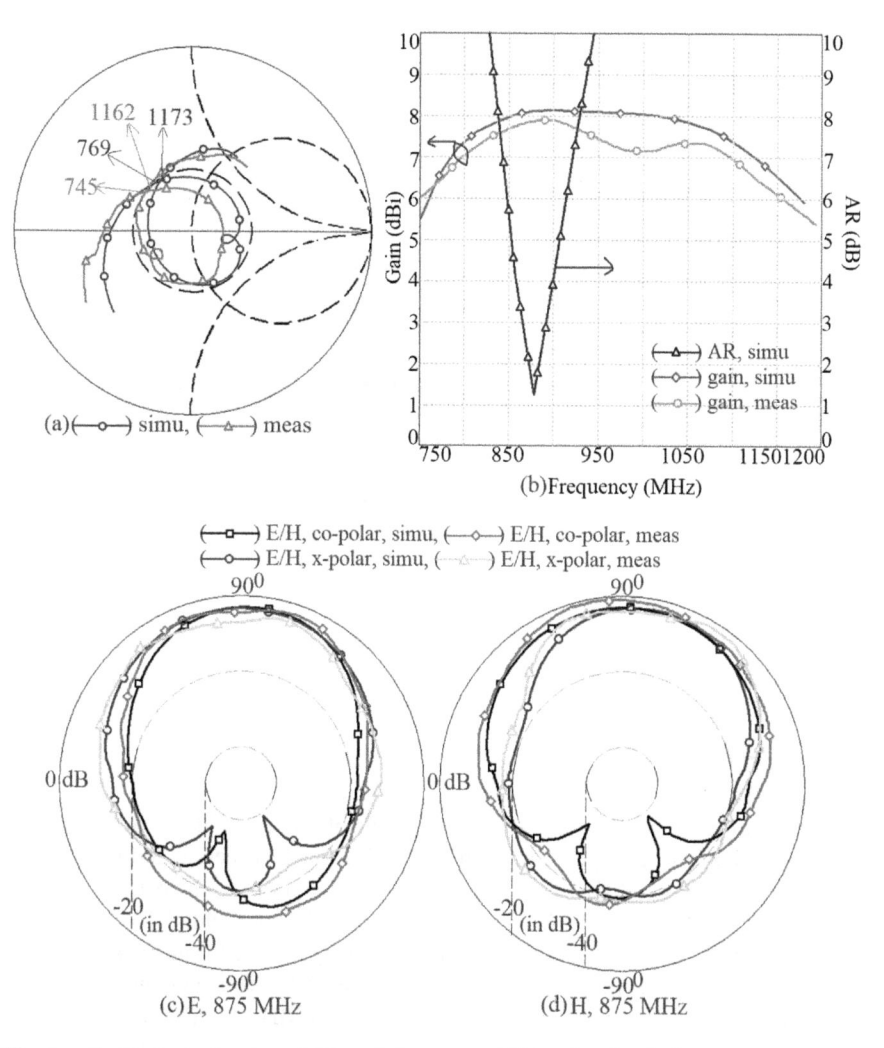

**Fig. 4** a Smith charts, b gain and AR variations across BW, and c, d radiation pattern at a center frequency of axial ratio BW for proximity-fed stub-loaded SMSA embedded with a pair of circular and rectangular slots

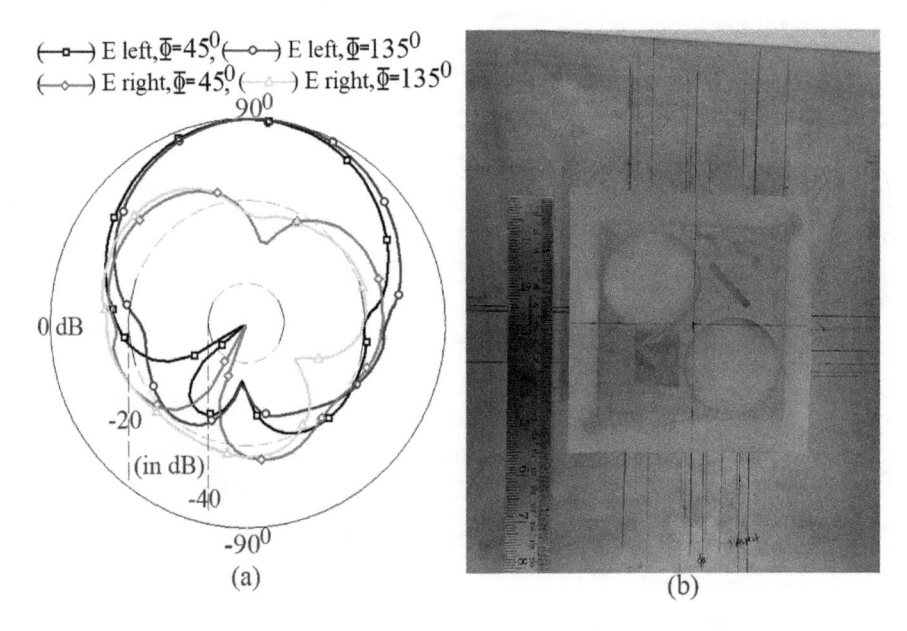

**Fig. 5** a Simulated polarization plot and **b** fabricated antenna picture for proximity-fed circular and rectangular slot cut stub-loaded SMSA

## 3 Conclusions

The novel design of SMSA embedded with a pair of circular and rectangular slots and which is loaded with open-circuit stubs towards patch corners is proposed. The combination of stubs and slots degenerates fundamental mode of the patch into dual orthogonal modes that yield CP response. Proposed design yields VSWR BW of more than 400 MHz (>40%) with AR BW of nearly 4%. The antenna gives an average gain of more than 6 dBi over complete BW. This antenna can find applications in mobile communication systems in 900 MHz frequency spectrum.

## References

1. James, J.R., Hall, P.S.: Handbook of Microstrip Antennas, vol. 1. Peter Peregrinus, London (1989)
2. Lee, H.F., Chen, W.: Advances in Microstrip and Printed Antennas. Wiley, New York (1997)
3. Balanis, C.A.: Antenna Theory, Analysis and Design. Wiley, New York (1997)
4. Kumar, G., Ray, K.P.: Broadband Microstrip Antennas, 1st edn. Artech House, USA (2003)
5. Wong, K.L.: Compact and Broadband Microstrip Antennas, 1st edn. Wiley, New York (2002)
6. Sharma, P.C., Gupta, K.C.: Analysis and optimized design of single feed circularly polarized microstrip antennas. IEEE Trans. Antennas Propag. **31**(6), 949–955 (1983)

7. Huang, C.Y., Wu, J.Y., Wong, K.L.: Broadband circularly polarized square microstrip antenna using chip-resistor loading. In: IEE Proceedings on Microwaves, Antennas Propagation, vol. 146, no. 1, pp. 94–96 (1999)
8. Sze, J.-Y., Hsu, C.I.G., Chen, Z.-W., Chang, C.-C.: Broadband CPW-fed circularly polarized square slot antenna with lightening-shaped feedline and inverted-L grounded strips. IEEE Trans. Antennas Propag. **58**(3), 973–977 (2010)
9. Deepukumar, M., George, J., Aanandan, C.K., Mohanan, P., Nair, K.C.: A peripherally fed broadband modified circular microstrip antenna. In: IEEE Antennas and Propagation Society International Symposium, vol. 1, pp. 37–40, AP-S. Digest, Baltimore, MD, USA (1996)
10. Yang, S.L.S., Lee, K.F., Kishk, A.A.: Design and study of wideband single feed circularly polarized microstrip antennas. Prog. Electron. Res. **80**, 45–61 (2008)
11. Tong, K.F., Wong, T.P.: Circularly polarized U-slot antenna. IEEE Trans. Antennas Propag. **55**(8) (2007)
12. Khidre, A., Lee, K.F., Yang, F., Eisherbeni, A.: Wideband circularly polarized E-shaped patch antenna for wireless applications. IEEE Antennas Propag. Mag. **52**(5) (2010)
13. Baudha, S., Dinesh Kumar, V.: Corner truncated broadband patch antenna with circular slots. Microw. Opt. Technol. Lett. **57**(4), 845–849 (2015)

# Wideband Designs of 60° Sectoral Microstrip Antenna Using Parasitic Angular Sectoral Patches

Amit A. Deshmukh and S. B. Deshmukh

**Abstract** Wideband designs of the novel and compact 60° Sectoral microstrip antennas are proposed. At fundamental mode frequency of 1500 MHz, single Sectoral patch yields bandwidth of greater than 550 MHz (>35%) with a peak gain of above 7 dBi. Enhancement in bandwidth and the peak gain of Sectoral configuration is obtained by gap-coupling angular Sectoral patches of the same angle with 60° Sectoral patch antenna. Design with additional Sectoral patch gives a bandwidth of nearly 750 MHz (>45%) whereas that with two angular Sectoral patches it yields a bandwidth of approximately 900 MHz (>50%). The peak gain of 10 dBi is realized in the gap-coupled design of 60° Sectoral patch to that with two angular Sectoral patches.

**Keywords** 60° Sectoral microstrip antenna · Angular Sectoral patch antenna Broadband microstrip antenna · Multi-resonator configuration

## 1 Introduction

The microstrip antenna (MSA) being planar in design, are more preferred in most of the modern day communication systems [1–3]. Their integration with signal processing circuitry that employs microwave integrated circuits, makes them more suitable for personal mobile communication applications as well as other wireless communication systems designs [1–4]. The gain of MSA is increased by modifying the effective area of MSA, and this is obtained in array configuration [5]. In arrays patches with an identical pattern, characteristics are coupled together in specific electrical and physical manner to realize a maximum of gain in the specific direction [5]. The use of reflector array has become very popular as it avoids the use of feeding power divider network that provides an input signal to the antennas [6, 7].

---

A. A. Deshmukh (✉) · S. B. Deshmukh
Department of Electronics and Telecommunication Engineering, SVKM's,
D J Sanghvi College of Engineering, Mumbai, India
e-mail: amit.deshmukh@djsce.ac.in; amitdeshmukh76@gmail.com

Using microstrip patches, in these variations namely, two designs have been implemented, reflect array and space-fed array [6–9]. The only disadvantage of using these arrays that they are too bulky in design. The gain of dipole antenna has been enhanced by suitably placing linear wires around the dipole wire, and this arrangement leads to the design of well-known Yagi–Uda antenna [5]. Due to the presence of reflector behind the dipole, Yagi structure yields one directional radiation characteristics whereas the presence of director, further enhances the directional properties. Using this concept, MSA variation of Yagi–Uda antenna has been realized [10]. It yields BW of 600 MHz (20%) with a peak gain of 8 dBi [10]. Using Sectoral patch which is fed to angular Sectoral patches, BW enhancement has been reported [11]. Using one director as an annular Sectoral patch, BW from 5100 to 5850 MHz (13.6%) is obtained. With two director BW increases from 5050 to 6020 MHz (17.6%) [11]. Design with two directors yields peak gain of 8.2 dBi. The enhancement in BW is reported because of the coupling between fundamental modes of Sectoral and annular Sectoral patches and gain improvement is the result of increase in aperture area.

In this paper, the first design of Sectoral Yagi antenna as reported in [11] is discussed. The configurations as reported in [11] are proposed in 5000 to 6000 MHz frequency band. Using single director, gap-coupled design yields BW of 13.6% which increases to above 17% of BW in case of two directors [11]. The peak gain of 8.2 dBi is observed for two director case. Further using proximity feed, Yagi design of printed Sectoral and annular Sectoral patches for fundamental mode frequency of 1500 MHz on substrate thickness of $0.1\lambda_0$, is presented. A single 60° Sectoral MSA gives BW of above 550 MHz (>35%) with a peak gain of more than 7 dBi. The BW of 60° Sectoral patch is now increased using multi-resonator technique in which, angular Sectoral patches are gap-coupled to fed MSA. Using single angular Sectoral patch, BW of more than 750 MHz (>45%) is observed. This BW value further increases to greater than 850 MHz (>50%) when second angular Sectoral patch as the director, is used. In the two director design, antenna gives peak gain of 10 dBi across the BW for which reflection coefficient over frequencies is less than 0.333. Thus using multi-resonator concept based upon Yagi–Uda antenna concept, a wideband antenna with a gain of 10 dBi is obtained. These antenna characteristics are better as compared to those obtained from the reported paper of [11]. The improvement in antenna parameters is the result of a suspended design that is fed using proximity feeding strip. The IE3D software is used here to initially study and optimize the design. Further, the antennas were fabricated and experimental verification was carried out. In experimentation, high-end instruments like, "ZVH–8", "FSC 6" and "SMA-100 A", were used. Using minimum far field distance, pattern and gain measurements were carried out in the Antenna Lab.

## 2 Design of Sectoral Patch Antenna Gap-Coupled with Angular Sectoral Patches

A multi-resonator gap-coupled design of Sectoral patch to that with angular Sectoral patches is explained in Fig. 1a, b [11]. The reported antenna is optimized on substrate thickness which has parameters as, dielectric constant ($\varepsilon_r$) = 2.55, substrate height ($h$) = 1.5 mm. The square substrate of length ($L$) 48 and 60 mm is selected in single director and dual director design of gap-coupled antenna [11]. Using single director, antenna offers BW of 13.6% and with the addition of the second director, it increases to 17.6% of BW as shown in Fig. 1c. The antenna offers a gain of 8.2 dBi in dual director design. In terms of resonant modes as provided in Fig. 1d, it is reported in [11] that BW is the result of coupling between $TM_{11}$ and $TM_{12}$ modes on Sectoral patches. Due to thinner substrate, BW reported for these antennas is less than 20%. To enhance the BW, multi-resonator designs of gap-coupled Sectoral patches are proposed using thicker substrate, as presented in Fig. 1.

Here the fundamental mode frequency of 1500 MHz is selected and the antenna is optimized using total substrate thickness of $0.1\lambda_0$. Radiating patch is etched on FR4 substrate which has substrate parameters as, dielectric constant = 4.3, loss tangent = 0.02 and height = 1.6 mm. Substrate is kept above ground plane with an air gap "$h_a$". For these substrate dimensions and frequency, radius for Sectoral antenna is found to be 8.0 cm as shown in Fig. 1e, f. By appropriately tuning the dimension of strip and its position with reference to null field line at fundamental mode in 60° Sectoral patch, wider BW in single element is achieved. Input impedance loci over 1000–2000 range of frequencies that is observed in simulation and that obtained in experimentation is given in Fig. 2a. Here using simulations BW is 555 MHz (35.7%) and that observed in experimentation is 571 MHz (36.3%). This Sectoral antenna yields peak gain of 8.5 dBi in the co-polar broadside direction as provided in gain plots in Fig. 3a.

For 60° Sectoral patch single resonant mode in its impedance curve is noticed around the frequency of 1500 MHz as given in Fig. 3b. Enhancement in gain and BW for Sectoral design is realized using its gap-coupled designs. Addition of single or dual angular Sectoral patch next to 60° Sectoral patch adds additional resonant modes in gap-coupled design as shown in Fig. 3b, which enhances the BW and gain. Tuning of air gaps between fed and parasitic elements yields optimum BW and Smith charts for additional single or dual angular Sectoral antenna are provided in Fig. 2b, c, respectively. For single parasitic element, respective BW's using simulation and experimentation are, 783 MHz (46.7%) and 800 MHz (48.4%). For dual Sectoral patches, two BWs are 888 MHz (48.3%) and 920 MHz (50.2%), respectively.

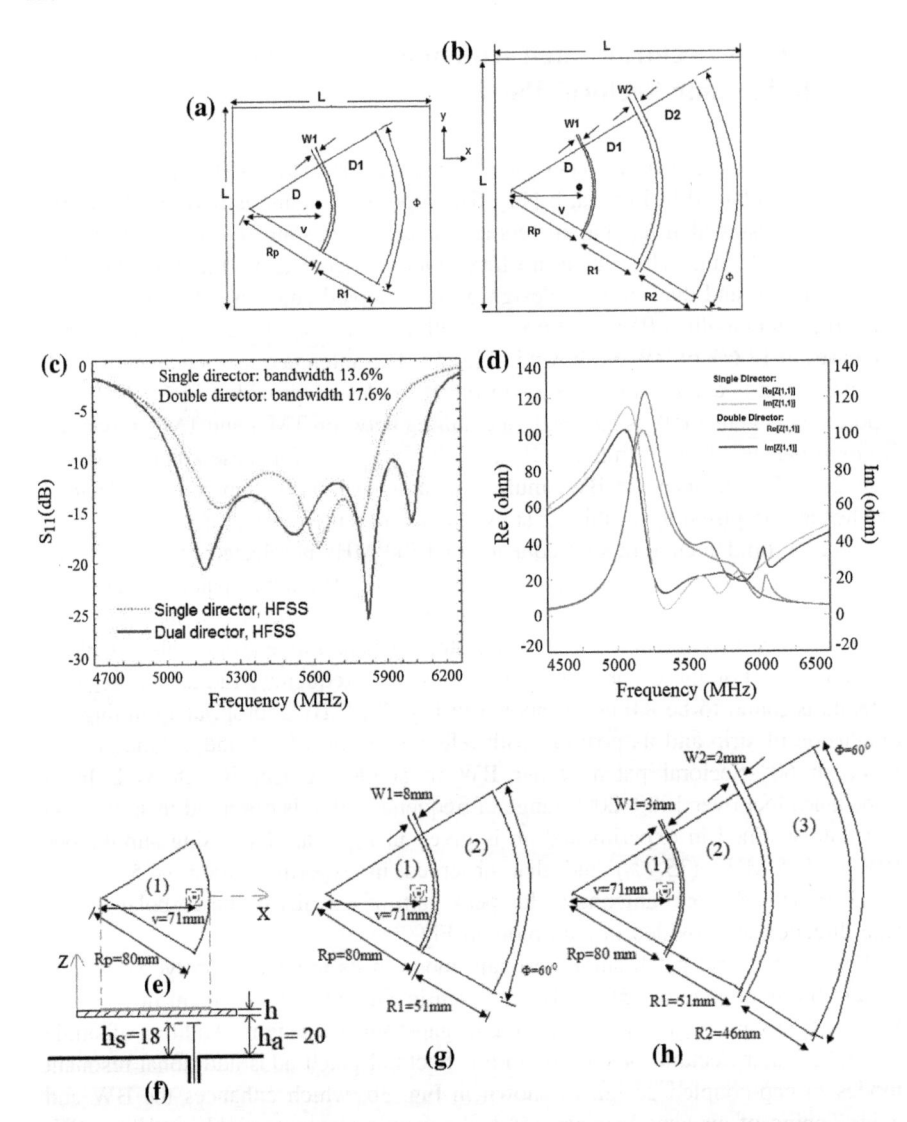

**Fig. 1** a Single and b dual director designs of Sectoral patches, and their c return loss and d resonance curve plots [11], designs of e, f 60° Sectoral MSA and its gap-coupled design with g single and h dual directors in 1500 MHz frequency range

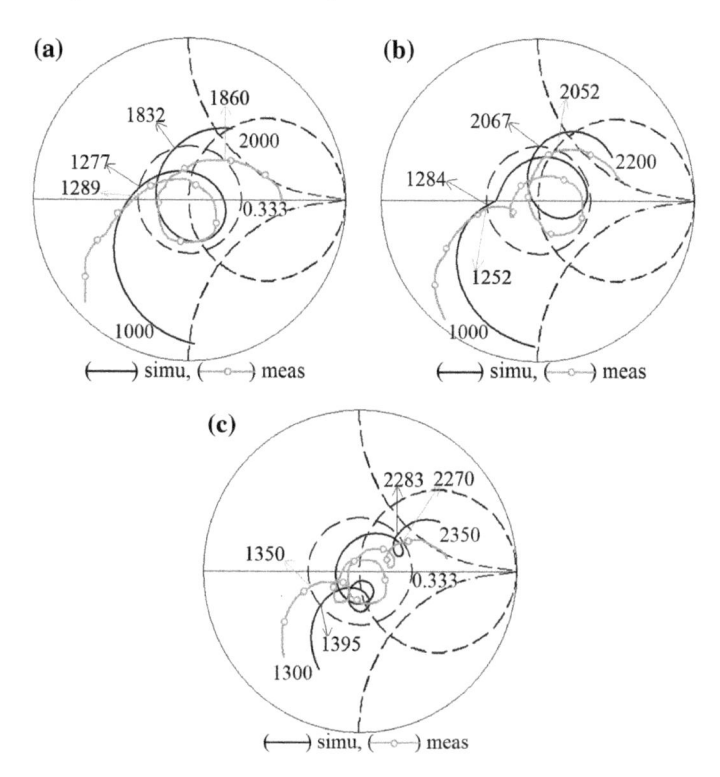

**Fig. 2** Smith charts for **a** 60° Sectoral patch and for its gap-coupled design with **b** single and **c** dual director designs

Simulated gain variations over BW for two gap-coupled designs and pattern across two frequencies over BW for the dual director case are provided in Fig. 3a, c–f. In these gap-coupled antennas, as phase and geometric centre of the composite structure is shifted towards first gap-coupled patch, maximum broadside gain is obtained in $\theta = 30°$ directions and it remains maximum in the same direction throughout the BW. Thus while in practical implementation, only modification in terms of antenna orientation is needed from 0° to 30° in the $\theta$ direction. Dual director design shows peak gain close to 10 dBi. Pattern across the BW remains in boresight direction with cross polar content less than 10 dB as compared with co-polar content. The picture of fabricated for dual angular Sectoral patch antenna is provided in Fig. 4a, b.

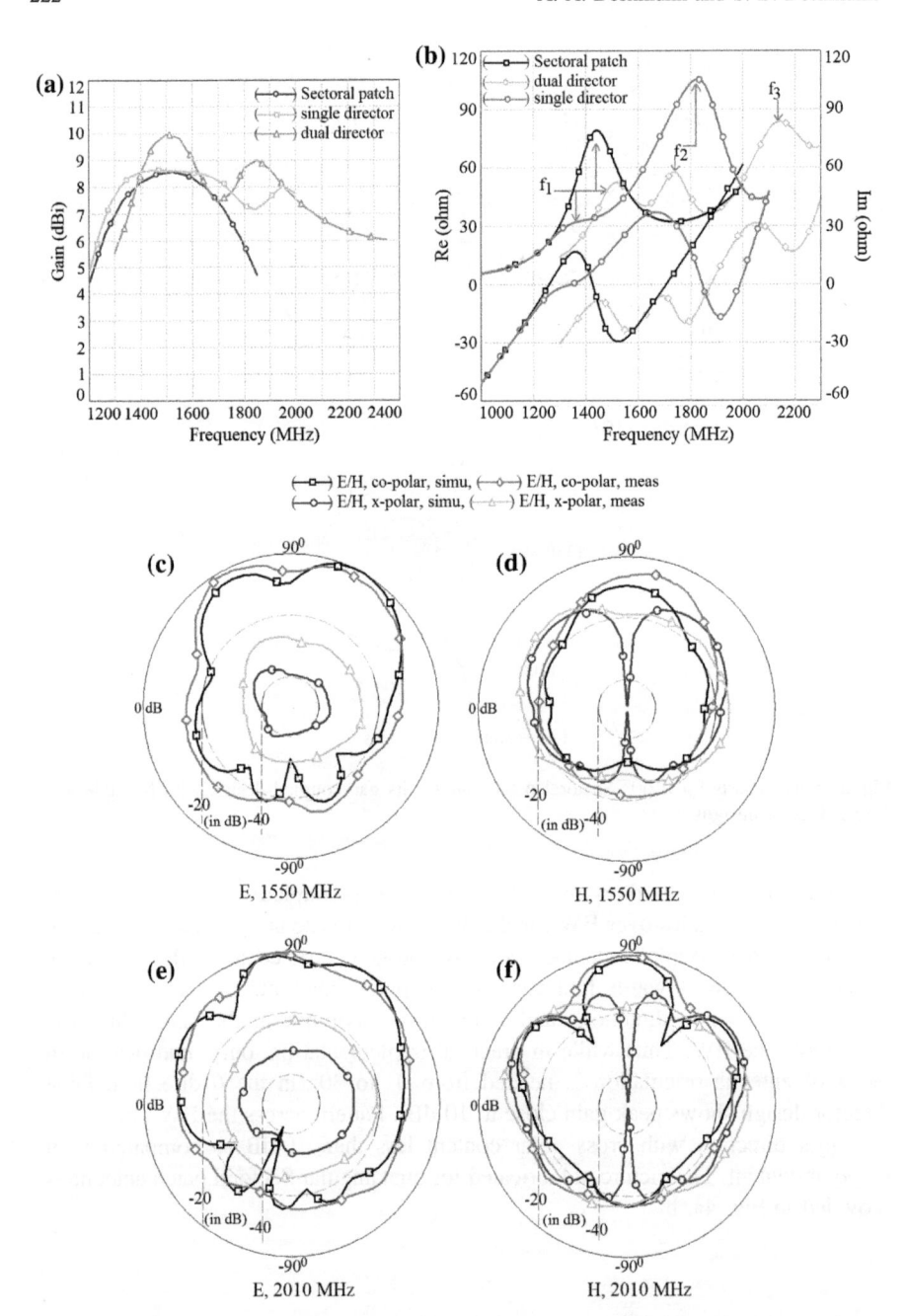

**Fig. 3 a** Simulated gain and **b** impedance curves for gap-coupled variations of the angular Sectoral patch with 60° Sectoral patch and **c–f** pattern at two frequencies across BW for gap-coupled dual directors design

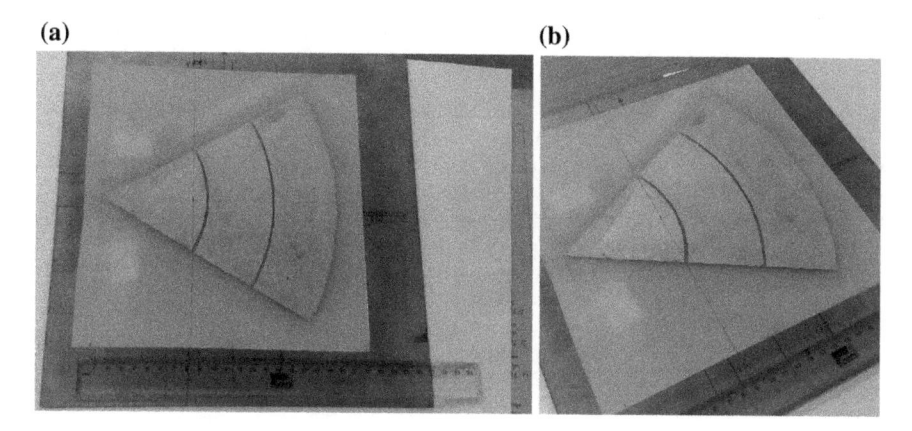

**Fig. 4 a, b** Fabricated pictures of gap-coupled variations of the dual angular Sectoral patch with 60° Sectoral patch

## 3 Conclusions

Design of gap-coupled antenna based on Yagi–Uda antenna concept in 5000 to 6000 MHz range is discussed. They yield BW in the range of 13–17% showing peak gain of 8.2 dBi. For BW and gain enhancement, gap-coupled design of 60° Sectoral MSA to that with angular Sectoral patch is presented in proposed work. An optimum response with BW of 900 MHz ($\sim$50%) with a peak gain of 10 dBi is observed in gap-coupled design with two angular Sectoral patches (dual director design) with 60° Sectoral patch. In proposed work angle for Sectoral and angular Sectoral patches was taken to be 60°. Further, in the extension of proposed work, a similar study will be carried out for other Sectoral variations with the varying angle. Formulations for designing patches at any other frequency will be developed. Also, above designs with slot embedded in the patch will be studied for further increment in BW.

## References

1. James, J.R., Hall, P.S.: Handbook of Microstrip Antennas, vol. 1. Peter Peregrinus, London (1989)
2. Lee, H.F., Chen, W.: Advances in Microstrip And Printed Antennas. Wiley, New York (1997)
3. Kumar, G., Ray, K.P.: Broadband Microstrip Antennas, 1st edn. Artech House, USA (2003)
4. Wong, K.L.: Compact and Broadband Microstrip Antennas, 1st edn. Wiley, New York (2002)
5. Balanis, C.A.: Antenna Theory, Analysis and Design. Wiley, New York (1997)
6. Yu, S., Li, L., Kou, N.: One-bit digital coding broadband reflectarray based on fuzzy phase control. IEEE Antennas Wirel. Propag. Lett. **16**, 1524–1527 (2017)

7. Yin, J., Wu, Q., Yu, C., Wang, H., Hong, W.: Low-sidelobe-level series-fed microstrip antenna array of unequal interelement spacing. IEEE Antennas Wirel. Propag. Lett. **16**, 1695–1698 (2017)
8. Bhide, R., Kumar, G.: Equivalence of space-fed microstrip antenna array with horn antenna. Microw. Opt. Technol. Lett. **52**(5), 1180–1183 (2010)
9. Bhide, R., Kumar, G.: Circularly polarized space-fed microstrip antenna arrays. Microw. Opt. Technol. Lett. **52**(10), 2221–2223 (2010)
10. Mandal, K., Sarkar, S., Sarkar, P.P.: Bandwidth enhancement of microstrip antennas by staggering effect. Microw. Opt. Technol. Lett. **53**(10), 2446–2447 (2011)
11. Liang, Z., Liu, J., Zhang, Y., Long, Y.: A novel microstrip quasi yagi array antenna with annular sector directors. IEEE Trans. Antennas Propag. **63**(10), 4524–4529 (2015)

# Part III
# Embedded Systems/Communication

# Performance Evaluation of Transform Domain Methods for Satellite Image Resolution Enhancement

## Mansing Rathod and Jayashree Khanapuri

**Abstract** Today satellite images are extensively considered in different fields of research. But the main problem associated with satellite images is their resolution. Hence, we propose a method to resolve the resolution problems associated with satellite images with transform domain methods such as Discrete Wavelet Transform, Dual-Tree Complex Wavelet Transform and Discrete Wavelet Transform with Stationary Wavelet Transform methods. Wavelet transform decomposes the low-resolution input image into four different subband images such as Low–Low, Low–High, High–Low, and High–High. Then Bicubic interpolation is applied on subband to resize the subband images and to get estimated images. All the estimated images and low-resolution images are combined by using Inverse Discrete Wavelet Transform to obtain a high-resolution image. All the methods are compared with different satellite images. It is observed that Discrete Wavelet Transform with Stationary Wavelet Transform maintains the high-frequency components due to interpolation applied to subband images. It preserves sharpness and details of high-frequency components in the images. Direction selectivity is also very good in Discrete Wavelet Transform with Stationary Wavelet Transform. This provides better results as compared to other methods. The results are evaluated for quantitative peak signal-to-noise ratio, Root Mean Square Error, Mean Square Error, Mean Absolute Error, and Time to prove the supremacy.

**Keywords** Resolution · Discrete wavelet transform · Stationary wavelet transform · Dual-tree complex wavelet transform · Enhancement

---

M. Rathod (✉)
Department of Information Technology, K.J.S.I.E./IT, Mumbai 400022, India
e-mail: rathodm@somaiya.edu

M. Rathod
Pacific Academy of Higher Education and Research University, Udaipur 313003, India

J. Khanapuri
Department of Electronics and Telecommunication, K.J.S.I.E./IT, Mumbai 400022, India
e-mail: jayashreek@somaiya.edu

© Springer Nature Singapore Pte Ltd. 2018
H. Vasudevan et al. (eds.), *Proceedings of International Conference on Wireless Communication*, Lecture Notes on Data Engineering and Communications Technologies 19, https://doi.org/10.1007/978-981-10-8339-6_25

# 1 Introduction

In the current scenario, a number of applications need satellite images for various research activities. Some important fields that use satellite images are astronomy, geosciences, and geography information. Application in image processing such as resolution enhancement improves the quality of satellite images. Some of the conventional methods using transform domain use interpolation to maintain the resolution of digital images. These methods provide sharper images because they directly work on pixel values of the images. There are three methods of interpolation such as nearest neighbor, bilinear, and bicubic interpolation. Bicubic interpolation is an effective method that provides sharper images and other two methods produce smoothened edges in the images. In interpolation, new pixel values are determined with neighboring pixels. The main disadvantage of interpolation is a computational problem which increases as the order of interpolation factor is increased. Wavelet domain is one of the techniques to improve the resolution of images. The wavelet transform is a quite new signal processing tool that has been successfully used in a number of areas. Wavelet transform resolves the limitation of Fourier transform because time information is lost in Fourier transform. Wavelet analysis provides time as well as frequency information together, i.e., which frequency at what time. The word wavelet means a small wave, with finite energy. Wavelet possesses the ability to allow simultaneous time and frequency analysis. Many researchers have used wavelet to improve the quality of satellite image. Wavelet improves the resolution of satellite images with interpolation. This is done by determining wavelets coefficient that increases the sharpness of the images. In this paper, we have compared transform domain methods such as DT-CWT, DWT, SWT, and DWT with SWT. These methods are tested with different satellite images. It concluded that DWT with SWT gives better resolution of the satellite images because of its good directional selectivity.

# 2 Literature Survey

A massive literature survey has been carried out in the transform domain and spatial domain methods to generate high-resolution satellite images. Paper [1] discusses about Discrete Wavelet Transform (DWT). It decomposes the low resolution of the input image into different subband images and all these subband images are downsampled images. The DWT gives sharpness of the image. The disadvantage is interpolation factor 2 is needed to resize the downsample image. Paper [2] has discussed about DWT with SVD algorithm. Singular Value Decomposition (SVD) is used to enhance the brightness of the image. Paper [3] has presented about the SWT, it overcomes the drawback of DWT and provides high frequency components. The main characteristic of DWT is the use of downsample images and SWT does not use downsample image. Paper [4] has presented transform domain and spatial domain methods to improve the resolution of satellite images. The comparison of both domains and suggests that transform domain is better than spatial domain. Paper [5] has worked out to improve the sharpness of satellite image. It is concluded that using interpolation in wavelet domain provides good resolution enhancement. Paper [6] has suggested removal of the blur and noise in the satellite image by adopting good shift invariance and directional property of DT-CWT transform. Paper [7] has presented about existing methods such as wavelet zero padding (WZP) cycle spinning (CS), complex wavelet transform (CWT) and discrete wavelet transform (DWT) for satellite images. The peak signal-to-noise ratio is improved with DWT algorithm. Paper [8] has compared the transform domain methods such as DWT, SWT, DWT with SWT, and DT-CWT. It is concluded that DT-CWT gives good quality for resolution enhancement, due to good directional selectivity. Paper [9] proposes noise removal, resolution enhancement, and Retinex methods. The BM3D+enhancement+retinex improves the PSNR. BM3D method gives better result of noise removal than another method. Paper [10] has presented contrast and resolution enhancement techniques for color and gray-level satellite images. It has used DWT and Singular Value Decomposition method to improve the resolution as well as the contrast of the image. Paper [11] has presented the improvement of the contrast and resolution of satellite images using DWT with Singular Value Decomposition (SVD). It has compared general histogram equalization and local histogram equalization. In this Paper [12], a model is presented for determining the silent object of the given images of the surface. After clarifying the image the model removes the deblurring and noise of the image. Paper [13] has compared discrete cosign transform (DCT) and DWT. It concludes that DWT gives the better resolution enhancement due to maintained high-frequency components. Paper [14] has presented SWT and DWT with SWT and both methods have been compared with PSNR, RMSE. It is suggested that DWT with SWT produces high resolution of satellite image. Paper [15] has studied literature survey on transform domain methods and compared some of the important algorithms such as DWT, SWT, DT-CWT, and SWT and DWT. The above-mentioned methods are compared and analyzed with the evaluation parameters such as Peak Signal-to-Noise Ratio, Root Mean Square Error, Mean Square Error, and Mean Absolute Error.

## 3 Resolution Enhancement Methods for Satellite Images

1. **Discrete Wavelet Transform (DWT)**: It is one of the frequency domain methods to increase the resolution of satellite images. This method decomposes the input image into four subband images such as Low–Low (LL), Low–High (LH), High–Low (HL), and High–High (HH) and all these subband images are downsampled images. After decomposition, bicubic interpolation factor 2 is applied on subband images for resizing the downsampled subband images. Low–Low (LL) subband image and low-resolution input image provide the

difference image and the difference image is combined with remaining subband images such as LH, HL, and HH to determine three estimated subband images. Bicubic interpolation with factor alpha/2 is applied on estimated images as well as low-resolution input image [1] (Fig. 1).

Inverse Discrete Wavelet Transform (IDWT) is used to generate high resolution of the input image. High-frequency components have not maintained because DWT is used to downsample image.

2. **Stationary Wavelet Transform (SWT)**: It is another method to enlarge the low-resolution image. Stationary Wavelet Transform to divide the low-resolution image into four subband images such as LL, LH, HL, and HH. Subband images and input image have the same size because does not use downsample image, so need not apply bicubic interpolation factor 2. Interpolation factor alpha/2 is applied on LH, HL, HH, and low-resolution input image [8]. Inverse Stationary Wavelet Transform (ISWT) is used to generate the resolution of low-resolution input image. High-frequency components have been maintained. To minimize the loss in high-frequency components, but redundancy is the drawback of SWT (Fig. 2).

3. **D-Tree Complex Wavelet Transform (DT-CWT)**: It is a very good method for super-resolution of satellite images because DWT and SWT are having very poor directional selectivity and shift variance of the input image. These two drawbacks are overcome with the help of DT-CWT due to good directional selectivity and shift variance of the image. DT-CWT is used to decompose the input image into six high-frequency subband image and two low-frequency subband images [6]. The six high-frequency contain real and imaginary parts with wavelet coefficients of the original input image in the degree of +15, +45,

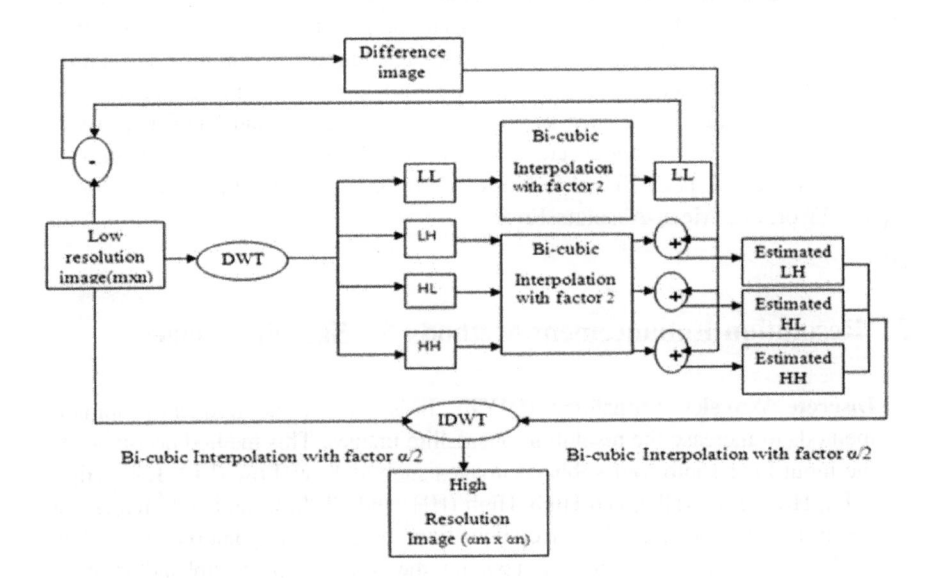

**Fig. 1** DWT-based resolution enhancement

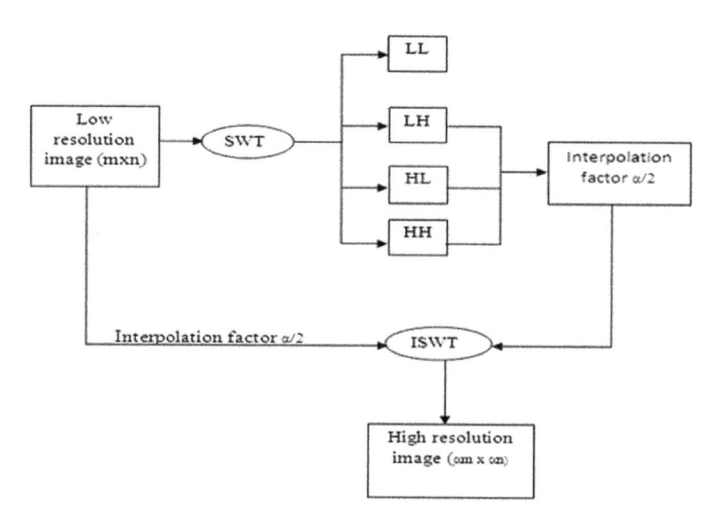

**Fig. 2** SWT-based resolution enhancement

+75, and −75, −45 and −15. These are the directional edges of the image. Interpolation factor alpha/2 is applied to six high-frequency subband image and the input image. Inverse DT-CWT is applied to both images such as interpolated subband images and interpolated input image to produce the high-resolution enhancement of the satellite image. It concluded that DT-CWT gives the better quality image compared to DWT and SWT (Fig. 3).

4. **Stationary Wavelet Transform (SWT) and Discrete Wavelet Transform (DWT)**: This method uses high-frequency components of DWT and SWT, in the first step, four subband images are obtained by DWT, i.e., LL, LH, HL, and HH. In the second step, four subband images are obtained by SWT as in the DWT. The size of DWT subband images are half of the input image. But SWT subband images and input image are of the same size. The third step is interpolation with factor 2 applied to high-frequency subband images of DWT. In the fourth step, the estimated high-frequency subband images such as LH, HL, and HH are obtained from interpolated subband images of DWT and subband images of SWT. Fifth step implements bicubic interpolation factor alpha/2 on estimated high-frequency subband images and low-resolution input image. Finally, Inverse Discrete Wavelet Transform (IDWT) is applied to produce the high-resolution image of the input image. To overcome the drawback of DWT and SWT, hidden information is exposed in edges of the image [3] (Fig. 4).

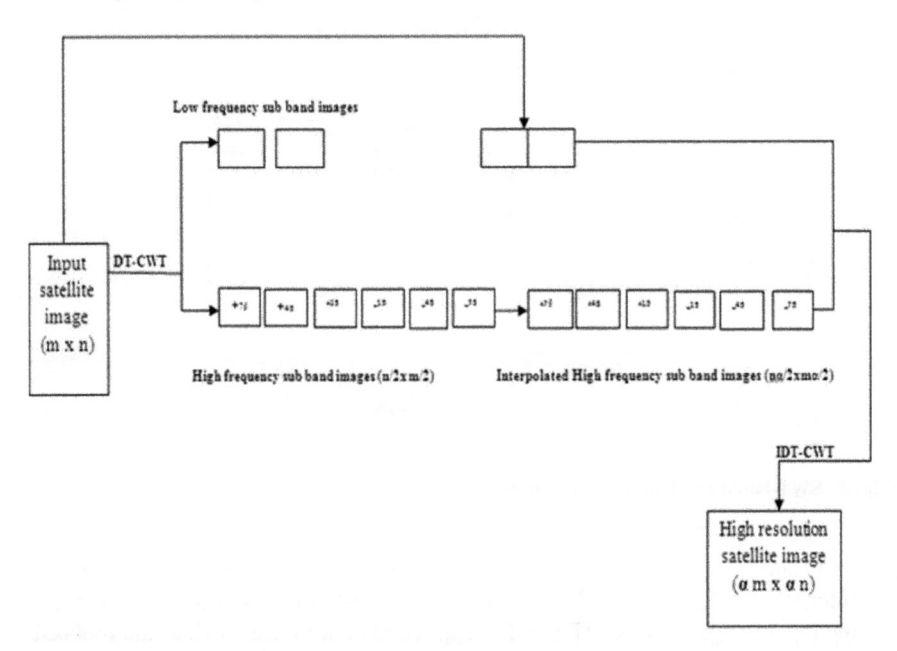

**Fig. 3** DT-CWT-based resolution enhancement

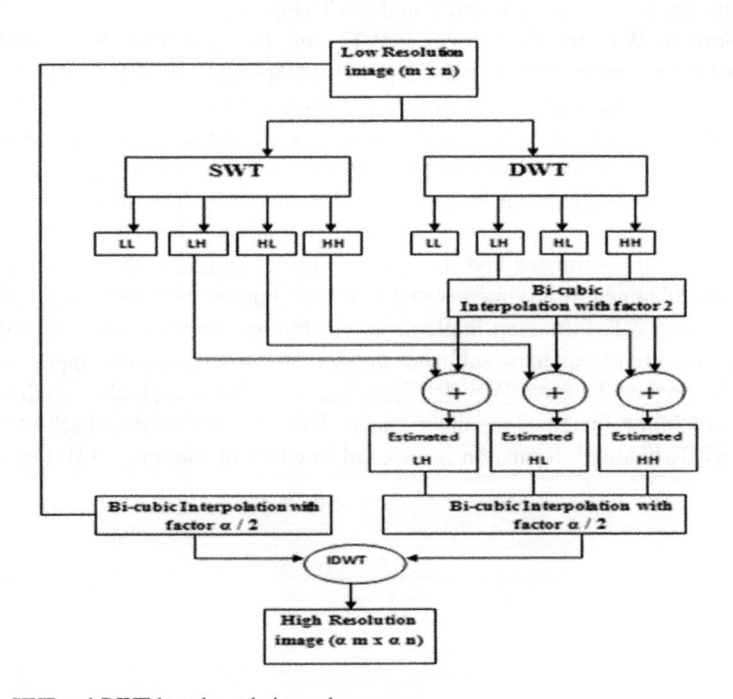

**Fig. 4** SWT and DWT-based resolution enhancement

## 4 Evaluation Methods for Resolution Enhancement Techniques

Different evaluation methods are used to determine the performance of implemented algorithms. The parameters which are considered for evaluating the resolution enhancement methods are Peak Signal-to-Noise Ratio (PSNR), Root Mean Square Error (RMSE), Root Mean Square Error (RMSE), and Mean Absolute Error (MAE). Frequency domain methods are compared with these parameters to shows the superiority of the methods over the satellite image.

1. **Mean Square Error (MSE)**: MSE formula is measuring the MSE between the input image ($I$) and the original image ($O$).

$$\text{MSE} = \frac{\sum_{i,j}(I\,\text{in}(i,j) - I\,\text{org}(i,j))^2}{M \times N} \quad (1)$$

where $M$ and $N$ are the size of the images.

2. **Root Mean Square Error (RMSE)**: The equivalent pixels in the reference image and the obtained high-resolution image ($H_r$) is calculated. The better quality image will always have less RMSE value.

$$\text{RMSE} = \sqrt{\frac{1}{MN}\sum_{i=1}^{M}\sum_{j=1}^{N}(H_r(i,j) - H(i,j))^2} \quad (2)$$

3. **Mean Absolute Error (MAE)**: Quantitative result is also measured in terms of Mean Absolute Error (MAE). MAE calculates the difference between of equivalent pixels in the reference image and high-resolution image ($H_r$).

$$\text{MAE} = \frac{1}{MN}\sum_{i=1}^{M}\sum_{j=1}^{N}|H_r(i,j) - H(i,j)| \quad (3)$$

4. **Peak Signal to Noise Ratio (PSNR)**: PSNR determines the ratio between the original image and reconstructed image. The high PSNR indicate better quality of the image. $R$ is the maximum fluctuation of the input image. The maximum fluctuation of input image is $R$.

$$\text{PSNR} = 10\log_{10}\left(\frac{R^2}{\text{MSE}}\right) \quad (4)$$

## 5 Result and Discussion

The medical and satellite image suffer with resolution. Many important transform domain methods such as DWT, SWT, DT-CWT, and DWT with SWT are implemented to overcome the problem in both the cases. The methods are tested and compared with satellite images. Various evaluation methods such as PSNR, RMSE, MSE, MAE and TIME are applied to determine the supremacy of the methods. These evaluation parameters are carried out to check the quality of the image. Here we have taken high resolution of image and down sampled that image. This downsampled image is considered as a low-resolution input (Fig. 5).

Tables 1, 2 and 3 shows the results obtained by DWT, SWT, DT-CWT, and SWT with DWT and comparison of various evaluation parameters. The resolution is enhanced from $128 \times 128$ to $512 \times 512$ ($\alpha = 4$).

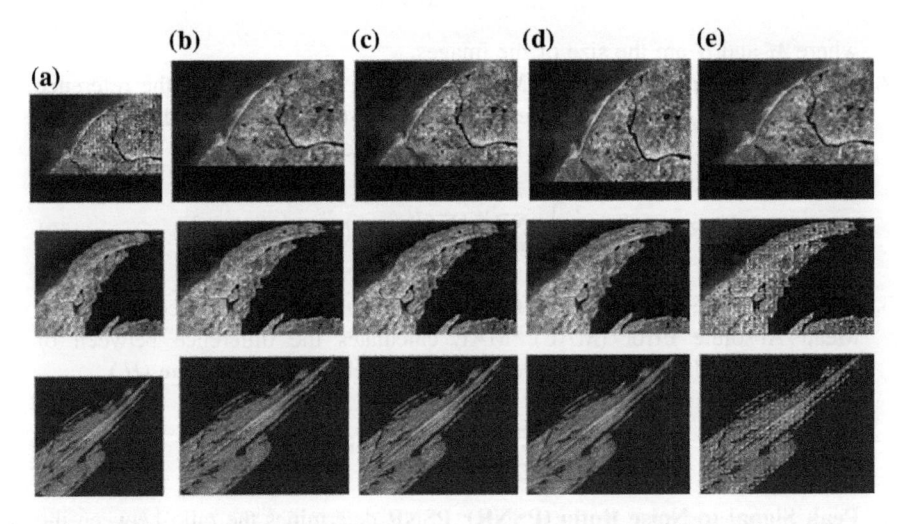

**Fig. 5** a Original low resolution (LR) image, **b** enhanced image using DWT, **c** enhanced image using SWT, **d** enhanced image using DT-CWT, **e** enhanced image using DWT and SWT

**Table 1** The result obtained for LR.1 Image

| Methods/parameters | PSNR | RMSE | MAE | TIME |
|---|---|---|---|---|
| DWT | 23.97 | 16.19 | 9.18 | 10.19 |
| SWT | 24.27 | 15.64 | 9.03 | 3.99 |
| CWT | 24.99 | 14.40 | 8.16 | 15.84 |
| SWT and DWT | 38.38 | 49.32 | 39.32 | 8.70 |

**Table 2** The result obtained for LR.2 Image

| Methods/parameters | PSNR | RMSE | MAE | TIME |
|---|---|---|---|---|
| DWT | 23.52 | 17.05 | 10.34 | 7.62 |
| SWT | 23.56 | 17.00 | 10.51 | 4.26 |
| CWT | 24.50 | 15.24 | 09.26 | 16.31 |
| SWT and DWT | 45.80 | 20.98 | 12.53 | 9.13 |

**Table 3** The result obtained for LR.3 Image

| Methods/parameters | PSNR | RMSE | MAE | TIME |
|---|---|---|---|---|
| DWT | 25.95 | 12.90 | 06.17 | 7.45 |
| SWT | 26.66 | 11.88 | 05.95 | 4.61 |
| CWT | 27.11 | 11.29 | 05.48 | 16.32 |
| SWT and DWT | 48.08 | 16.15 | 07.53 | 9.17 |

## 6 Conclusion

The resolution enhancement of satellite images has been implemented by using transform domain methods such as DWT, SWT, DT-CWT, and DWT with SWT and compared in terms of evaluation parameters such as PSNR, RMSE, TIME, and MAE. The resulting parameters determine the superiority of methods. It is concluded that the resultant image obtained by using DWT with SWT gives a sharper image than other methods. Tables show the results of RMSE, PSNR, TIME, and MAE. The PSNR graph shows the supremacy of the method.

## References

1. Demirel, H., Anbarjafari, G.: Discrete wavelet transform-based satellite image resolution enhancement. IEEE Geosci. Remote Sens. **49**, 1997–2004 (2011)
2. Mathew, A.A., Kamatchi, S.: Brightness and resolution enhancement of satellite images using SVD and DWT. Int. J. Eng. Trends Technol. (IJETT) **4**(4), 712–718 (2013)
3. Hasan, D., Gholamreza, F.: Image resolution enhancement by using discrete and stationary wavelet decomposition. IEEE Trans. Image Process. **20**(5), 1458–1460 (2011)
4. Shekokar, R.U., Pawar, Y.S.: Resolution enhancement of image captured by satellite using DWT. Int. J. Multidiscip. Res. Dev. **2**(7), 238–243 (2015)
5. Gupta, A., Sonika, M.: Image resolution enhancement technique by interpolation in wavelet domain. Int. J. Prog. Eng. Manag. Sci. Humanit. (IJPE) **1**(3), 2395–7794 (2015). ISSN: 2395-7786
6. Jayanthi, P., Jagadeesh, P.: Image resolution enhancement based on edge directed interpolation using dual tree-complex wavelet transform. In: IEEE-International Conference on Recent Trends in Information Technology (ICRTIT), pp. 759–763. IEEE (2011). ISBN: 978-1-4577-0590-8/11/$26.00
7. Kole, P.S., Patil, N.: Satellite image resolution enhancement using discrete wavelet transform. Int. J. Eng. Sci. Comput. (IJESC) **6**(4), 3719–3721 (2016). ISSN: 2321-3361

8. Bala Srinivas, P., Venkatesh, B.: Comparative analysis of DWT, SWT, DWT & SWT and DT-CWT-based satellite image resolution enhancement. Int. J. Electron. Commun. Technol. (IJECT) 5(4), 137–141 (2014)
9. Sontakke, M.D., Kulkarni, M.S.: Multistage combined image enhancement technique. In: International Conference on Recent Trends in Electronics Information Communication Technology, pp. 212–215. IEEE (2016). ISBN: 978-1-5090-0774-5
10. Bidwai, P., Tuptewar, D.J.: Resolution and contrast enhancement techniques for grey level, color image and Satellite image. In: International Conference on Information Processing (ICIP), pp. 511–515. IEEE (2015). ISBN: 978-1-4673-7758-4
11. Sharma, A., Khunteta, A.: Satellite image contrast and resolution enhancement using discrete wavelet transform and singular value decomposition. In: International Conference on Emerging Trends in Electrical, Electronics and Sustainable Energy System (ICETEESES), pp. 1–5. IEEE (2016). ISBN: 978-1-5090-2118-5
12. Mazhar, A., Hassan, F., Anjum, M.R, Maria, S., Muhammad, A.S.: High resolution image processing for remote sensing application. In: International Conference on Computing Technology (INTECH), pp. 302–305. IEEE (2016). ISBN: 978-1-5090-2000-3
13. Rathod, M., Khanapuri, J.: Resolution enhancement of satellite image using discrete cosign transform (DCT) and discrete wavelet transform (DWT). Asian J. Converg. Technol. (AJCT) 3(3), 67–71 (2017). ISSN: 2350-1146
14. Rathod, M., Khanapuri, J.: Satellite image resolution enhancement using stationary wavelet transform (SWT) and discrete wavelet transform (DWT). In: International Conference on Nascent Technologies in the engineering Field (ICNTE), pp. 1–5. IEEE (2017). ISBN: 978-1-5090-2794-8
15. Rathod, M., Khanapuri, J.: A comparative study of transform domain methods for resolution enhancement of satellite image. In: International Conference on Intelligent System Control (ISCO), pp. 287–291. IEEE (2017). ISBN: 978-1-5090-2718-7

# Test Case Analysis of Android Application Analyzer

**Bushra Almin Shaikh**

**Abstract** An Android Application Analyzer (AAA) is a tool that helps user to detect malware in Android devices. In addition, it also provides provision to remove detected malware. The system detects various types of existing malware using a novel approach by learning behavior of Android Apps based on its permissions. Android Apps are generally using combination of permissions to perform their desired actions on the device such as reading from or writing to internal/external storage, accessing network or Bluetooth etc. In most of the cases, the malicious Apps use combination of permissions that may pose a risk to user. The implemented system learns these malicious permissions of Apps by using data mining algorithm such as clustering and classification and classifies them either into harmful or safe. The paper discusses the test cases that show how the system is accurately able to classify malware by increasing the number of clusters size.

**Keywords** Apps · Android malware · Permissions · Benign

## 1 Introduction

Android being an open-source continues to be one of the leading mobile operating systems that is led by Google. However, the security issues in Android still needs to be discussed. Most of the Android devices are exposed to one or the other vulnerability. The most common threats are due to installation that leads to malicious codes installed; the second one is dynamic code loading where an existing app downloads new malicious codes, and finally, the last threat is injection where an attacker injects malicious codes directly into the device's existing system. Each of

---

B. A. Shaikh (✉)
SIES Graduate School of Technology, Navi Mumbai, India
e-mail: bushra.sk@siesgst.ac.in

© Springer Nature Singapore Pte Ltd. 2018
H. Vasudevan et al. (eds.), *Proceedings of International Conference on Wireless Communication*, Lecture Notes on Data Engineering and Communications Technologies 19, https://doi.org/10.1007/978-981-10-8339-6_26

these threats requires regular updates and patches from both the manufacturers and the software companies. However, Android devices receive an average of only 1.26 updates per year to fix these evolving threats. The bottom line is that Android users must be more careful about how they use their devices, and take extra precautions in securing their personal data [1].

## 1.1 Android Security Threats

The real threat to the Android is mobile malware. A malware is a malicious software designed to damage or disrupt the mobile devices. These malicious programs are installed unknowingly by users or get installed by themselves and then perform functions without user knowledge. Once a user grants the permissions to an application while installing, it does not ask for any further permission and performs their task in the background. The malware can be distributed through the Internet via a mobile browser, downloaded from play store or even installed via device messaging functions. The objectives of mobile malware are stealing device details such as IMEI or IMSI, downloading and installing further hidden malware apps. Spyware and Adware, Trojans and Viruses, and Phishing Apps are some of the categories of the malware.

## 1.2 Android Permission Mechanism

Android has its own security mechanism which lies in its permission model, with the help of which it tries to restrict an access to the system's or user's information. However, it nevertheless provides a way to use only certain allowed permission according to the user. A user who wishes to install and use any application does not understand the significance and meaning of the permissions requested by an application, and thereby simply grants all the permissions. As a result of this, harmful applications also gets installed and perform their malicious activity behind the scene. The user's inability of analyzing the risk of any application results in compromised security and privacy. Thus, a system is needed which can identify and remove such harmful applications.

An Android Application Analyzer is implemented in order to protect the user by identifying and removing malware applications and thereby it provides a security to user's personal information and android device. The implemented system identifies as well as removes harmful applications using clustering and classification algorithm.

## 2 Android Application Analyzer

An Android Application Analyzer [2] is a secure system that determines which installed apps are malicious and needs to be uninstalled. The decision to remove a malicious app is based on the analysis of the permissions of installed applications. The implemented system aims to reduce the burden of analyzing and removing those harmful applications from the user's phone and thereby provides security.

The implemented system uses techniques such as clustering and classification to analyze the permissions of installed apps. In Fig. 1, we give the architecture of proposed system "Android Application Analyzer".

### 2.1 Working of Android Application Analyzer

The system consists of five major modules such as identification of installed applications, permission extraction, clustering of known permissions into categories, classification of benign and malicious apps and removal of malicious apps [2]. The modules thus are explained in detail below.

The PackageManager API is used in the first module to identify the list of apps installed on the phone. The Permission Extraction module extracts the permissions of installed applications along with their related information such as app's name, package name, version, required features, etc. The third module assigns an app into one of the malicious permission's clusters or safe permission's cluster. It uses $k$-means clustering algorithm [3]. In order to create a cluster, a particular family of malware has to be considered. Every cluster that is created consists of the list of permissions represents one family of malware. For a particular cluster, the individual values of all the permissions that it defines are extracted from the database of malicious permissions lists and are added to represent its centroid. If the distance is

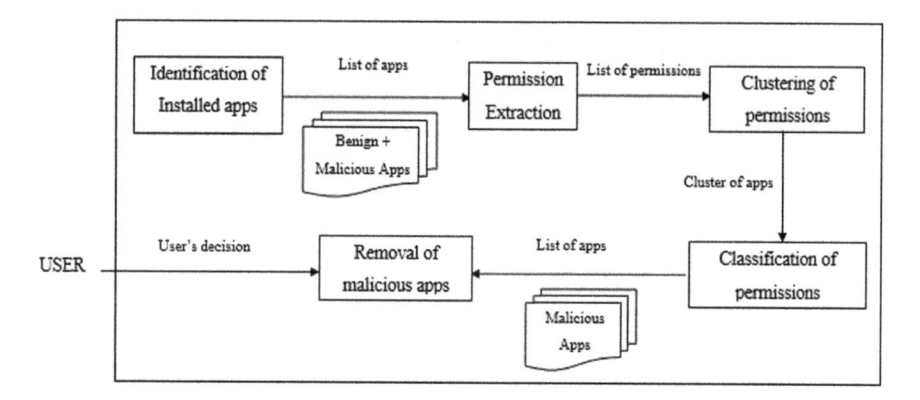

**Fig. 1** System architecture of android application analyzer [2]

of an app's overall permissions is found to be closer to any one of the cluster's centroid, then it is assigned to that particular cluster. The objective of clustering is to group similar applications into a particular malicious cluster based on the closest distance of an app to its cluster. The result of clustering is the list of applications which are malicious in nature. The clusters and its permissions are then used together to form the attributes for classification. The fourth module is needed to accurately classify an app into the benign or malicious category. This is achieved using Naïve Bayesian classification algorithm [4]. For a given an application and its cluster, the probability for each set of permission's combination occurring in a particular cluster is calculated. The combination of the permissions is derived from the list of the permissions that are defined by a particular cluster. Each possible permission's combination has been assigned to one of the two classes such as "malicious" and "benign" considering its behavior. If a new app to be installed is requesting a permission that matches with the permissions declared in a dataset, then its corresponding probability of being benign and malicious are calculated. If the probability of a given application being malicious is found greater than the probability of being benign, then that application is declared as malicious else benign. The fifth module allows the user either to delete the harmful app from the phone or to retain it.

## 3 Test Case Analysis

The following samples of malicious applications were already used for testing: DogWar (Trojan), ICalendar (Premium SMS), and SuperSolo (DroidDream) [5]. These malware were successfully detected and its test results were already discussed in [2]. However, we have performed the testing of our implemented system to detect 15 samples of known/unknown malware by increasing the number of clusters. The most widespread malicious objects detected on Android smartphones can be divided into three main groups: SMS Trojans, advertising modules and exploits to gain root access to smartphones [5].

We have collected around 15 malware out of which 6 are Trojans, 3 are SMS Trojans, 2 are ad SMS and 1 is of type root exploit. Among 6 unique Trojans, 2 samples of same characteristics and family are chosen. The malware samples that are considered for testing are collected from [5]. The test cases of the selected malware are given in the following section.

### 3.1 Test Case Scenarios

Following are some of the test case scenarios discussed. The same way different test cases for different categories of malware are found to be successfully detected. The test case analysis of all the samples of malware detected is shown in the next section.

For instance, we started with the first malware known as DroidKungFu [6] which gains and exploit root privilege and send device specific information to a remote server. DroidKungFu comes in many forms such as DroidKungFu1, DroidKunFu2, etc. We have selected a sample malware DroidKungFu2 for testing purpose and it has been categorized in Cluster 1 as shown in Fig. 2a, since the sum of permissions requested by DroidKungFu2 is very close to the centroid of Cluster 1.

We have also tested the family of malware known as Geinimi [7]. Geinimi is the most sophisticated and is the first Android malware in the wild that displays botnet-like capabilities. Geinimi usually comes in many variants such as Geinimi A, Geinimi B, etc. and we have selected two samples such as Geinimi Shoppers (Geinimi B) and Geinimi GoldMiner (Geinimi D). Both the samples have been categorized in Cluster 2. For example, we have Geinimi B as shown in Fig. 2b and detected as malware in classification stage in Fig. 2c. Apart from Geinimi, we have found that the Trojan DogWar [8] has also been detected by Cluster 2. This Trojan was already detected in [2]. Figure 2c also shows detection list of the rest of malware.

## 3.2 Test Result Analysis

In our earlier work [2], we have created only few sample clusters considering few sample malware's family and we proposed to create more clusters in order to detect more sample of malware's family. This test result analysis in Table 1 aims to prove that our proposed method of increasing clusters helps us to detect more samples of known and unknown malware as well.

Table 1 depicts that out of 15 malware only 1 malware, i.e., KMHome (B) [9] could not get detected since it has been clustered under the safe category. The reason behind KMHome (B) being clustered under safe category is that the sum of its permissions is far away from the centroids of all the clusters that are created and also there was no cluster created separately for the family of KMinn [9].

In our implemented system, each cluster is created for a particular known family of malware so that if other unknown malware of the same family is installed then they also get detected. However, we have found out that KMinn (B&C) and HippoSMS has been detected in Cluster 3 that was created considering ICalendar malware since the sums of KMinn as well as HippoSMS [10] are very close to Cluster 3.

This indicates that if the sum of a particular malware or app is closest to any of the cluster's centroid, then they are put inside that cluster. Hence the major advantage of our system is that the lesser number of clusters are required to detect other unknown malware. For instance, we have created only 7 clusters that are

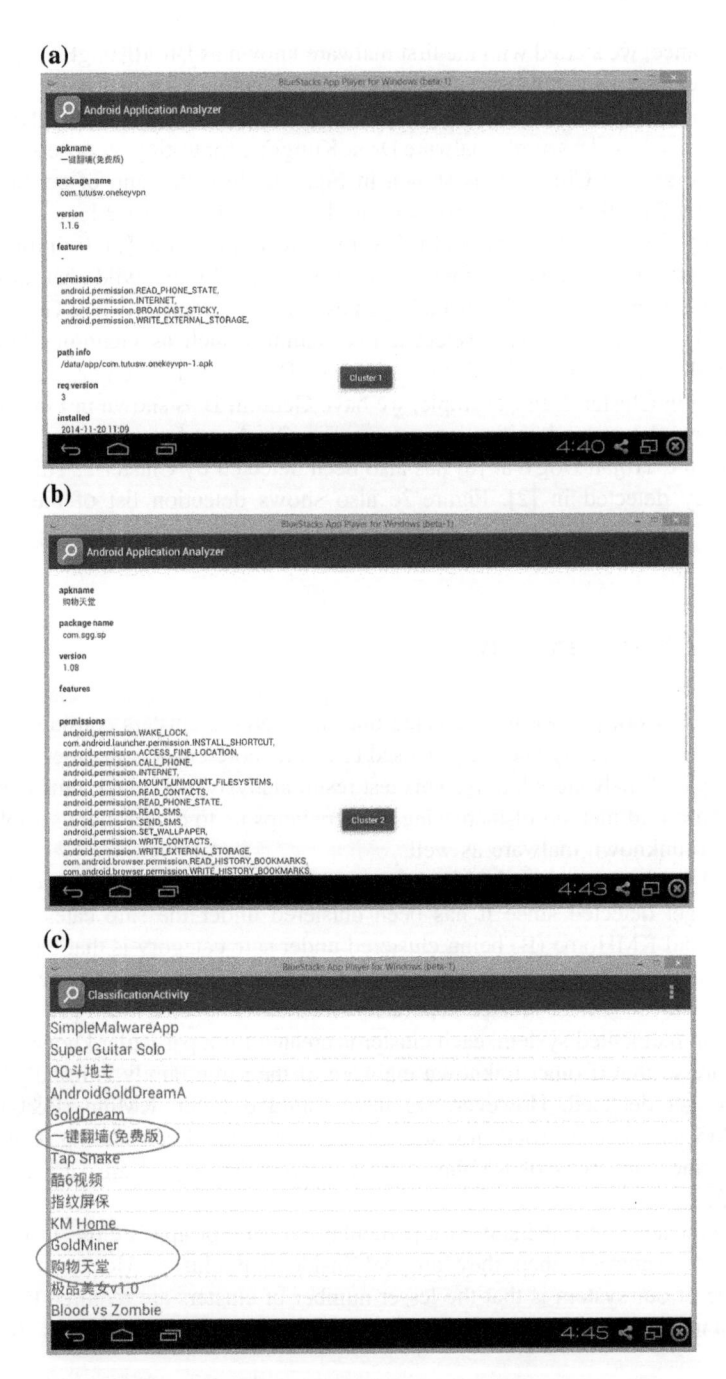

**Fig. 2 a** Clustering of DroidKungFu in Cluster 1, **b** clustering of Geinimi B in Cluster 2, **c** classification of DroidKungFu, Geinimi B and D, and other malware

**Table 1** Summary of test result analysis

| S. No. | Malware | Cluster # | Safe cluster | Correctly classified | Not classified | Result |
|---|---|---|---|---|---|---|
| 1 | Geinimi-Shoppers | 2 | | ✓ | | Correct |
| 2 | Geinimi-GoldMiner | 2 | | ✓ | | Correct |
| 3 | GoldDreamA | 5 | | ✓ | | Correct |
| 4 | GoldDreamB | 5 | | ✓ | | Correct |
| 5 | DroidDream-SuperSolo | 4 | | ✓ | | Correct |
| 6 | PJApps | 6 | | ✓ | | Correct |
| 7 | DogWar | 2 | | ✓ | | Correct |
| 8 | ICalendar | 3 | | ✓ | | Correct |
| 9 | JimmRussia | 7 | | ✓ | | Correct |
| 10 | TapSnake | 7 | | ✓ | | Correct |
| 11 | HippoSms | 3 | | ✓ | | Correct |
| 12 | KMHome(C) | 3 | | ✓ | | Correct |
| 13 | KMHome(B) | | ✓ | | ✓ | **Incorrect** |
| 14 | DroidKungFu | 1 | | ✓ | | Correct |
| 15 | QQTencent | 4 | | ✓ | | Correct |

capable of detecting around 14 malware out of 15 malware that are used for testing. Also if a cluster is created for one type of known malware's family, then other unknown malware samples that belong to the same family are also being detected.

## 4 Comparative Analysis

Table 2 shows a comparative study of our system with the well-known and popularly used anti-malware applications such as 360 Mobile Security [11], Avast Anti-virus [12], Lookout Security and Antivirus [12], Kaspersky [12] McAfee [13], Norton Security and Antivirus [14], ESET [15], NetQin Mobile Security [16] and Zoner [17] etc. to find out how accurately our system distinguishes the malware samples that are considered for testing from the benign applications.

The advantage of our system that it classifies most of the malware applications correctly which is also detected as malware by the other anti-malware apps was already discussed in [2]. However, all the anti-malware apps require an updated database of known malware's signatures without which they cannot detect any known or unknown malware whereas our system can detect unknown malware of a particular family once its predecessor has been clustered and classified [2].

**Table 2** Comparative analysis of android application analyzer with anti-malware

| S. No. | Malware | 360, Avast, Norton, McAfeeESET | Lookout | Karspersky | NetQin | Zoner | Android application analyzer |
|---|---|---|---|---|---|---|---|
| 1 | Geinimi-Shoppers | ✓ | ✓ | ✓ | ✓ | ✓ | ✓ |
| 2 | Geinimi-GoldMiner | ✓ | ✓ | ✓ | ✗ | ✓ | ✓ |
| 3 | GoldDreamA | ✓ | ✓ | ✓ | ✓ | ✗ | ✓ |
| 4 | GoldDreamB | ✓ | ✓ | ✓ | ✓ | ✓ | ✓ |
| 5 | DroidDream-SuperSolo | ✓ | ✓ | ✓ | ✓ | ✓ | ✓ |
| 6 | PJApps | ✓ | ✗ | ✓ | ✓ | ✓ | ✓ |
| 7 | DogWar | ✓ | ✗ | ✗ | ✓ | ✓ | ✓ |
| 8 | ICalendar | ✓ | ✓ | ✓ | ✓ | ✓ | ✓ |
| 9 | JimmRussia | ✓ | ✗ | ✓ | ✓ | ✗ | ✓ |
| 10 | TapSnake | ✓ | ✓ | ✗ | ✓ | ✓ | ✓ |
| 11 | HippoSms | ✓ | ✗ | ✗ | ✓ | ✓ | ✓ |
| 12 | KMHome (C) | ✓ | ✗ | ✓ | ✓ | ✓ | ✓ |
| 13 | KMHome (B) | ✓ | ✗ | ✓ | ✓ | ✓ | ✗ |
| 14 | DroidKungFu | ✓ | ✓ | ✓ | ✓ | ✓ | ✓ |
| 15 | QQTencent | ✓ | ✗ | ✗ | ✓ | ✓ | ✓ |

## 5 Conclusion

We have created more number of clusters for other malware's families so that unknown malwares could also get detected. However, it has been found in testing that some of the already created clusters are also able to detect the malware of other family. This is a major advantage of our system since no cluster was created for other malware's family, but still our system is able to detect it. This detection has been made possible due to the similarity with the already created clusters. Thus we are able to detect the malware that belongs to the same or known family as well as some of the malware that belongs to the unknown family by increasing clusters.

## References

1. Protecting Android Devices and Why They're so Vulnerable. https://us.norton.com/internetsecurity-malware-android-malware.html
2. Almin, S.B., Chatterjee, M.: A novel approach to detect android malware. Procedia Comput. Sci. **45**, 407–417 (2015)
3. KardiTech's Page: K-means clustering algorithm: tutorial—how k-means clustering algorithm works? http://people.revoledu.com/kardi/tutorial/kMean/Algorithm.htm
4. Scribd: Classification algorithms used in data mining. http://www.scribd.com/doc/11352378/Classification-algorithms-used-in-Data-Mining-This-is-a-lecture-given-to-Msc-students
5. Contagio: Take a sample leave sample mobile malware. http://contagiodump.blogspot.in/2011/03/take-sampleleave-sample-mobile-malware.html
6. Security Alert: New DroidKungFu Variant—AGAIN!—Found in Alternative Android Markets. http://www.csc.ncsu.edu/faculty/jiang/DroidKungFu3/
7. The Official Lookout Blog: Security Alert Geinimi, Sophisticated New Android Trojan Found in Wild. https://blog.lookout.com/blog/2010/12/29/geinimi_trojan/
8. Virus Bulletin: Covering the Global Threat Landscape: Droiddream Malware. https://www.virusbtn.com/virusbulletin/archive/2012/03/vb201203-DroidDream
9. Networks: Juniper Networks, Mobile Signatures. www.trapezenetworks.com/us/en/local/pdf/additional-resources/jnpr-mobile-signatures.pdf
10. Security Alert: New Android Malware—HippoSMS—Found in Alternative Android Markets. http://www.csc.ncsu.edu/faculty/jiang/HippoSMS/
11. Forensic blog: Mobile Phone Forensics and Mobile Malware. http://forensics.spreitzenbarth.de/android-malware/
12. Look into Mobile: 5 Best Android Antivirus and Security Apps. http://lookintomobile.com/android-antivirus-security-apps/
13. Tom's guide: Best Android Antivirus Software 2014. http://www.tomsguide.com/us/best-android-antivirus,review-2102.html
14. Google Play: Norton Security and Antivirus. https://play.google.com/store/apps/details?id=com.symantec.mobilesecurity&hl=en
15. Digital Trends: Top 5 Android Security Apps: Do they protect you. http://www.digitaltrends.com/mobile/top-android-security-apps/
16. Google Play: NQ Mobile Security and Antivirus. https://play.google.com/store/apps/details?id=com.nqmobile.antivirus20&hl=en
17. Google Play: Zoner Antivirus. https://play.google.com/store/apps/details?id=com.zoner.android.antivirus&hl=en

# Effect of Windowing in the Performance of OFDM Systems

Ranjushree Pal

**Abstract** Orthogonal Frequency-Division Multiplexing is the key broadband technology, which is the basis for 4G (Fourth Generation) wireless communication systems. Undistorted OFDM carries many un-filtered Quadrature Amplitude-Modulated (QAM) subcarriers. Thus, the out-of-band spectrum decreases very slowly. This out-of-band spectral characteristic of each symbol interferes with the next OFDM symbol. In this paper, several windowing techniques are used to reduce the out-of-band spectrum of each OFDM symbol. The effects of different windows in Peak to Average Power Ratio (PAPR) and Bit Error Rate (BER) of OFDM signals are also observed. The results are summarized in the concluding section.

**Keywords** Orthogonal frequency-division multiplexing (OFDM)
Quadrature amplitude-modulated (QAM) · Windows · Peak to average power ratio (PAPR) · Bit error rate (BER) · Out-of-band spectral characteristics

## 1 Introduction

Orthogonal Frequency-Division Multiplexing (OFDM) is the basis of fourth-generation wireless systems. It is a radical broadband technology which can support data rates above 100 mbps [1]. In OFDM, the high-rate input data stream is divided into several low-rate output data streams and these are transmitted parallelly by a number of low-frequency carriers known as subcarriers. As the frequency of the sub-carriers decreases, the symbol duration increases, which decreases the time domain dispersion which is caused by the delay spread in multipath channel. This time domain dispersion is known as Intersymbol Interference (ISI) which can be completely removed by adding a guard-interval with every OFDM symbol. This guard-interval consists of cyclically extending the OFDM symbol which is called

---

R. Pal (✉)
Department of Electronics and Telecommunication Engineering, Dwarkadas J. Sanghvi College of Engineering, Mumbai, India
e-mail: ranjushreepal@gmail.com; ranjushree.pal@djsce.ac.in

© Springer Nature Singapore Pte Ltd. 2018
H. Vasudevan et al. (eds.), *Proceedings of International Conference on Wireless Communication*, Lecture Notes on Data Engineering and Communications Technologies 19, https://doi.org/10.1007/978-981-10-8339-6_27

cyclic prefix [2]. The frequency of the sub-carriers is chosen so that they are orthogonal to each other which ensures no interference among the sub-carriers even when they overlap [3]. Thus the spectral efficiency increases tremendously. OFDM is exploited for data communications in wideband radio channels, in high bit rate DSL (HDSL) technology supporting up to 1.6 mbps, Asymmetric DSL (ADSL) which supports data rate up to 6 mbps over existing Cu lines, Very-High-Speed-DSL supporting data rate up to 100 mbps, Digital Audio Broadcasting and High-Definition television [3].

## 2 Basic OFDM System Model

Figure 1 shows the basic OFDM transmitter–receiver model [4].

The data input is first mapped to a constellation mapper either BPSK, QPSK, QAM. OFDM symbol is then formed by obtaining IFFT and adding a cyclic-extension. The reverse process is carried out in the demodulator. An example of the OFDM signal is shown in Fig. 2.

Figure 2 shows the sharp phase transitions at the symbol boundaries caused by modulation. Actually, an OFDM signal carries a number of un-filtered generally QAM sub-carriers. As a result, the out-of-band spectrum decreases rather slowly according to a sinc function [2]. This out-of-band characteristic should be reduced in order to reduce interference with the next OFDM symbol. To make the OFDM symbol decrease more rapidly, each OFDM symbol can be passed through different windows. After windowing in the time domain, the amplitude of the OFDM symbol at the symbol boundaries decreases to zero very smoothly [2].

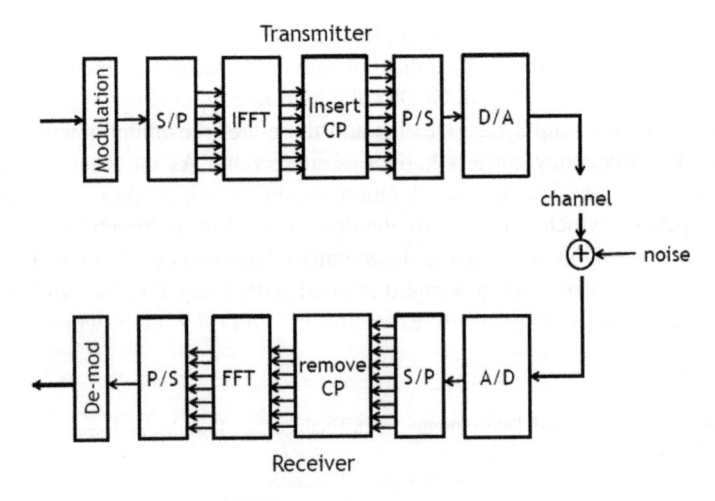

**Fig. 1** Block diagram of basic OFDM transmitter–receiver

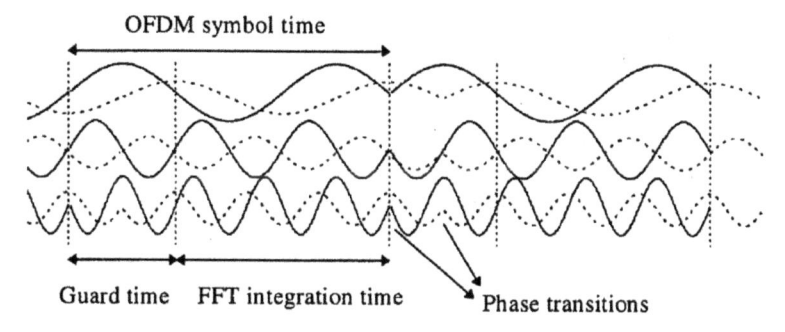

Fig. 2 OFDM signal with three sub-carriers in two-ray multipath channel

## 3 Implementation

Simulink is used to implement the OFDM model [5]. The OFDM parameters used during implementation are partly according to IEEE 802.11a standard with 16 QAM. The specifications used are given in Table 1.

Figure 3 shows the Simulink model of OFDM transmitter–receiver.

The flow of the model is as follows:

Figure 4 shows Random Bernoulli binary generator generating zeros and ones with equal probability with sample time 1/Fs. It is converted to 16 QAM signals.

Figure 5 shows the generation of OFDM signals. It is a subsystem, which includes implementation of 64-point IFFT of the 16 QAM signals followed by cyclic prefixing and transmitter gain.

The OFDM signal so generated is passed through three types of windows:

(a) Hamming Window

$$W(n) = 0.54 - 0.46\cos(2\pi n/N), \quad 0 \le n \le N \quad (1)$$

(b) Hanning Window

$$W(n) = 0.5 - 0.5\cos(2\pi n/N), \quad 0 \le n \le N \quad (2)$$

Table 1 OFDM parameters used in implementation

| | |
|---|---|
| Symbol data rate Fs (uncoded in Hz) | 20e6 |
| Number of data-carriers | 48 |
| Number of pilot-carriers | 4 at (−21, −7, 7, 21) |
| FFT length | 64 (48 data + 12 null + 4 pilot) |
| Modulation of 52 carriers | 16 QAM |
| Cyclic prefix length | 16 |

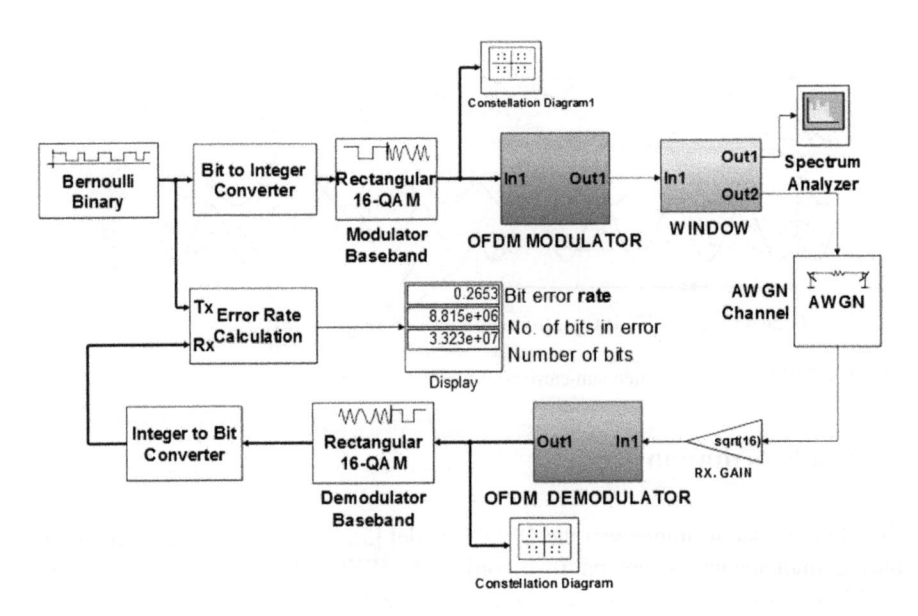

**Fig. 3** Simulink model of OFDM transmitter–receiver

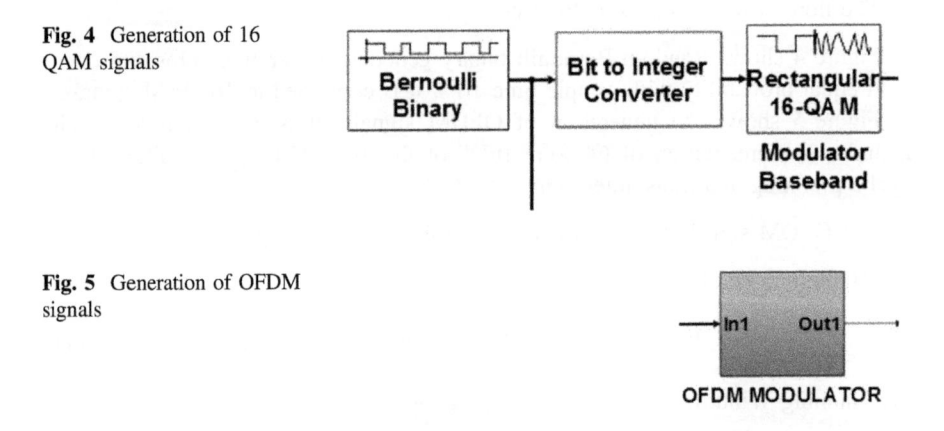

**Fig. 4** Generation of 16 QAM signals

**Fig. 5** Generation of OFDM signals

(c) Kaiser Window (beta = 10), $\alpha$ is attenuation in dB

$$\beta = 0.1102(\alpha - 8.7), \quad \alpha > 50 \qquad (3)$$

$$= 0.5842(\alpha - 21)^{0.4} + 0.07886(\alpha - 21), \quad 50 \geq \alpha \geq 21 \qquad (4)$$

$$= 0 \quad \alpha < 21 \qquad (5)$$

These windows are generated in the subsystem as shown in Fig. 6 and compared with the undistorted OFDM signal in the Spectrum Analyzer.

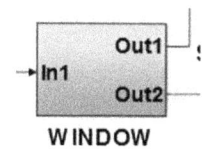

Effect of Windowing in the Performance of OFDM Systems 251

**Fig. 6** Windowing of OFDM signals

After windowing OFDM signal is mixed with AWGN as shown in Fig. 7 and the SNR in AWGN block is varied from 0 to 80 dB.

The demodulator block is shown in Fig. 8.

OFDM demodulator block is a subsystem which consists of amplifying the received signal, removal of the cyclic prefix and carrying out FFT. After FFT, the signal is demodulated by the 16-QAM demodulator and converted into bits for calculation of the Bit Error Rate.

Figure 9 shows the BER calculation of the transmitted and received signal with various values of the SNR in AWGN block.

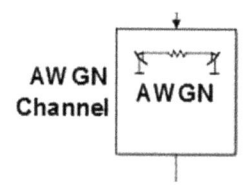

**Fig. 7** OFDM signals mixed with AWGN

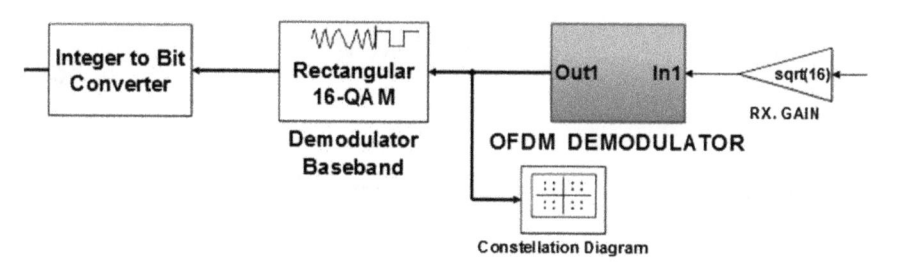

**Fig. 8** OFDM demodulator block diagram

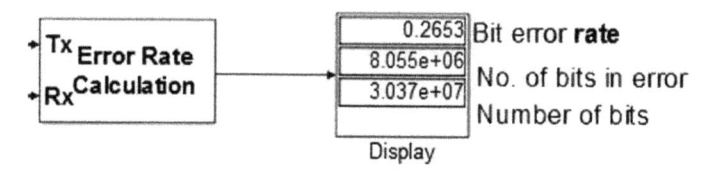

**Fig. 9** Bit error rate calculation

## 4 Results

The OFDM signal is passed through three different windows, e.g., Hamming, Hanning, Kaiser (beta = 10) and their spectrum is compared with the spectrum of the undistorted signal in the Spectrum Analyser. Figure 10 shows the comparison of the different spectra.

The OFDM spectra for undistorted (black), Hamming (yellow), Hanning (blue) and Kaiser (red) plotted against dBm in $y$ axis versus Frequency (in MHz) in $x$ axis. The undistorted OFDM spectra has more out-of-band spectral characteristics. The out-of-band spectral characteristics decrease when passed through Hamming, Hanning window and become minimum in Kaiser window.

The plot of Peak to Average Power Ratio (PAPR) of OFDM signal after passing through different windows Undistorted (blue), Hamming (black), Hanning (red), Kaiser (magenta) as plotted against Probability (%) versus dB above average power is shown in Fig. 11.

The Peak to Average Power Ratio (PAPR) of OFDM signal increases as it is passed through different windows which can be seen from Fig. 11. The undistorted spectra have minimum PAPR while PAPR increases as the OFDM signal is passed through Hamming, Hanning, Kaiser window respectively.

The Bit Error Rate versus SNR in OFDM for different windows is shown in Fig. 12.

When BER is plotted against SNR, the undistorted signal gives the lowest BER. BER increases when the OFDM signal is passed through Hamming, Hanning, Kaiser window, respectively, as shown in Fig. 12.

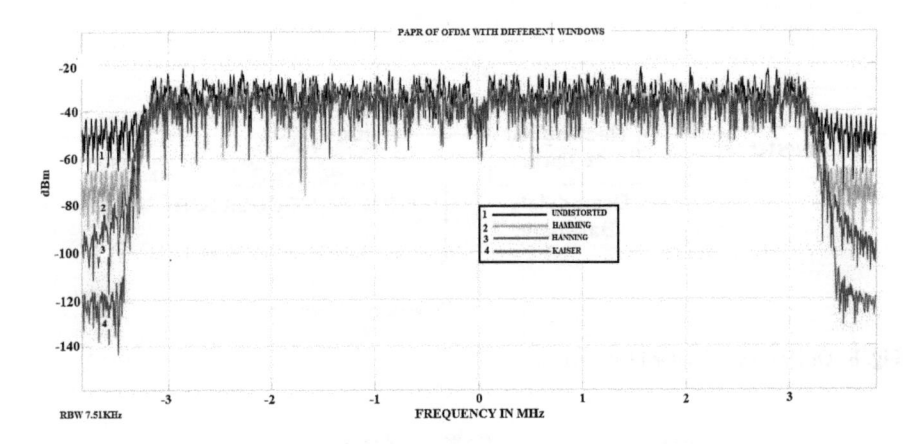

**Fig. 10** OFDM spectra through different windows

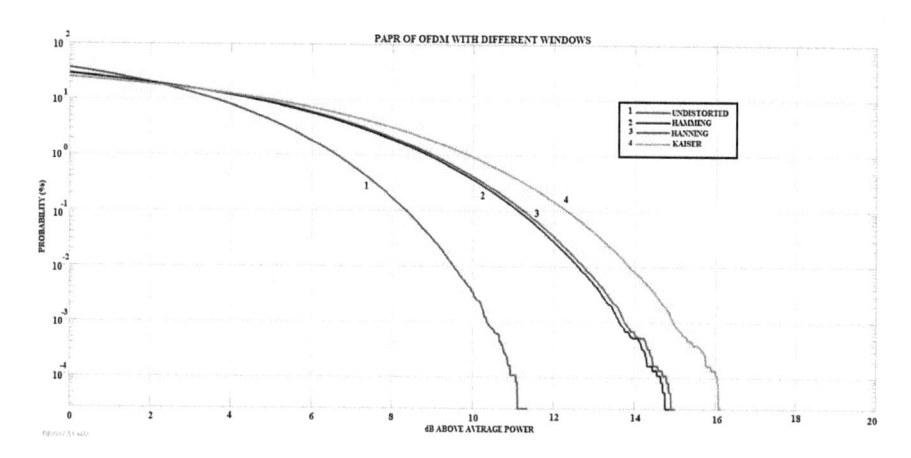

Effect of Windowing in the Performance of OFDM Systems

**Fig. 11** PAPR of OFDM spectra through different windows

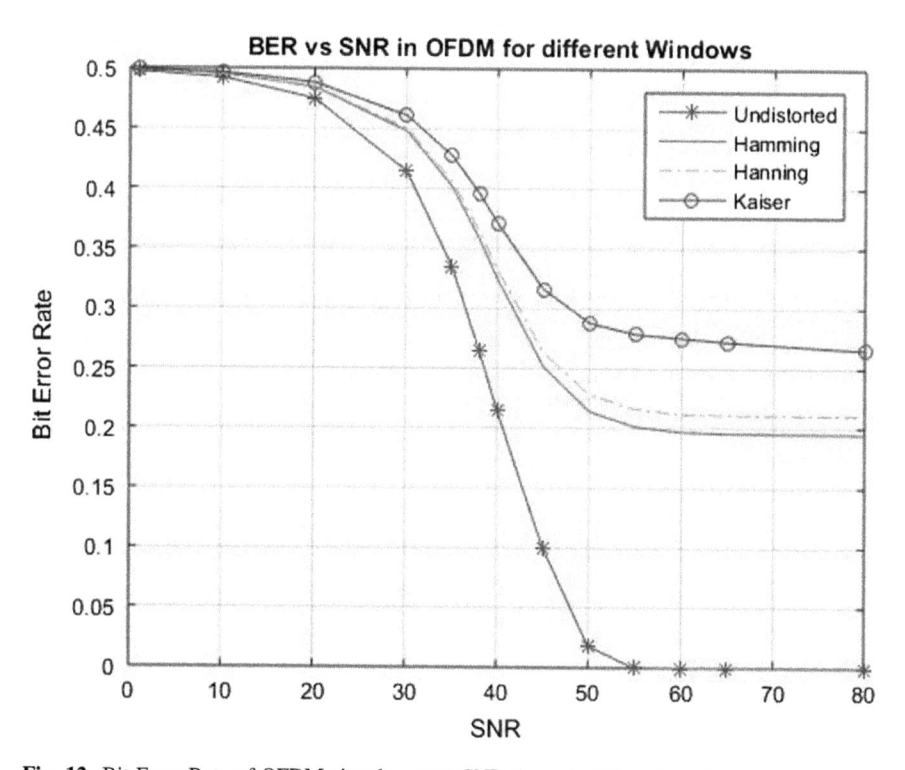

**Fig. 12** Bit Error Rate of OFDM signals versus SNR through different windows

## 5 Conclusion

The undistorted OFDM signal consists of many un-filtered QAM modulated carriers. Thus, the phase of the carriers changes abruptly across the signal boundaries. When these abrupt changes in time domain visualized in frequency domain gives large out-of-band spectral characteristics. The out-of-band spectral characteristics of one OFDM symbol will interfere with the next OFDM symbol. Thus, before transmission, each OFDM symbol can be passed through a window. Here Hamming, Hanning and Kaiser windows are used. It can be seen from Fig. 10 that the out-of-band spectral characteristics are decreasing when passed through Hamming, Hanning and Kaiser windows, respectively, but the Peak to Average Power Ratio (PAPR) as shown in Fig. 11 and Bit Error Rate as shown in Fig. 12 of OFDM signal increases as it is passed through Hamming, Hanning and Kaiser window respectively. Thus, one has to compromise between time and frequency domain in choosing the type of window. As the response in frequency domain becomes compact, there is an increase in the Peak to Average Power Ratio (PAPR) and Bit Error Rate of the OFDM signal in time domain. Thus, it is up to the designer to choose the type of window.

## References

1. Jagannatham, A.K.: https://onlinecourses.nptel.ac.in, Advanced 3G 4G wireless communications, IIT Kanpur
2. Van Nee, R., Prasad, R.: OFDM for Wireless Multimedia Communications. Artech House, Boston (2000)
3. Prasad, R.: OFDM for Wireless Communication Systems. Artech House Inc., Boston (2004)
4. https://www.complextoreal.com
5. https://www.mathworks.in/help/matlab/

# Telemedicine: Making Health Care Accessible

Akshita V. Nichani, Shruti T. Pistolwala, Amit A. Deshmukh and Manali J. Godse

**Abstract** Telemedicine is the confluence of telecommunication technology and medicine. The objective is to develop a low-cost proof of concept of the same, to be implemented in rural areas with minimum access to healthcare facilities. The proposed system consists of biomedical sensors interfaced with embedded technology to read vital information of the patient and store it in a database. This is available to be accessed by a specialist based in a city hospital for remote diagnosis and suggested further treatment. This setup also has provisions for a live video session with the medical practitioner, along with data security features and a unique identification system for each individual based on fingerprint scanning.

**Keywords** Telemedicine · ECG · Biomedical sensors · Fingerprint identification Health care · Solar energy · Server

## 1 Introduction

India is a vast country with a population of about 1.3 billion, with a rural population of 67.25% [1]. Due to the extensive geographic terrain, several rural areas are isolated from hospitals and basic healthcare systems. Moreover, because of logistic

---

A. V. Nichani · S. T. Pistolwala (✉) · A. A. Deshmukh
Department of Electronics and Telecommunication Engineering,
SVKM's, D J Sanghvi College of Engineering, Mumbai, India
e-mail: shruti19996@gmail.com

A. V. Nichani
e-mail: akshita1596@gmail.com

A. A. Deshmukh
e-mail: amit.deshmukh@djsce.ac.in

M. J. Godse
Biomedical Engineering, SVKM's, D J Sanghvi College of Engineering,
Mumbai 400056, India
e-mail: manali.godse@djsce.ac.in

© Springer Nature Singapore Pte Ltd. 2018
H. Vasudevan et al. (eds.), *Proceedings of International Conference on Wireless Communication*, Lecture Notes on Data Engineering and Communications Technologies 19, https://doi.org/10.1007/978-981-10-8339-6_28

and infrastructural difficulties, adequate facilities cannot be extended to these remote regions. According to World Health Organization (WHO) statistics, the total expenditure on health care in India is only 4.7% of the Gross Domestic Product (GDP). There are 8 hospital beds per 10,000 population [2] and 6.49 physicians per 10,000 population [3]. The impact of the above shortcomings is particularly profound in case of rural areas, where people from these areas need to travel long distances to go to the main city hospital. If they are daily wage workers, not only do they have additional expenses of traveling and consultation fees, but their wage for the day is also compromised. Thus, there is a need for an implementation of Telemedicine. Telemedicine is defined as "The delivery of healthcare services, where distance is a critical factor, by all healthcare professionals using information and communication technologies for the exchange of valid information for diagnosis, treatment and prevention of disease and injuries, research and evaluation, and for the continuing education of healthcare providers, all in the interests of advancing the health of individuals and their communities" [4].

The proposed system uses the concept of Telemedicine and allows for remote diagnosis in case of villages at a considerable distance from a hospital. A booth will be set up at a central location in the village. It will consist of medical sensors to sense basic parameters of each patient. These sensors send the patient readings via embedded technology set up to a server, that is accessible by a specialist at a hospital in the city. The specialist then sends back feedback along with the diagnosis. This is a compact and mobile system that can be set up at any remote location. Hence, preventive measures can be provided at initial stages, to reduce the need for invasive action later. Early detection can reduce a great amount of resources, time, as well as suffering for the patient.

## 2 Existing Technology Review

The most relevant systems that have been implemented previously along with their features are as listed as follows. In "Design of advanced Telemedicine system for remote supervision" [5], commercial off-the-shelf medical devices have been used to monitor patient health and integrate it with consumer electronics which store the data on to a database. It provides login for patients and doctors through a web application and carries out analysis on patient data. The limitation of this system is that, it was predominantly implemented for COPD (Chronic Obstructive Pulmonary Disease) and since it uses only commercial devices, the overall investment is higher. The "Patient Monitoring System based on e-health sensors and web services" [6] focuses on real-time monitoring of patient parameters using the e-health senor shield. It does not consist of a database set up for storing medical records, and the use of the e-health sensor shield makes it expensive.

The "Patient Monitoring System using LabVIEW" [7] monitors ECG, Pulse, Galvanic Skin Response (GSR), and Body Temperature using the respective sensors interfaced with Arduino. It provides real-time monitoring over the Internet

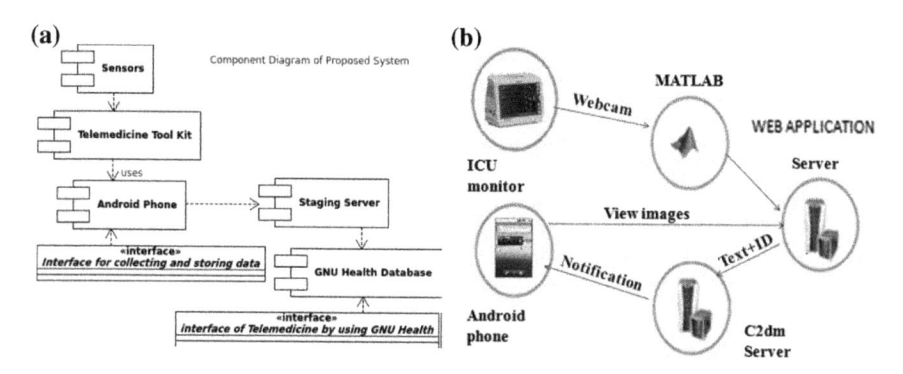

**Fig. 1 a** Component diagram of "development of a telemedicine model with low cost portable tool kit for remote diagnosis of rural people in Bangladesh" [9], **b** block schematic representation of "remote patient monitoring system" [8]

using LabVIEW interface. However, it does not have a database system to store patient information and medical records. "Remote Patient Monitoring System" [8] takes images using a webcam, of the ECG waveform from the display screen of the ECG machine. This image is then sent to MATLAB for performing image processing to detect any abnormalities. If any problems are detected, an alert is sent to the hospital as shown in Fig. 1b. This system uses a database to store the above information, however setup and maintenance of an ECG machine can be expensive and it also requires a reliable supply of electricity.

The "Development of a Telemedicine model with low cost portable tool kit for remote diagnosis of rural people in Bangladesh" [9] uses e-health sensor shield for monitoring ECG and Blood Pressure as shown in Fig. 1a. It utilizes an open source software, GNU for database and patient information management. The limitations are that the e-health sensor shield is expensive and using GNU makes the system reliant on a third-party software. Overall, the mobile Telemedicine system proposed in this paper overcomes the limitations of the above-mentioned systems as it has provisions not only for real-time monitoring using a webcam, but also a database system to store information for the specialist to refer to. It uses medical sensors to make it into a low-cost alternative and has additional features including a solar-powered setup and a fingerprint identification system and provisions for a live consultation session.

## 3 Proposed System

The overview of the proposed system is shown in Fig. 2. It comprises of a Telemedicine unit set up in a rural area. This unit consists of medical sensors used to monitor basic health parameters of a patient. These sensors are interfaced with a laptop (or Raspberry Pi) via an Arduino. The signals can be amplified using a

**Fig. 2** Overview of proposed system

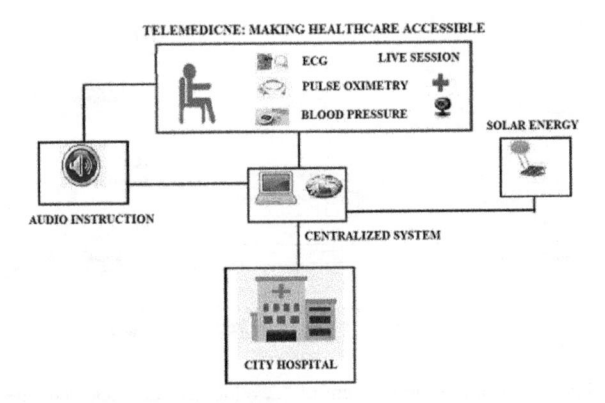

preamplifier and noise can be removed by using an ADC noise reduction code in Arduino. The readings of these sensors are then transferred to the laptop using a software such as LabVIEW. A centralized user application system is set up that consists of a web application for login and is linked to a database with all the patient details along with vital parameters, which are protected by data security features. Login options are available for the patient to enter information while monitoring and view medical practitioner's feedback, as well as for the medical practitioner (based at the city hospital) to send across the diagnosis. Patient login can also be carried out by fingerprint identification since literacy levels are low in rural regions and it is a faster process. The user interface also has an option to play audio instructions in the regional language of the village to allow the user to navigate the system and prepare for medical tests correctly. Due to scarcity of electricity in villages, solar energy will be used to power the system.

## 4 System Architecture

### 4.1 Hardware

**ECG module**: Electrocardiogram (ECG) is a measure of the electrical activity of the heart. The heart has four chambers, the two upper auricles and two lower ventricles. The contraction of any muscle is associated with depolarization, followed by relaxation with repolarization. ECG is recorded using 12 different lead configurations by picking up potentials from different location on the body, giving information of various parts of the heart. Among these 12 leads, LEAD II configuration is used in this system where P, Q, R, S, T waves with R wave of maximum amplitude can be measured. The contraction of the atria is represented by the P wave. Due to greater density of the ventricular mass as compared to the atria, the ECG is sharper during depolarization of ventricles and is represented by the QRS complex. The T wave indicates the repolarization of the ventricles [10].

Cardiac arrhythmias can be diagnosed based on the measurement of amplitudes and time intervals of individual waves in the recorded waveform. Timely diagnosis can help save a patient's life which is why ECG is a very important parameter. The controller used in this system is an Arduino Uno. The Arduino has been chosen for this system due to its ability to read the analog output of the medical sensors through its analog pins, as it supports serial communication and displays the same on to the computer screen using its serial plotter functionality. Overall, due to its ease of use, versatility and economical cost factor, it is the preferred choice. To measure ECG signal, the Sparkfun AD8232 Single lead heart rate monitor is used. The Sparkfun board brings out 9 pins from the AD8232 IC which are—Ground, 3.3 V, LO+ (Leads on detect), LO− (Leads off detect), Output, SDN (shut down pin) along with provisions to connect to LA (Left arm), RA (Right arm), LL (Left leg) that is the LEAD II configuration. There is also a jack for connecting all three electrodes directly [11]. A power supply module MBV2 is used to provide power to the AD8232 module. It works on an input of 6.5–12 V and provides an output of 3.3 V or 5 V. LO+ and LO− are connected to pin 10 and 11 of Arduino, respectively. Output of the module is connected to analog read pin A2 of Arduino for reading the ECG values. The three electrodes are connected via the jack (Fig. 3). The electrodes sense the biopotential and send it to the IC after which these values are sent to the Arduino and displayed on the serial plotter on the laptop or Raspberry Pi.

**Solar power module**: Since power supply is unreliable in case of rural areas, solar energy is used to power the setup. The sun's rays are incident on a solar panel; however, the solar panel cannot produce energy at night or during cloudy periods. For this purpose, we use rechargeable batteries, to store electricity. The photovoltaic panels charge the batteries during the day and this power can be drawn upon in the evening. This battery is connected to a solar inverter that converts variable DC output of the photovoltaic solar panel into a utility frequency alternating current

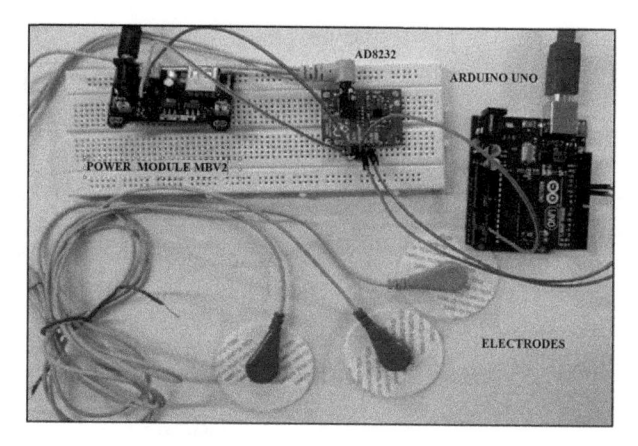

**Fig. 3** ECG measurement module

**Fig. 4** Solar power module

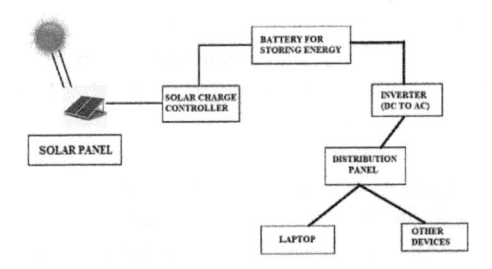

[12]. This can then be used for powering up the system which includes medical devices, laptop, and the embedded technology (Fig. 4).

## 4.2 User Application

The user application consists of a front-end web application and a database that can be accessed using the laptop, to which the embedded technology and sensors are connected as shown in Fig. 5. The front-end web application shown in Fig. 6 has been designed using Hypertext Markup Language (HTML), Cascading Style Sheets (CSS), and JavaScript. It consists of a home page and sections for patient login, hospitals (which consists of the doctor login page), and an NGO section in case the proposed system is being used by a nongovernment organization in villages and a Help section. In the patient login section, a new patient can sign up or an existing one can sign in using the respective ID and password assigned to proceed with medical tests and update and send relevant information to the database. Under the hospital section, a doctor can login using the medical registration number and assigned password to view patient data and give feedback. Audio instructions are provided on the patient login page to give guidance in a regional language regarding login. Instructions will also be provided for carrying out necessary

**Fig. 5** Block diagram of user application

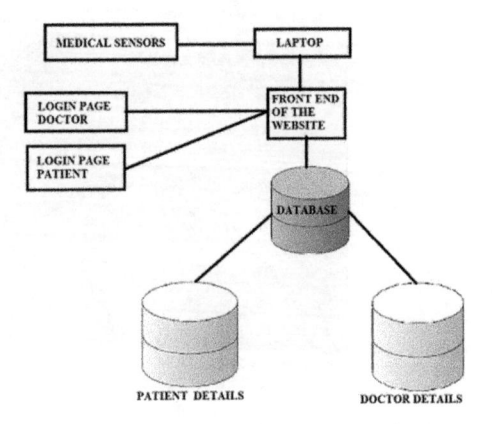

Telemedicine: Making Health Care Accessible

medical tests and connecting the equipment. It will also have a provision to conduct a live session with a specialist using a webcam.

The database is linked to the web application and has been created using MYSQL and Apache. This centralized database basically allows for two-way communication between the patient and doctor. It comprises of patient details and doctor details. In case of patient details, there are 14 fields which include Patient ID, Name, Age, Gender, Weight, Height, Address, Chronic Illness, Blood group, ECG, Blood Pressure, ENT Details, Phone number, and Doctor's diagnosis as shown in Fig. 7b. Doctor details include Name, Contact details, Registration Details, Specialization, Hospital visited, and City as shown in Fig. 7a. LabVIEW software is a platform for a visual programming language and is compatible with Arduino, where it can be used to view the ECG waveform graphically. It allows for graphs and images to be exported and saved to the database created, under the ECG field. It can similarly be used for exporting readings of other medical sensors interfaced with Arduino.

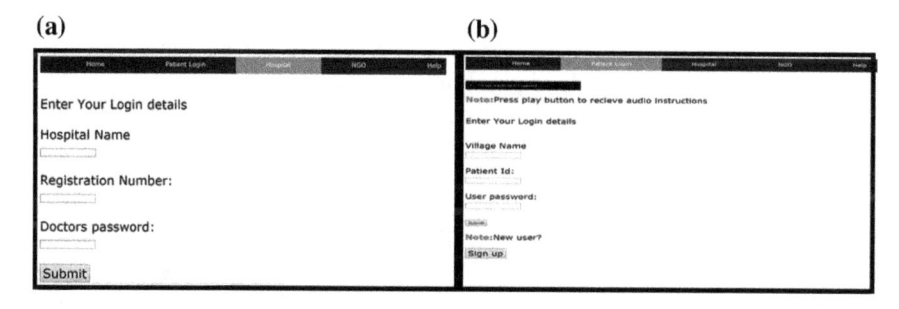

**Fig. 6** **a** Doctor login page, **b** patient login page

**Fig. 7** **a** Database of doctor's details, **b** database of patient's details

## 5 Conclusion and Scope of Future Work

The proposed system demonstrates the use of the concept of Telemedicine effectively by using ECG as a parameter. Further, pulse rate can be found out using this ECG signal. To this setup, additional medical parameters including measurement of Blood Pressure, pulse oximetry can be added in the future for expansion. The AADHAR is a unique 12-digit identification number that is issued to every resident of India based on biometric and demographic data. This data is indexed with the Unique Identification Authority of India (UIDAI). About 99% of the Indian population, that is, 1.17 billion people have AADHAR cards [13]. In the future, if this system is expanded all over the country, by collaborating with the UIDAI, a fingerprint identification system can be indexed on to a database with the AADHAR number that stores biometric information. By this method, it is possible to develop a robust, fully centralized patient identification and monitoring system, by using a Standardization, Testing and Quality Certification (STQC) approved fingerprint scanner. Thus, this eliminates the need for issuing a separate set of patient ID numbers.

LiFi is a wireless communication system, it uses LEDs (Light emitting diodes) to transmit information using visible light. It allows data transfer at very high speeds of up to 224 gigabits per second under lab conditions. At the transmitter, the information is sent by varying the brightness of the light signal and the receiver consists of a photodetector that detects this signal and converts it into binary format [14]. Since the LEDs are already widely used, it will be more convenient to develop LiFi systems on the existing framework. LiFi has advantages over Wi-Fi as, since it has higher speed and lower cost of implementation. Moreover, since it uses the visible light spectrum, it is less hazardous as compared to Wi-Fi which operates on radio frequency. The range of the visible spectrum is 10,000 times that of the radio wave spectrum. Since it does not contribute to adverse effects caused by electromagnetic radiation, it can safely be used in hospitals and near medical equipment such as MRI and CT Scan machines, and can be used along with the proposed system in the future for cost and energy efficiency.

Based on the data collected, diagnosis can be carried out to observe general healthcare patterns and most common health problems in various regions. This diagnosis can then be used to organize medical camps in the rural areas and come up with solutions to reduce the occurrence of diseases. The concept of the mobile Telemedicine system can also be used to promote skill development of the youth in villages for operation of the unit and consequently lead to overall progress. It can be further developed to fulfill applications in the field of distance education and training for medical interns, emergency consultation, disaster management where medical aid is not available immediately.

**Acknowledgements** The authors would like to thank Mr. Ashim Purohit, Business Consultant, Director-Advisory Board at AGD, Chairman-Advisory Board India at ChemoTech, for his invaluable guidance and support throughout the course of this project.

# References

1. Rural population of India, Trading Economics, https://tradingeconomics.com/india
2. World Health Statistics 2013, World Health Organisation
3. World Health Organisation Country Office for India, http://www.searo.who.int/india/en/
4. Telemedicine-Report on the second global survey on eHealth. Global Observatory for eHealth series, vol. 2 (2010)
5. Görs, M., Albert, M., Schwedhelm, K., Herrmann, C., Schilling, K.: Design of an advanced telemedicine system for remote supervision. IEEE Syst. J. **10**(3), 1089–1097 (2016)
6. Hameed, R., Mohamad, O., Hamid, O.: Patient monitoring system based on e-health sensors and web services. In: 8th International Conference on Electronics, Computers and Artificial Intelligence, pp. 17–22. https://doi.org/10.1109/ecai.2016.7861089
7. Mohanraj, T., Keshore Raj, S.N.: Patient monitoring system using LabVIEW. Int. J. Emerg. Technol. Comput. Sci. Electron. **24**(4) (2017)
8. Sebastian, S., Jacob, N., Manmadhan, Y., Anand V.R., Jayashree, M.J.: Remote patient monitoring system. Int. J. Distrib. Parallel Syst. **3**(5) (2012)
9. Prodhan, U., Rahman, M., Abid, A., Bellah, M.: Development of a telemedicine model with low cost portable tool kit for remote diagnosis of rural people in Bangladesh. In: International Conference on Innovations in Science, Engineering and Technology. https://doi.org/10.1109/iciset.2016.7856513
10. Hampton, J.R.: The ECG Made Easy
11. Sparkfun electronics, https://sparkfun.com
12. Bucci, J., Fang, R.: Project Green Stations: Prototype Zero
13. Unique Identification Authority of India official website, https://uidai.gov.in/
14. Forget Wi-Fi. Meet the new Li-Fi Internet by Harald Haas, TED Global (2015), https://www.ted.com/talks/harald_haas_a_breakthrough_new_kind_of_wireless_internet

# Virtual Piano

**Aditi Patel, Abhishek Satpute, Mital Pattani
and V. Venkataramanan**

**Abstract** Music is a limitless form of art. Accessibility is an essential part of its emergence. Traditional musical instruments like piano, drum, etc., are expensive and arduous which reduces its accessibility and portability. To make it universally available and efficient, this is an attempt to demonstrate the use of an alternate medium for creating music using gestures. This makes it available to large number of music enthusiasts. Devices, nowadays, use touch-based and key-array-based controls, limiting the interface to two dimensions. For this purpose, image processing has been used to demonstrate a gesture-controlled piano, giving it another dimension. Existing image processing systems use shadow detection methods, IR detection methods, etc., whereas here, colour detection method for gesture recognition has been used to identify the fingertips and simulate the pressing of keys. The sensitive colour detection and processing method eliminate the use of coloured gloves/fingertip covers/nail paints/drumsticks, etc., making it more user-friendly and autonomous.

**Keywords** Music · Detection · Gesture recognition

---

A. Patel · A. Satpute · M. Pattani · V. Venkataramanan (✉)
Electronics and Telecommunication Engineering, DJSCE, Mumbai, India
e-mail: Venkataramanan.V@djsce.ac.in; rvvenkat.mtech@gmail.com

A. Patel
e-mail: aditispatel20@gmail.com

A. Satpute
e-mail: abhishek@7pute.com

M. Pattani
e-mail: mitalpattni@gmail.com

© Springer Nature Singapore Pte Ltd. 2018
H. Vasudevan et al. (eds.), *Proceedings of International Conference on Wireless Communication*, Lecture Notes on Data Engineering and Communications Technologies 19, https://doi.org/10.1007/978-981-10-8339-6_29

# 1 Introduction

Of late, human–computer interaction (HCI) has found natural and comfortable effectuations. Gesture control is one such process which has contributed to this cause. Techniques like shadow detection, colour detection, etc., for gesture recognition, have gained reputation in recent times. Vision-based gesture control has applications that employ image processing to facilitate the process of remotely controlling the piano. Segmentation and binarization are the basic techniques used for similar applications. Image processing has gained a broader scope in computer vision due its rising demand in gaming industries and scientific visualization.

This project is an attempt to simulate a mechanical piano using gestures, eliminating the need of complex and heavy piano components. In a Virtual Piano, gestures are given as input and we get the desired sound as output hence, eliminating the need of arduous components.

Several algorithms are used to recognize hand gestures based on hand segmentation based on wavelet network as well as supervised feedforward neural network algorithm for gesture recognition [1]. Attempts have been made to make HCI a comfortable experience using speech and body language. Digital image processing used for gesture recognition is implemented using different procedures for HCI. It refers to the techniques in which image is taken as an input to either extract meaningful information from it or to manipulate. A real-time command system uses general purpose hardware and low-cost sensors for hand gesture recognition, making it plausible for implementing the real-time command system on household as well as industrial level [2].

For better quality of image, Poisson–Gaussian is used for reducing the noise [3]. The Poisson component accounts for the signal-dependent uncertainty, while the Gaussian mixture component accounts for the other signal-independent noise sources is used for denoising the image and for fish localization in underwater images. Computer vision is being widely used in object detection, and vision-based gesture recognition system is also being operated in automotive environment [4]. Image processing having wide range of intuitive applications in interaction of human with machines is also used for entertainment purpose. In such an application, the calibrated camera makes the computer screen work like a touchscreen by sensing the presence of an opaque object, which, when detected, can sense the position of the touch [5].

In a mechanical piano when a key is struck, a chain reaction occurs wherein, a string is struck by a hammer which makes wire resonate. When the key is released, the damper dampens the vibration, producing the required sound. Thus, it consists of various components which makes it bulky and requires maintenance.

Normally, hand gestures are referred to as micro and macro. The dynamic gestures of hands and fingers are called as 'macro' gestures whereas gestures which refer to the relative position of the fingers are called as 'micro' gestures [6].

Virtual Piano

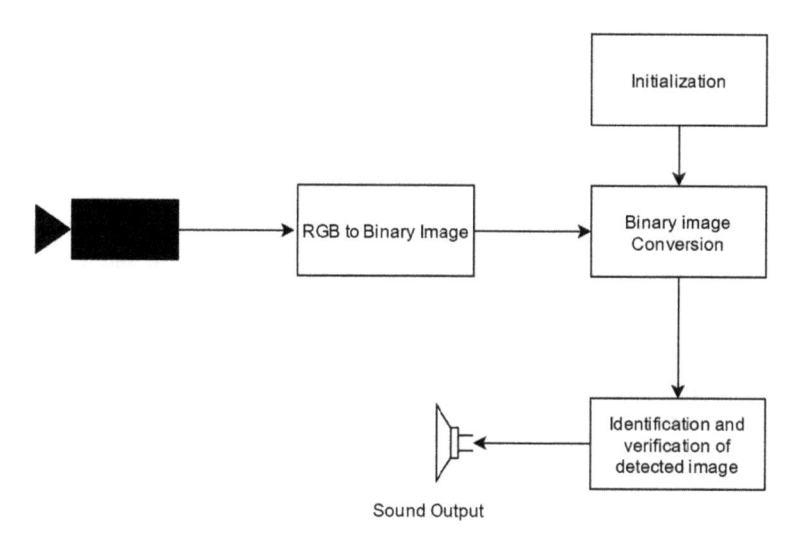

**Fig. 1** Block diagram explaining the working of virtual piano

## 2 Block Diagram

See Fig. 1.

## 3 Working Principle

A digitally generated image is a 2D array of pixels (picture elements), the position of which is defined by the Cartesian coordinate and the colour is defined by different RGB values [7]. The camera used to capture the image is the acquisition device, which senses wavelength of the light and then converts it into a digital image. It encodes this information captured by the lens to digital image defining the RGB values for each pixel. The input images are divided into two domains, upper 70% of the image and lower 30% of the image. Out of which the lower domain is used to sense and display the array of keys, hence it is the main working domain (named as 'pressed' domain). The lower part of the images is further divided into multiple subdomains which define individual keys for the Virtual Piano. These subdomains have been assigned the audio frequencies in accordance with their symphonic hierarchy. This division of domains can be seen in Fig. 6.

The code is divided into four scripts with routines defined in accordance to the flow of the algorithm. The initialization of the key notes is done in the first script, wherein, musical notes from the two major octaves and their audio files are loaded into the workspace. The notes can be changed according the user's preference. The user gives variables, 'colour (col)' as an input (i.e. colour of the nail or finger which is supposed to be detected) and 'sensitivity (sens)' as two mandatory inputs. The program then checks for colours of each pixel in the image within the range (determined from the sensitivity variable 'sens') from the given colour (col) and defines the pixels following the above condition as 'True' and rest of the pixels as 'False'. It then creates a new binary image according to this new information. This is how the program tracks the position of the finger.

There are two major cases of the position of hand that the program is able to recognize, first being a 'key pressed' case as seen in Figs. 2 and 4, and the second being 'key not pressed' case as seen in Figs. 3 and 5. These two cases determine if the virtual key, defined by the domains, is pressed or not (Fig. 6).

After this, the code calculates a mean position of all the 'True' pixels. This mean decides the position of the user's finger with reference to the two-dimensional Cartesian coordinates. Figure 7 shows the lower 30% of the screen the 'working

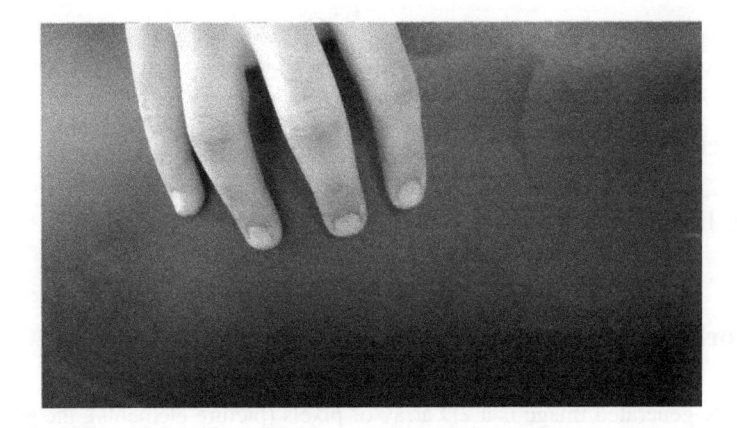

**Fig. 2** Raw RGB image of hand when the key is not pressed

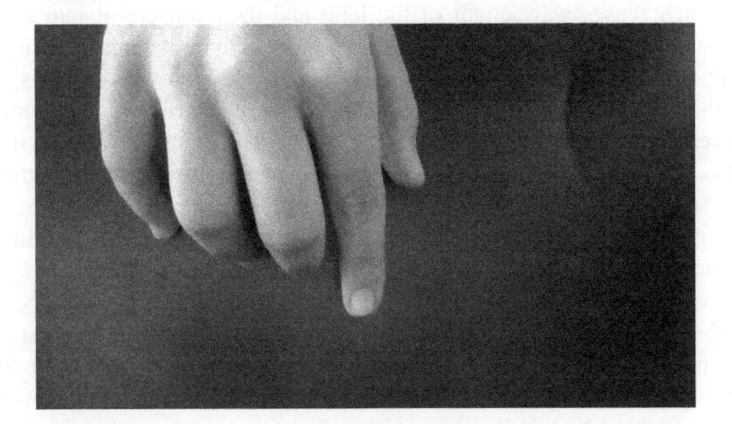

**Fig. 3** Raw RGB image of hand when the key is pressed

Fig. 4 Binary image of hand when key is not pressed

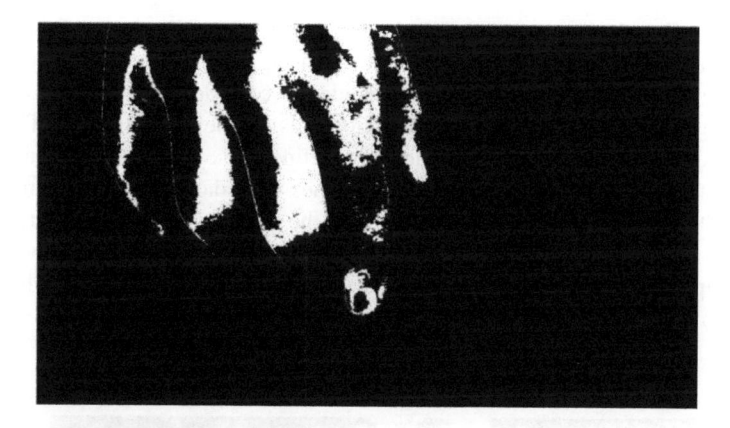

Fig. 5 Binary image of hand when key is pressed

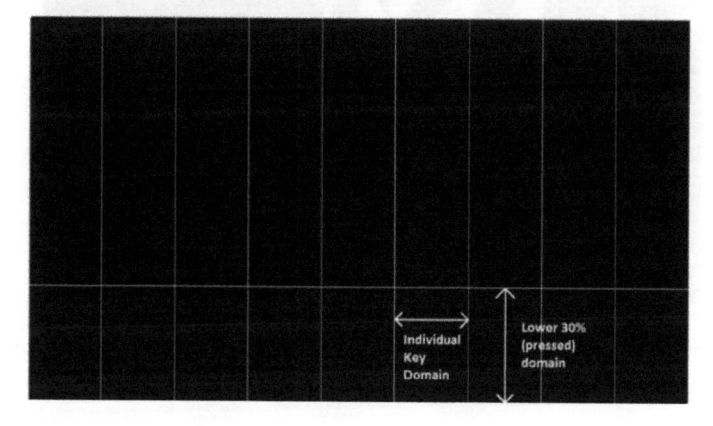

Fig. 6 Domain boundaries are set

**Fig. 7** The part of the user's finger visible in the working domain. These true pixels decide the keys to be played

**Fig. 8** The mean values from Fig. 7 are represented in this figure by two white lines parallel to $x$- and $y$-axes

domain'. In Fig. 7, the part of the user's finger is visible, the mean of these true pixels is taken to position the key press shown in Fig. 8. The key corresponding to the subdomain over which the mean coincides is considered as pressed and the pressed key glows white for visual stimulation, this can be seen in Fig. 9. The note assigned to the key is then played hence completing one cycle of the Virtual Piano Simulation. This cycle is repeated multiple times to update the changes in user's gestures. It continues to track the gestures until the user terminates the program. An additional feature has been added, so that the same key is not played again when the cycle is repeated. As the cycle continues, the user can play multiple notes, like in a mechanical piano. The final visual output can be seen in Fig. 9.

**Fig. 9** This is the final visual output of the interface visible to the user, with the help of which they can decide the position of their hand

## 4 Simulation Environment

The key domain that the 'avg' falls under is selected and the corresponding sound is played.

Let $x_1, x_2, x_3, \ldots x_{10}$ define the upper limits of the domain,

$$x_1 = x, \quad x_2 = 2x, \quad x_3 = 3x \ldots x_{10} = 10x \tag{1}$$

Each domain can be defined as

$$d_1, d_2, d_3, \ldots d_{10}, \quad d_n = (x_n) - (x_{n-1}), \tag{2}$$

where

$$n = 1, 2, 3 \ldots 10$$

Position of the finger is calculated using the average formula,

$$a = \frac{\text{Number of foreground pixels}}{\text{Total number of pixels}}$$

If $a \subseteq d_n$, the finger is in domain $n$. Hence, $n$th key is selected.

Here, each domain is assigned a key note. When $n$th domain is selected, the keynote defined by that domain is played.

## 5 Result and Conclusion

In the demo implementation of Virtual Piano, experimental results are presented to show the working of digitally simulated musical instrument. A new approach for detecting user's finger from a constant background has been used wherein average position of the true pixels (which includes the values of skin colour of user) is calculated for HCI. Giving the skin colour as input for recognizing the true pixels makes the system to respond to only the skin colour of the user trying to access the keys. Use of constant background eliminates the intervention of external factors such as background noise in the live feed taken by the camera. Construction of this effective system for determining the position of user's finger on piano keys and playing the audio files associated with corresponding keys was done using MATLAB on Windows platform. The live feed taken by the camera is the mirror image of the actual image, but the system has been made such that the problem of mirror image is eradicated. Thus, it succeeded in detection and simulation of piano keys in accordance to the gestures of the users. In the cases where gestures of the user try to simulate the same key for a longer time, the playing of the same note is avoided by keeping a check on the time for which the key is being simulated.

However, simulation and synchronization of multiple musical keys can be the advancements which can help the user to play different motion-based musical instruments. By adding the more number of musical keys, this system can be used as a launch pad. The proposed project detects the skin colour in a constant background, but recognition of gesture in a dynamic background can make the system robust for playing the air instruments efficiently.

# References

1. Jalab, H.A.: Static hand gesture recognition for human computer interaction. Inf. Technol. J. **11**, 1265–1271 (2012)
2. Martina, J., Nagarajan, P., Karthikeyan, P.: Hand gesture recognition based on real time command system. IJCSMC **2**(4), 295–299 (2013)
3. Boudhane, M., Nsiri, B.: Underwater image processing method for fish localization and detection in submarine environment. J. Vis. Commun. Image Represent. **39**, 226–238 (2016)
4. Akyol, S., Canzler, U., Bengler, K., Hahn, W.: Gesture Control for Use in Automobile. The University of Tokyo, Japan (2000)
5. Li, C., Hu, R.: Piano Hero Using Virtual Keyboard. Cornell University ECE (2013)
6. Oliveira, P.C., Moura, J.P., Fernandes, L.F., Amaral, E.M., Oliveira, A.A.: A non-destructive method based on digital image processing for calculate the vigor and the vegetative expression of vines. Comput. Electron. Agric. **124**, 289–294 (2016)
7. Cisneros, A.H., Sarmiento, N.V.R., Delrieux, C.A., Piccolo, C., Perillo, G.M.E.: Beach carrying capacity assessment through image processing tools for coastal management. Ocean Coast. Manag. **130**, 138–147 (2016)

# Automatic Garbage Collector Bot Using Arduino and GPS

Niharika Mehta, Shikhar Verma and Shivani Bhattacharjee

**Abstract** This project's objective is to build a robot, Automatic Garbage Collector Bot (AGCB), which will be able to collect garbage from a designated geographical area without the need for human intervention. NI LabVIEW accepts Google Maps coordinates and plots an area for the bot to cover. This information is then transmitted to the bot via the ESP8266 module. The robot then proceeds to sweep the entire area. Ultrasonic sensors are used for obstacle detection and are used under the I2C protocol. The encoder calculates the distance travelled by the bot and the gyroscope gives it a sense of direction. The aim of the project is to first have one fully functioning robot that can collect garbage efficiently. As a future scope of innovation, we have explored the domain of Swarm robotics. The concept is based on decentralization between a group of robots and the external environment. This reduces total computational load of the system by dividing the process, improves efficiency, and reduces processing time.

**Keywords** ESP8266 · NI LabVIEW · AGCB · I2C protocol · Swarm robotics

## 1 Introduction

Garbage is a major problem worldwide. But cleanliness is something we as human beings strive for. But we see litter everywhere. Public places get littered every day, especially in India. People can be very insensitive towards public hygiene. There is also dust that settles down. Therefore, any property eventually starts looking dishevelled. In places like malls, where thousands of people visit every day, the floor gets especially dirty. The authorities then have to hire cleaners and sweepers to maintain the cleanliness of the property. Just one sweeper cannot sweep an entire

N. Mehta · S. Verma · S. Bhattacharjee (✉)
Electronics and Telecommunication Department, SVKM's,
Dwarkadas J. Sanghvi College of Engineering, Vile Parle (W),
Mumbai 400056, India
e-mail: shivani.bhattacharjee@djsce.ac.in

© Springer Nature Singapore Pte Ltd. 2018
H. Vasudevan et al. (eds.), *Proceedings of International Conference on Wireless Communication*, Lecture Notes on Data Engineering and Communications Technologies 19, https://doi.org/10.1007/978-981-10-8339-6_30

mall or a huge parking lot, forcing the authorities to hire more people. This is taxing on the organization financially as well as logistically. This problem can be solved by automating the entire process. The sweeping process was first automated when the vacuum cleaner was invented by the British engineer, Hubert Cecil Booth, in 1901. The sweeper now just had to roll around a machine which did his work for him. Our idea is to make the need to have a sweeper redundant and have a robot clean up the given geographical area. In our approach to providing this solution, our robot will require Google Earth coordinates to recognize what area needed to be cleaned and the robot shall then proceed to sweep. The idea is to ultimately have a team of robots cleaning fractions of the area and further reducing the sweep time.

## 2 Hardware Structure

### 2.1 Encoder

A rotary encoder is a variant of position sensor. It determines the angular position of a rotating shaft or wheel by generating an electrical signal, either analog or digital, with respect to the rotational movement. The encoder has two switches. For working of the encoder, we generally consider three pins namely A, B and C. These switches connect pin A to pin C and pin B to pin C, respectively. The open or closed status of the switch modified by the encoder's click is integral to compute the rotation. The diameter of the wheels determines the circumference of the wheel. The code for interfacing the encoder with the Arduino is uploaded into the microcontroller. In accordance with the code, each encoder ticks amount to a particular angular rotation. This relation is predetermined in the code. The distance travelled is the product of rotations of wheels and circumference of each wheel. On detecting obstacle, the ultrasonic sensor sends interrupt to the encoder. The redundant distance traversed due to the obstacle is put in different variables as per the code. The logic of the code included identification of the useful distance versus the redundant distances.

### 2.2 Ultrasonic Sensor

The ultrasonic sensor calculates the time it takes for a high-frequency sound pulse to reflect from an obstacle and the echo to reach back to the sensor. The sensor has one opening which transmits the ultrasonic waves and one other opening for receiving the echo of the transmitted wave. We have used three ultrasonic sensors and keeping in mind the limited number of serial ports available to us on the Arduino Mega, we interfaced the three sensors with an ATtiny85 development board which has a 16-bit microcontroller IC. This microcontroller is interfaced with the Arduino Mega using I2C protocol as this allows us to connect multiple components on the same port leaving the serial ports free for components which are not easily interfaced using I2C protocol. The readings from all three sensors are simultaneously transmitted to the Arduino Mega for the obstacle avoidance operation.

## 2.3 Gyroscope

The gyroscope we used employs micro-electro-mechanical system (MEMS). This type of gyroscope is small and cheap but it is usually used to measure angular velocity for objects that are not spinning very fast. They provide stability and precision in calculation of the rotation along the three axes, $x$, $y$, $z$. This is achieved as a result of the high sensitivity of the gyroscope as it rotates a few degrees on each of the three axes. Our Bot uses the gyroscope to get the bearing angle of the Bot when it rotates on the spot. The angle from the gyroscope is compared with the angle calculated by our NI LabVIEW CalculatePoints virtual instrument for that particular slope. The spot rotation of the Bot is stopped when the two values match and the Bot continues on its path.

## 2.4 Wi-Fi Module

The ESP8266 Wi-Fi module wirelessly communicates with the PC via Wi-Fi and accepts the Google Maps coordinates generated by LabVIEW. This data is accepted from a ThingSpeak server. The ESP8266 communicates with the Arduino board via UART. The information is passed onto the Arduino which then proceeds to travel along the designated path. The sole purpose of the ESP8266 is to establish a communication channel in between the bot and the computer device.

## 2.5 Servo and DC Motor

A servomotor consists of a potentiometer, DC motor, and control circuit. The gears connect the motor with the wheels. The control circuit acts as a regulator of the motion of the motor and of direction control. The motor's neutral position is defined as the position where the servo has the equal potential rotation in the clockwise or counterclockwise direction. To control the servo, an electrical signal of variable pulse width is used. The length of the pulse determines the distance the motor will transverse. The servo motors can be AC or DC. In this bot, the servo motor controls the broom bristles and periodically opens the garbage bin to empty the tray. A DC motor converts electrical energy into mechanical energy. It contains of a current carrying armature placed between an electromagnet. This armature is connected to the output end via commutator brushes and segments. On supplying power, the electromagnetic effect of the electromagnet generates mechanical force which rotates the shaft. The speed of the motor is controlled using pulse width modulation (PWM) technique. It requires a driver IC L298N to operate.

## 2.6 GPS and Arduino

The GPS is used to increase the accuracy of the Bot position even though it is not required as such as the Arduino mainly requires the distance and the slopes of the designated areas. The GPS is not as accurate as it has a margin of error in metres which causes problems for our Bot operation. We have tried to take that margin into consideration and code the Arduino based on that but the margin is too high.

## 3 Software Structure

By entering four Google Earth coordinates into NI LabVIEW, we were able to successfully generate an enclosed figure with those coordinates as the end points. The graphical programming language based on C is used for NI LabVIEW. The coordinates are fed to the bot using Wi-Fi module. The coordinates are mapped to the most suitable polygon shape. The gyroscope is used to compute the slopes long the boundaries of the area. It also helps in adaptive rotational motion. The AGCB then follows a designated path to travel all over that area. The encoder will calculate the distance travelled by the bot. Using ultrasonic sensors, the encoder will be prompted of redundant obstacle distance. Thus, the actual distance calculation remains precise. The gyroscope and ultrasonic sensors are interfaced using I2C protocol on a common I2C bus. Using Arduino Mega, various components are interfaced using embedded C programming language.

NI LabVIEW and Google Earth Interface: This interface required a .kml file to be used for the interfacing resulting in compatibility issues. Also, there were compatibility issues such as the NI LabVIEW virtual instrument which we created for interfacing with the Google Earth was not able to find the path to the Google Earth application and hence the program was not executing. This was solved by adding a 'config' file in the Google Earth program files which enabled the virtual instrument to find the application and run the program.

NI LabVIEW and ThingSpeak Server: In order to send multiple data through ThingSpeak, the data file needs to be in a specific format of the .csv spreadsheet type. We had to add a sub-virtual instrument program after manipulating different array structures and give provisions for appropriate column and row headers. This was then transferred to the ThingSpeak server. The esp8266 NodeMCU Wi-Fi module read this data and transferred it to the Arduino Mega. We have used I2C protocol to interface ultrasonic sensors with the Arduino. This enabled us to connect multiple devices sing only two wires. This protocol simplifies the operations and reduces processing time by performing faster computation (Figs. 1 and 2).

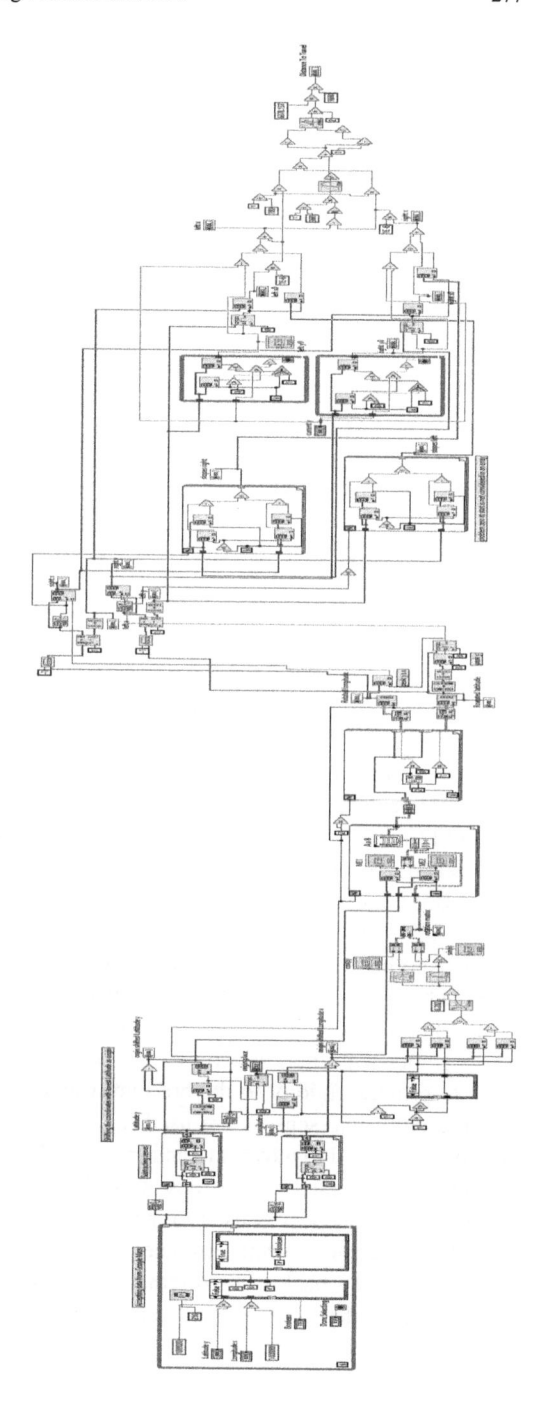

**Fig. 1** CalculatePoints virtual instrument code

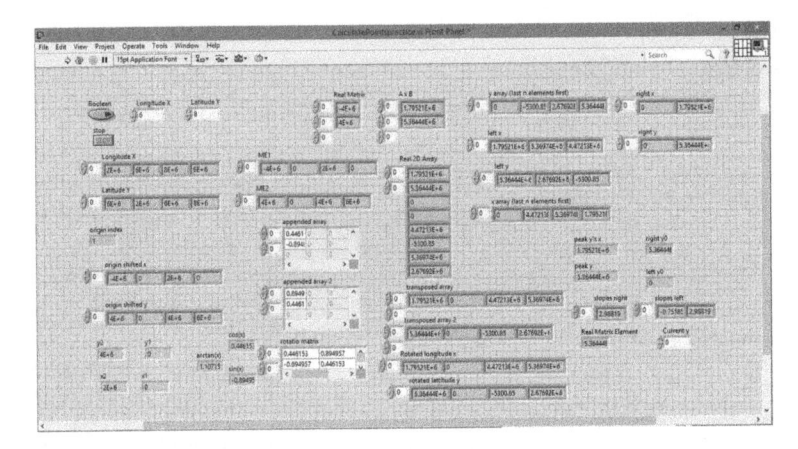

**Fig. 2** Front panel of CalculatePoints virtual instrument

## 4 Working Principle

First, the operator designates the points of the polygon which forms the area the Bot is supposed to traverse and clean. These points are marked on Google Earth. Using kml files, the Google Earth is interfaced with NI LabVIEW which computes the slopes and lengths of the sides of the marked polygon. This data is then sent wirelessly to the Arduino Mega on board the Bot using esp8266 wireless module and ThingSpeak server. The Arduino stores this data in various variables which are later used to aid the Bot's movements. The Bot starts its operation from one of the vertices and goes along the horizontal side. DC motors and Omni wheels are used for the Bot's movement. There are broom bristles attached in the front which are used to collect the garbage on a tray. The bristles are rotated on a shaft using servo motors. Another servo motor is used to lift the tray up. There is a garbage collector box which forms the bulk of the Bot structure. The opening of this bin is at the top and levers are used to open the flap of the bin simultaneously tilting the garbage tray to empty it in the bin. Trusses are added as the primary structural reinforcements to provide rigidity to the bot over rough surfaces.

Three ultrasonic sensors are used, one is mounted on a truss in the front and servo motor is used to move it to and fro in the horizontal direction as to cover more area. Two are mounted on the sides of the Bot. These are used for obstacle detection and avoidance. Encoder is used to measure the distances at various sub-operations of the Bot especially during obstacle avoidance since it is the encoder which aids the Bot in coming back to its original path. The bearing angle of the Bot and the spot rotation of the Bot are aided using a gyroscope (Fig. 3).

**Fig. 3** Diagrammatic representation of how the Bot traverses the area

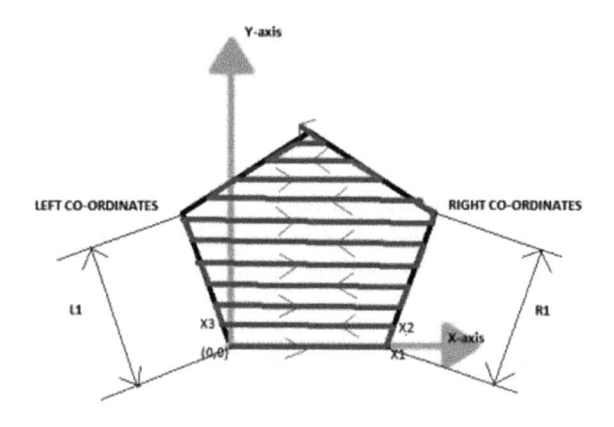

## 5 Future Scope

Swarm mechanism is adopted from the self-organizing mechanisms of the ant community. There are two integral parts of the Swarm mechanism: Software and Hardware. Software handles the processing complexity and computing time. Hardware performs the mechanical operations of assembling the robots and implementing the actions instructed by the software. The concept behind our integration of Swarm robotics is that is to first use GPS modules on each bot to determine the number of Bots present in the designated area of operation. After this, the NI LabVIEW CalculatePoints VI will be modified automatically to divide the area of operation in equal sections for each Bot present. Each Bot will have its dedicated ThingSpeak server and will receive the data on distances and slopes from there. Once this data is transmitted to each bot, the general operation of our single bot described in this report will be followed. This provides physical independence of topology. Using dynamic routing, the sectors demanding extra effort will be autonomously interrupted to the respective bots. Thus, adding the flexibility quotient to the system. A decentralized connection between the group of robots and the governing environment removes the need of a central control. This provides contingency measure against failure of the system due to central control. Incorporating self-organization mechanisms, feedback control can be implemented. Feedback can provide error correction control and improve functional capability of the system. Through training and learning algorithms namely reinforcement learning and genetic algorithms, the bots can operate autonomously, independently in dynamic environments. Overall, Swarm robotics provides multiple advantages of energy efficient, higher efficiency, and faster processing time. With automation as the upcoming necessity of everyday life, Swarm robotics offers a promising future.

## 6 Results and Conclusion

The implementation of this bot involved using multiple components. Every component incorporates an effective feature into the bot. We have implemented the obstacle avoidance using interfacing of ultrasonic sensor and Arduino. We have implemented the rotation of the dc motor using driver IC L298N to control the motion of the wheels. The .vi files of NI Labview are interfaced with Google Earth successfully. The coordinates of the area are successfully transmitted using Wi-Fi module to the Arduino. Thus, data reception was successful.

This project presents the new future for humans. Using automation to ease manual effort has been the major predicate of this project. Automation not only reduces manual effort but also provides improved efficiency by elimination human error. It involves high computational complexity

# Cellulose Acetate Substrates for Design and Calibration of Strain Gauges in Angle Measurement

Megh Doshi, Maitri Fafadia, Charmi Gandhi and Sunil Karamchandani

**Abstract** Mechanical sensors for project and industrial processes are expensive as well as nonbiodegradable. Thus, the use of biodegradable and more economical materials for manufacturing of sensors has become a growing requirement. In this paper, we have proposed strain gauges made of cellulose acetate (paper) substrates with a constant gsm and compared their data for angle measurements. We have designed and manufactured two different sensors of varying length as well as conductor thickness with an organic substrate and correlated the sensor output with the corresponding industrial sensors and therefore determined that they can be used to obtain optimum outputs for smaller lengths of sensors.

**Keywords** Biodegradable · Cellulose acetate · Organic substrates

## 1 Introduction

Strain gauges are sensors which change its resistance in accordance with the physical stimuli applied on it, the stimuli may be force, pressure, tension, weight, etc. [1]. On application of external forces, stress and strain are produced. Stress can

---

M. Doshi · M. Fafadia · C. Gandhi · S. Karamchandani (✉)
Department of Electronics and Telecommunication, DJSCE,
University of Mumbai, Vile Parle (W), Mumbai 400056, India
e-mail: skaramchandani@rediff.com; sunil.karamchandani@djsce.ac.in

M. Doshi
e-mail: doshimegh@yahoo.co.in

M. Fafadia
e-mail: maitrifafadia15@gmail.com

C. Gandhi
e-mail: charmigandhi@gmail.com

© Springer Nature Singapore Pte Ltd. 2018
H. Vasudevan et al. (eds.), *Proceedings of International Conference on Wireless Communication*, Lecture Notes on Data Engineering and Communications Technologies 19, https://doi.org/10.1007/978-981-10-8339-6_31

be defined as the internal resistive forces of a body to deformation and strain is the change in the physical structure of an object when under the action of a force. This paper deals with the methodology of selection and calibration of a strain gauge for better accuracy of angle values [2, 3]. The experimental data is based on a 4.5″ flex sensor, a 2.5″ flex sensor, and a 4.5″ proposed bend sensor [4].

*Strain Gauge*: It is a resistance-type passive transducer that works on the principle that the resistance of a conductor or semiconductor changes when subjected to a tensile or compressive force. The initial step in preparing for any application of strain gauges is the selection of an appropriate strain gauge for the task. On first glance it might appear that the selection of a gauge is a simple task, but in reality it is quite the opposite. The selection of the gauge depends on characteristics of the operating conditions, accuracy of the measurements, ease of installation, and minimal cost. Thus, before selecting a gauge, we need to know the operating parameters.

*Flex Sensor*: It is a strain gauge specifically used for the purpose of measuring angles. The sensor consists of four parts, a substrate, a conducting film, an insulating film, and contact leads.

## 2 Need of Calibration

We propose the design of Cellulose Acetate Substrates for the design of strain gauges for angle measurements. These gauges need to be in the form of flex sensors so that they can be bent and also show variation in potential in such conditions Thus, there is a need to calibrate our design with the strain gauges which are robust and provide accurate readings.

## 3 Working of Flex Sensor

A flex sensor is a strain gauge which changes the resistance values according to the angle at which it is bent. We have tried and tested four different flex sensors for experimental data out of which two of them are proposed. It is a variable printed resistor which gets great form factor depending on the thickness and flexibility of the substrate. When the substrate is bent, the flex sensor changes its resistance output proportional to the **bend** radius—the lesser the radius, the greater the value of resistance.

## 4 Proposed Flex Sensors

We have made two flex sensors using aluminium sheets as the conducting material, cardboard as the resistive material, and acetate paper as the substrate. The wires are soldered to the two sheets of flexible aluminium. The proposed flex sensors were made with sheets of 1-mm-thick aluminium as the conducting material and paper with a thin layer of graphite on both sides as a substrate material, and the other sensor with a 0.1 mm thickness aluminium and paper as a substrate material. The criteria of selection of the conducting material and the substrate depend on the amount of current you can supply to the sensor as well as the cost-effectiveness, the substrate of the sensor determines the resistance of the sensor as the sensor resistance depends on the compatibility of the substrate with the conducting material as well as the actual resistance of the substrate. The resistance of the sensor changes in two ways, it can change as the contact of the substrate with the conducting surface changes or it can have a varying resistance according to the amount of bending it is under. Using both of these factors, we select a substrate and a conducting material depending on the substrate compatibility. The thickness of the conducting material also plays a part in the output of the sensor as the flexibility as well as conductivity of the sensor changes, the thickness of the conducting material is a compromise between sensor flexibility and the repeatability and conductivity of the material. The repeatability and conductivity of the sensor are directly proportional to the thickness of the sensor material. We have used 80–90 gsm paper covered over a thin layer of graphite on both sides as the substrate as varying thickness of the substrate can also affect the sensor characteristics (Figs. 1, 2, 3, 4 and 5).

**Fig. 1** Structure of flex sensor

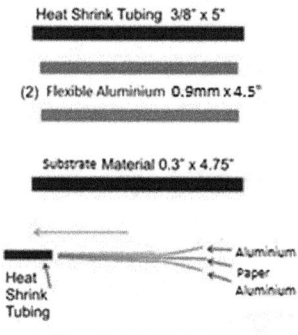

**Fig. 2** Proposed 4.5" 0.9 mm thickness Al flex sensor

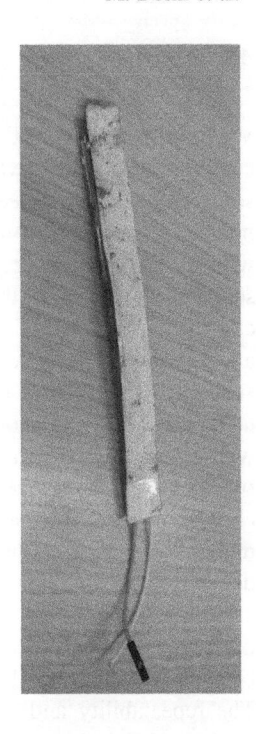

## 5 Designed Data Acquisition System

The proposed system comprises of an Arduino UNO microcontroller, a Micro SD card module with a digital low pass filter to reduce noise, a 16 * 2 LCD display which is used to display the voltage readings, the readings are stored on a SD card as a text file connected to the Arduino board. The readings can be later viewed for further analysis (Fig. 6).

## 6 Display System

The proposed system was made after the calibration was performed and the equation for each sensor was obtained. For this purpose, we used an ATMEGA 32A PU microcontroller along with a 16 * 2 LCD display and a constant current source giving 160 mA using 2 BC-547 so that the voltage output is within the range of the ADC of the microcontroller (Fig. 7; Table 1).

Cellulose Acetate Substrates for Design and Calibration ...

**Fig. 3** Proposed 2.5″ 0.1 mm Al flex sensor

**Fig. 4** Proposed 2.5″ spectra symbol flex sensor

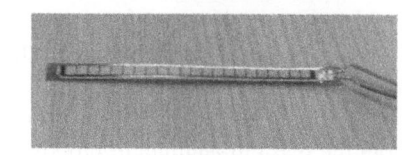

**Fig. 5** Proposed 4.5″ spectra symbol flex sensor

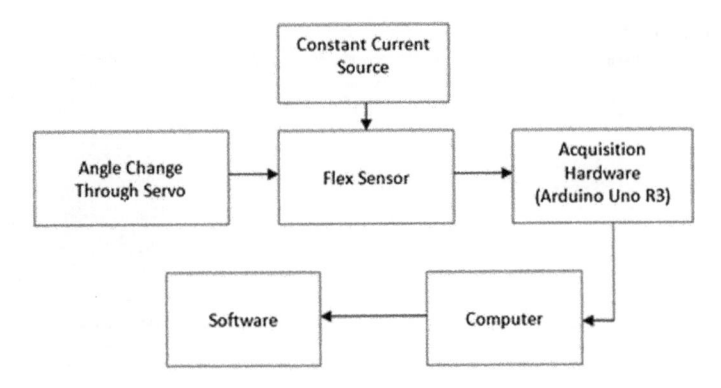

**Fig. 6** Block diagram of proposed DAQ system

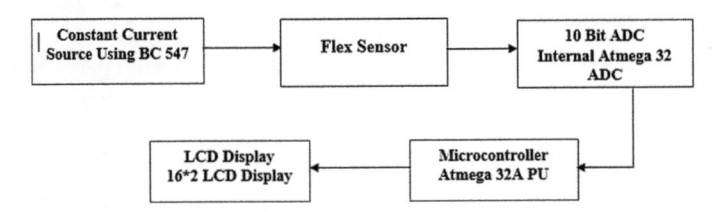

**Fig. 7** Proposed block diagram for sensor calibration

**Table 1** Constant current source for proposed sensors

|   | Current values for current source |         |
|---|-----------------------------------|---------|
| 1 | 4.5″ flex sensor                  | 200 mA  |
| 2 | 4.5″ 0.9 mm AL flex sensor        | 200 mA  |
| 3 | 2.5″ flex sensor                  | 70 mA   |
| 4 | 2.5″ 0.1 mm AL flex sensor        | 70 mA   |

## 7 Calibration Results

The calibration process for the proposed sensors was plotted as a measure of the variation in the angle measurement to the obtained electrical potential developed across them. The graphs areas are visualized in Figs. 8 and 9. The observations obtained with the spectra symbol flex sensors are compared with those obtained from the 0.9 mm aluminium proposed flex sensor.

Flex sensors are shown in Fig. 8. The proposed sensor exhibits prominent nonlinear characteristics and cannot be useful in practical applications.

The readings recorded with the 2.2″ spectra symbol flex sensor were plotted against the 0.1 mm aluminium paper proposed flex sensors in Fig. 9b. It can be concluded that the observations in the proposed sensors exhibit lesser nonlinear characteristics as compared with the sensor readings shown in Fig. 9b.

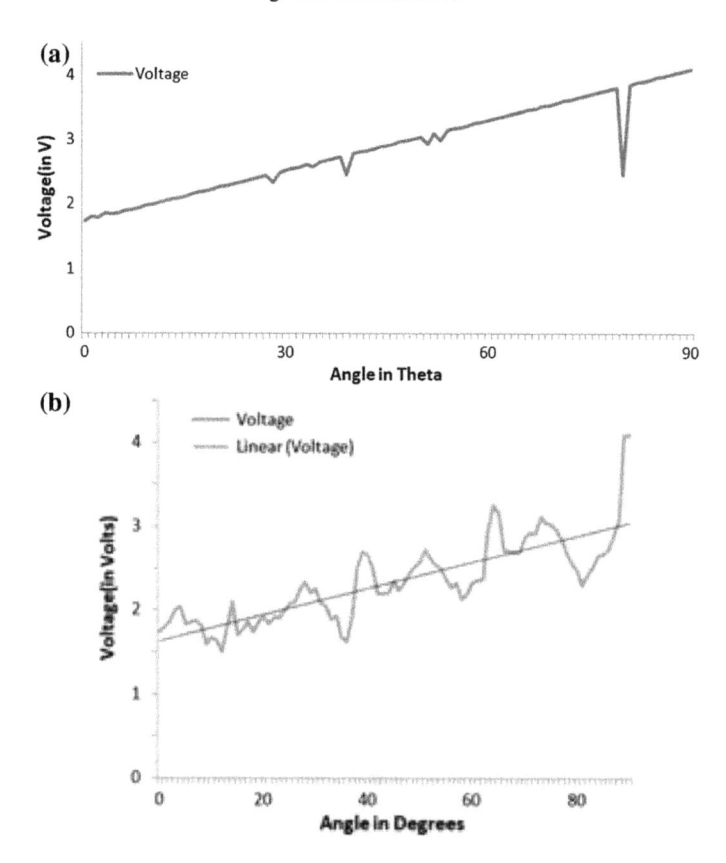

**Fig. 8** a Characteristics of 4.5″ spectra symbol flex sensor (angle in degrees), **b** Characteristics of 4.5″ 0.9 mm AL flex sensor

The nonlinearity is exhibited disappears around the angle between 30° and 60° else it shows sufficiently linear characteristics. Thus, organic material as substrate flex sensors can be designed with lower width sizes as they exhibit sufficient nonlinearity. We have plotted all the graphs by interpolation of the data obtained.

The results were satisfactory with 4.5″ and 2.5″ flex sensors which appeared to have linear characteristics as compared to the two self-made sensors which had irregular characteristics as we can observe from the graphs. As compared to the self-made sensors, the flex sensors had a resistance which varied linearly throughout, whereas the proposed sensors had a resistance which increased up to a point and then decreased.

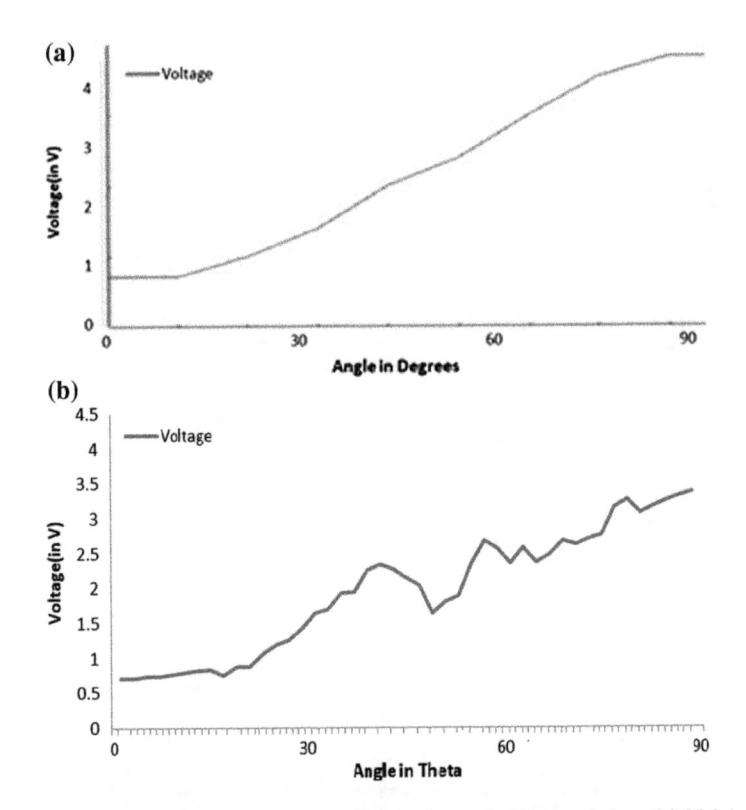

Fig. 9 a Characteristics of 2.2″ spectra symbol flex sensor, b Characteristics of 2.2″ 0.1 mm Al flex sensor (angle in degrees)

## 8 Conclusions

The design is the basic step to incorporate organic material as substrates in flex sensors. Though nonlinear characteristics majorly overshadow our design, we need to further allow chemical process to substrates so that they can provide conductivity for a large range of input parameters. This technology can further be used to make sensors for lesser lengths which show comparatively linear characteristics and are biodegradable as well as cost-effective.

## References

1. Zimmerman, T.G.: Optical flex sensor. US Patent 4542291, 1985
2. Saggio, G.: A novel array of flex sensors for a goniometric glove. Sens. Actuators A: Phys. **205**, 119–125 (2014)

3. Saggio, G., Bocchetti, S., Pinto, C., Orengo, G.: Wireless data glove system developed for HMI. In: 3rd International Symposium on Applied Sciences in Biomedical and Communication Technologies (ISABEL) (2010)
4. Ponticelli, R.: Full perimeter obstacle contact sensor based on flex sensors. Sens. Actuators A: Phys. **147**(2), 441–448 (2008)

# Segmentation Techniques for Differential Variations in Fingerprint Images

Aniruddha Garge, Sunil Karamchandani and Sweta Suhasaria

**Abstract** The fingerprints are the widely used human parameters for identification systems. The fingerprints are of various kinds depending upon the type of acquisition equipment, surrounding conditions, way of presenting the fingerprint sample to the system, etc. Image enhancement plays an important role in identification system as it reduces the noise in the image and intensifies the true fingerprint features (minutiae, Ridge endings, Ridge bifurcations). Segmentation is the initial and important part in fingerprint image enhancement as it decides the region of interest in which the fingerprint features are present. In this simulation, we have to adapt various segmentation algorithms while dealing with different types of fingerprints. The simulation results show the dilation method provides best results for dry types of fingerprints, but at the same time, the decision of threshold is very difficult. The manual segmentation is simple and very effective but cannot be used for evaluation with large databases and there is also a possibility of human error particularly for low-quality fingerprints. The complexity of decision of threshold is very less in variance threshold method and can also be used for evaluation with large datasets which makes it a best suitable segmentation method for fingerprints for our simulation which comprises of Traditional and Modified Gabor filters (TGF and MGF) as the filtration algorithms.

**Keywords** Fingerprint image enhancement · Segmentation · Variance threshold
Minutiae · Dilation

---

A. Garge · S. Karamchandani (✉)
Department of Electronics and Telecommunication,
DJSCE, University of Mumbai, Vile Parle (W), Mumbai 400056, India
e-mail: sunil.karamchandani@djsce.ac.in; skaramchandani@rediffmail.com

A. Garge
e-mail: aniruddhagarge007@gmail.com

S. Suhasaria
Centre for Development of Advanced Computing (CDAC), Mumbai, India
e-mail: Swetas@cdac.in

# 1 Introduction

The uniqueness of a fingerprint can be determined by the pattern of ridges and furrows (valleys), minutiae, and attributes of ridges, such as sweat pores, edge contours features, etc. [1, 2]. Minutiae points are local ridge characteristics that occur at either a ridge bifurcation or a ridge ending. Fingerprint identification methods are required for forensic applications. These methods can be incorporated in Deep Learning, AI-based applications wherein it is required to identify, classify the fingerprints among databases. Such methods can be conjoined with different biometric data and aid in forensics [3, 4].

The performance of AFIS depends on feature extraction and matching performance which totally depends on fingerprint quality. The fingerprint features [5, 3] are easily distinguished, but in Fig. 1b, c which represents the dry and wet fingerprint, respectively, the ridges and valleys are partially discriminated from each other and in Fig. 1d which represents the bad quality fingerprint, the ridge-valley structures are highly distorted. It is impossible to always have a good quality fingerprint image and thus, there is requirement of some image processing technique [3, 6] that can enhance the fingerprint features. The enhancement mechanism [3, 7] should extract the information from the corrupted ridge-valley structures and improving the matching results [4] for good, average (dry and wet), and bad quality fingerprint image. The initial and the main part in the image enhancement is the segmentation process which separates the foreground region from the background region. As the fingerprints are of various types (some of them as above), we cannot use a single segmentation algorithm for all kinds of fingerprints but we have to use multiple fingerprints and evaluate the segmentation process best suitable for each kind of fingerprint. It is also important to note that more effective the segmentation

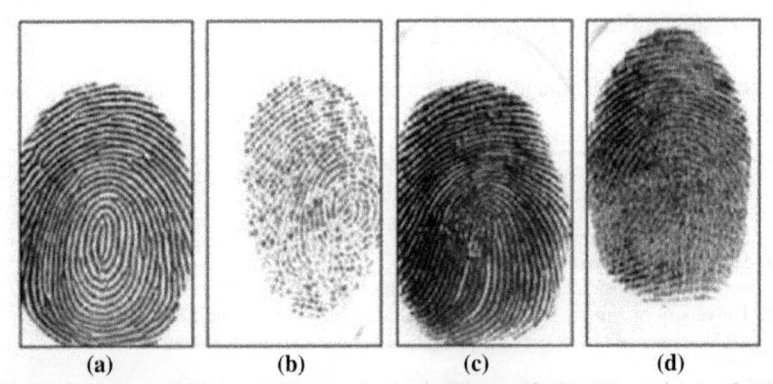

Figure. Examples of fingerprint images acquired with an optical scanner: a) a good quality fingerprint; b) a fingerprint left by a dry finger; c) a fingerprint left by a wet finger, d) an intrinsically bad fingerprint.

**Fig. 1** Variation in fingerprints

process, more is the probability of noise removable and it effectively leads to better enhancement of the fingerprint features (Minutiae) using filtration process [6] (in this simulation, TGF and MGF).

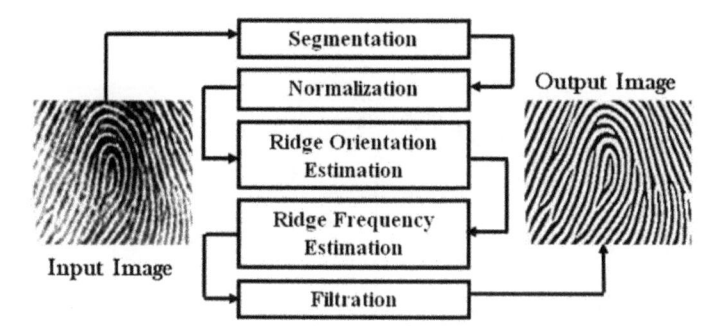

## 2 Fingerprint Image Enhancement

Noise corrupts the ridge-valley structures in fingerprint acquired images and thus leads to generation of spurious minutiae (false minutiae) adversely affecting the extraction and matching performances. In order to reduce the effect of spurious minutiae, the fingerprint image enhancement algorithm as shown in Fig. 2 is proposed which enhances the fingerprint features leading to improvement in the data extraction from the fingerprint and thus the matching performance. The algorithm is divided into two subprocesses: preprocessing steps and filtration. The process of image enhancement is performed by the filtration part but it requires prior information before performing the image enhancement and this prior information is obtained through preprocessing part.

**Preprocessing:**
The preprocessing steps consist of:

1. Segmentation and Normalization
2. Estimation of Ridge Orientation
3. Ridge Frequency Estimation.

The initial process in preprocessing is the Segmentation [7] and Normalization process. If this part is not evaluated effectively then it will leads to addition of contaminated components in the further processes and thus leading to inefficient filtration process which will in turn increase the probability of presence of spurious minutiae in the enhanced image degrading the extraction and matching performances respectively.

**Fig. 2** Fingerprint Image enhancement process

## 3 Segmentation and Normalization

Segmentation is a process of extracting the region of interest which consists of the fingerprint ridge-valley structures [8, 9]. It separates the foreground region from the background region which may consist of the noise depending upon the type of acquisition equipment, surrounding conditions, way of presenting the fingerprint sample to the sensor, etc. In addition to this, as the background part of fingerprint does not contain any of the fingerprint features, there is no need of applying enhancement procedure to this part as it will just increase the computation time boosting the background noisy features. Thus, it is very important to remove this background part from the foreground one and this is achieved using the segmentation process. For this simulation, the Adaptive Histogram Equalization [10] process is used for the normalizing the gray levels, as it gives closest flat response for the input image. In this simulation, we will be evaluating three segmentation methods:

1. Dilation Method
2. Manual Method
3. Variance threshold Method

### 3.1 Dilation Method

In this method, the morphological structured element is created with specific shape and size (examples are shown in Fig. 3) followed by the image erosion process which tries to blurred the gray level components with the selected parameters. Due to erosion process, the dark regions and light regions tend to spread with the larger area and hence by determining the proper threshold, the background region and foreground region can be separated. Finally, Segmented and Normalized image is obtained by multiplying the normalized image with the logical mask.

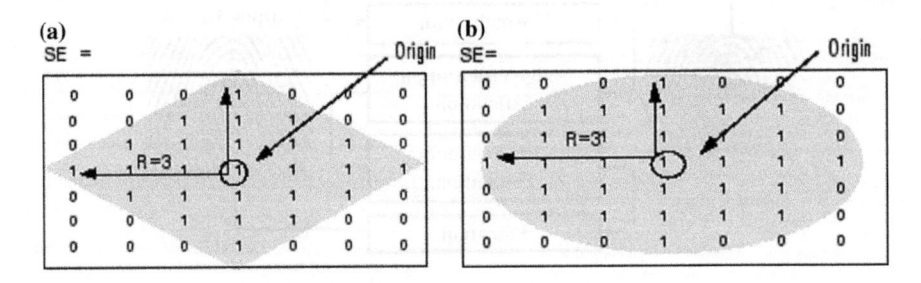

**Fig. 3** Structuring element: **a** Diamond ($R = 3$). **b** Disk ($R = 3$)

The dilation method extracts fingerprint edges with smooth approximation and can be used for evaluation of large fingerprint datasets. But the disadvantage of this technique is determination of threshold which is very complex which varies image to image and this can be used for large fingerprints datasets with purely white background only that maximally founds to be in dry types of fingerprints.

## 3.2 Manual Method

Manual method allows user to select region of interest freely and using the Canny edge detection, the edge contour is created for the selected region. Within the region, the logical mask is created by appending logical 1 within the selected region and logical 0 in the remaining. Finally, the mask is multiplied with the normalized image to obtain the Segmented and Normalized image.

This method is very easy and effective for all kind of fingerprints as it allows the user to choose the region of interest. But this method cannot be used for evaluation purpose with larger datasets as we have to select the region manually of each of the sample separately which is very tedious and time-consuming process. In addition to this, there is a high possibility of selection of noise in the background for low-quality fingerprints due to human error as the discrimination of ridge-valley structures in these types of fingerprints is very difficult with naked eyes.

## 3.3 Variance Threshold Method

This method uses the statistical parameter for the segmentation. In this technique, the image is first divided into number of blocks and then the image is converted into zero mean and unit standard deviation form. The variance is calculated for each block and using proper threshold value, the blocks are extracted for segmentation. Division of image into blocks provides the better way of separating the region of interest from the background. The blocks which have greater variance value than the threshold are converted into logical mask and finally multiplying mask with normalized image gives Segmented and Normalized Image. This method has less complexity in terms of threshold determination as standard deviation varies from 0 to 1. This method can be used for evaluating larger datasets with one particular value of the threshold.

## 4 Results and Discussion

The results are evaluated on fingerprints samples from various datasets, i.e., FVC2000, FVC2002, FVC2004, etc. This simulation is performed on the MATLAB 7.8.0 (R2009a).

### 4.1 Dilation Method

See Table 1.

**Table 1** Results for dilation segmentation method

| Original Image | Dilated Image | Segmented and Normalized Image |
|---|---|---|
| | | Threshold-22 to 140 |
| | | Threshold-65 to 160 |
| | | Threshold-17 to 125 |

## 4.2 Manual Method

See Table 2.

**Table 2** Results for manual segmentation method

| Original Image | Marked Image | Segmented and Normalized Image |
|---|---|---|
| | | |
| | | |
| | | |

## 4.3 Variance Threshold Method

See Table 3.

**Table 3** Results for variance threshold segmentation method

| Original Image | Zero Mean Image | Segmented and Normalized Image |
|---|---|---|
|  |  | Threshold-0.7 |
|  |  | Threshold-0.4 |
|  |  | Threshold-0.5 |

## 5 Conclusion and Future Work

The simulation results indicate that the dilation method is effective for fingerprints with uniform background variation specially for dry fingerprints with white background and it is also observed that there is vast and dynamic variation in the threshold value range which makes it complicated for the determination of threshold within a single dataset as well. We witness a very effective and useful segmentation results for manual method as it provides a user a capability to choose a region of interest but while evaluation of larger fingerprint datasets, it proves to be very time-consuming process as we have to select the region of interest for each and every fingerprint individually and in addition to this, due to human error, there is a high probability of noisy region getting selected in the segmented image. For variance threshold method, we observe an effective segmentation process with a very little variation in the threshold value which makes it a better choice for the evaluation of larger fingerprints datasets. The advantage of this technique is that, if the value for 1 sample of the database is determined then it is applicable for a whole dataset. In future, our objective is to evaluate the performance of Traditional and Modified Gabor Filters (TGF and MGF) for all types of fingerprint datasets and for this, the variance threshold method proves to be better choice for Good quality, Average (dry and wet), and Low qualities of fingerprints.

## References

1. Paper, P.: Position paper for an R&D project under cyber security. DIT, Govt. of India, no. 9, pp. 1–115 (2009)
2. Maltoni, D., Maio, D., Jain, A.K., Prabhakar, S.: Handbook of Fingerprint Recognition. Springer, New York (2003)
3. Bhargava, N., Bhargava, R., Mathuria, M., Dixit, P.: Fingerprint minutiae matching using region of interest. Int. J. Comput. Trends Technol. **4**, 515–518 (2013)
4. Chaudhari, A.S., Patil, S.S.: A study and review on fingerprint image enhancement and minutiae extraction. IOSR J. Comput. Eng. **9**(6), 53–56 (2013)
5. Hong, L., Wan, Y., Jain, A.: Fingerprint image enhancement: algorithm and performance evaluation. IEEE Trans. Pattern Anal. Mach. Intell. **20**(8), 777, 789 (1998)
6. Greenberg, S., Aladjem, M., Kogan, D., Dimitrov, I.: Fingerprint image enhancement using filtering techniques. In: Proceedings 15th International Conference on Pattern Recognition, vol. 3, pp. 322, 325 (2000)
7. Fleyeh, H., Davami, E., Jomaa, D.: Segmentation of fingerprint images based on bi-level processing using fuzzy rules. In: 2012 Annual Meeting of the North American Fuzzy Information Processing Society, NAFIPS 2012, vol. 4, no. 3 (2012)
8. Zhao, S., Hao, X., Li, X.: Segmentation of fingerprint images using support vector machines. In: Second International Symposium on Intelligent Information Technology Application, vol. 2, pp. 423–427 (2008)

9. Bazen, A.M., Gerez, S.H.: Segmentation of Fingerprint Images. In: ProRISC 2001 Workshop on Circuits, Systems and Signal Processing (2001)
10. Pizer, S.M., Johnston, R.E., Ericksen, J.P., Yankaskas, B.C., Muller, K.E.: Contrast-limited adaptive histogram equalization: speed and effectiveness. In: Proceedings of the First Conference on Visualization in Biomedical Computing, pp. 337–345 (1990)

# Smart Traffic Density Management System Using Image Processing

Jeet D. Sanghavi, Alay M. Shah, Saurabh S. Rane and V. Venkataramanan

**Abstract** With the population increasing manifold, the existing resources and infrastructure are not able to cope up with it for obvious reasons. Over the last few years, advancement in technology has led to growth in production of automobiles. More and more efforts are being made to ensure comparatively cheaper availability of vehicles for common man to afford it. On one hand, where the availability of vehicles is being facilitated, there are no outstanding improvements in the infrastructure In a city like Mumbai, heavy traffic is a major issue and it has become a growing concern as it is causing great distress to the citizens here. This is not just the story of Mumbai; major cities across the globe are facing such issues and are finding ways to mend them. As the famous saying goes, 'Rome was not built in a day', expecting radical changes in the present infrastructure overnight is impractical and a very demanding task in itself. A better solution would be bringing in sophistication in the present infrastructure using technology and making the con-

---

J. D. Sanghavi · A. M. Shah · S. S. Rane · V. Venkataramanan (✉)
Electronics and Telecommunication Engineering, DJSCE, Mumbai, India
e-mail: venkataramanan.v@djsce.ac.in

J. D. Sanghavi
e-mail: jeetsanghavi48@gmail.com

A. M. Shah
e-mail: alaymiteshshah@gmail.com

S. S. Rane
e-mail: saurabhrane.rane446@gmail.com

© Springer Nature Singapore Pte Ltd. 2018
H. Vasudevan et al. (eds.), *Proceedings of International Conference on Wireless Communication*, Lecture Notes on Data Engineering and Communications Technologies 19, https://doi.org/10.1007/978-981-10-8339-6_33

ditions better. Considering the most vital element of the traffic system, the traffic signal; this project aims at bringing the necessary sophistication in the way signals work with the help of image processing. A basic camera mounted on the top of existing traffic signals can be used for this purpose. Further, using MATLAB coding and basic image processing principles, the desired sophistication can be achieved. In the present system, each signal has predefined signal timing irrespective of the density of traffic in the lanes. A less populated lane will have comparatively the same signal time as that of more congested lane. This is where the present system's drawback lies. Equal signal timing for each lane without considering the factor of density of traffic results in inefficiency and is something which needs improvement. In addition to this, there is a great inconvenience caused to ambulance service due to the present working of the traffic signal. There have been several cases where the ambulance cannot move because the cars ahead of the ambulance are stuck due to the signal. A system where detection of ambulance in a particular lane will give immediate green signal access to that particular lane is developed. Such a system can prove of great importance during incidents where the need of ambulance is inevitable and the patient is in a critical condition.

**Keywords** Traffic · Density · Image processing · Ambulance · MATLAB Fire detection

## 1 Introduction

Eliminating the present anomalies and inefficiencies in the existing traffic system is the need of the hour. Here in this project, we have aimed to remove this inefficiency and have successfully implemented it [1]. The project is divided into two parts namely density-based traffic management system and ambulance detection. The first part of the project that is density-based traffic management system was implemented using two methods. The first method is a naïve method. Improvising on this method, we developed a better sophisticated method to meet the needs. In the first method, a basic approach of counting the intensity of a particular colour pixel is used. While in the second method, a well-devised algorithm involving complex MATLAB functions is used. In both the methods, the processing is done on the image captured by the camera mounted on the top of traffic signals. The information extracted from the images is then sent over to the Arduino UNO microcontroller which controls the signal timing. Signal timing is varied according to the density of traffic [2, 3]. A lane with more traffic will have more green signal time whereas a lane with less traffic will have a comparatively shorter green signal timing. Same is the approach in the case of ambulance detection. As soon as an ambulance is detected in a lane with the help of processing done on images captured by the overhead camera, the Arduino UNO makes the signal for that particular lane green and keeps the other lanes on halt till the ambulance passes the junction. All of this is achieved with the help of MATLAB software [4, 5].

## 2 Methodology

The traffic system as a whole is controlled and given directions by counting the number of vehicles that are present in a given lane at that particular instant [6]. We have considered a traffic light system at a four-lane intersection. So, we now have a given area, which can be divided into four parts. Each of the four parts can further be divided symmetrically by a line into two more parts. For our convenience, we denote the two parts as IN lane and OUT lane, depending upon the moving direction of a particular vehicle in that part. The part of the system which has vehicles moving towards the four-lane junction is denoted as IN lane and that part of the junction which has the vehicles moving away from the four-lane intersection is termed as OUT lane. Now, the algorithm is so developed that the number of vehicles in the IN lane determines the priority of a given lane, whereas the number of vehicles in the OUT lane determines the number of seconds for which the signal is green. Now, considering an example, say we have a dozen cars in a given part's IN lane, and the rest of the parts have a couple of cars less in their IN lanes. Then, the part with a dozen cars has the highest priority. But the duration for which the signal would be ON, i.e. the signal would be green, depending upon the number of cars in its corresponding OUT lane. Greater the number of cars OUT lane, lesser is the duration of the green signal.

Now, let us consider a possibility of an ambulance, or any emergency vehicle in a particular lane. Highest range of priority is given to the lane having such vehicles. The signal is immediately turned green for such lanes, and is turned red only after the emergency vehicle has made its way out. Similar will be the case in case of fire detection in any of the lanes. After this, the algorithm starts functioning normally until further disruption.

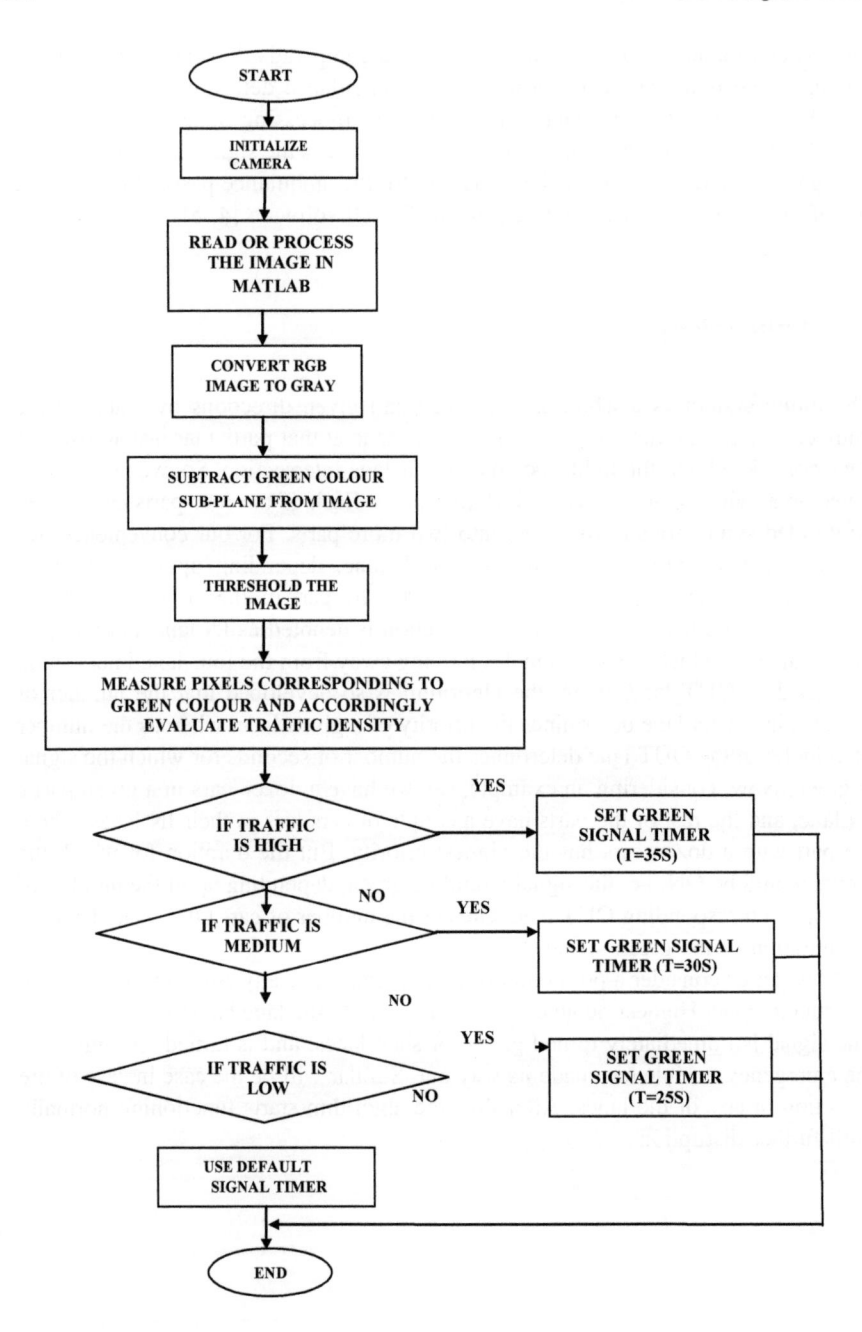

```
START
  ↓
INITIALIZE CAMERA
  ↓
READ OR PROCESS THE IMAGE IN MATLAB
  ↓
CONVERT RGB IMAGE TO GRAY
  ↓
SUBTRACT GREEN COLOUR SUB-PLANE FROM IMAGE
  ↓
THRESHOLD THE IMAGE
  ↓
MEASURE PIXELS CORRESPONDING TO GREEN COLOUR AND ACCORDINGLY EVALUATE TRAFFIC DENSITY
  ↓
IF TRAFFIC IS HIGH —YES→ SET GREEN SIGNAL TIMER (T=35S)
  ↓ NO
IF TRAFFIC IS MEDIUM —YES→ SET GREEN SIGNAL TIMER (T=30S)
  ↓ NO
IF TRAFFIC IS LOW —YES→ SET GREEN SIGNAL TIMER (T=25S)
  ↓ NO
USE DEFAULT SIGNAL TIMER
  ↓
END
```

## 2.1 Existing System

The present system employed in majority of the streets in Mumbai is not automated. So, a sudden increase in traffic in one of the channels does not result in increase in the duration of green light in that lane. This leads to long hours of traffic snarls, which has become synonymous with the city itself. A predetermined system is presently implemented, where the duration for green signal is fixed for each of the lanes. Also, there is not any special consideration for emergency vehicles like ambulances or emergency situations like fire. An ambulance stuck in a traffic snarl has to wait for the traffic ahead to clear. This results in losing of valuable time, which can be extremely critical in some cases. Thus, the existing system has proven to be extremely inefficient and has been unable to upgrade itself with the increase in the number of automobiles on the streets. Thus, an upgradation of the traffic system is the need of the hour. Moreover, there are a few disadvantages of the existing system like long traffic snarls, wastage of fuel, air pollution, emergency vehicles not getting priority, etc.

## 2.2 Proposed System

Two ways are proposed to solve this problem. They are as follows:

In the first system designed, using a camera (pre-installed), an image is received in the RGB colour format. This image is converted into grey image using MATLAB. The colour image can be classified into three different monochrome colour bands. Here, each band corresponds to a unique colour. Green colour subtraction is now performed on the image so as to count the pixel count. The edges of the different shapes that are formed are sharpened using different image processing techniques like using a threshold value, etc. Thus, the entire image now consists of sharp images. This image is now converted into black and white image. The white area represents the green strip area and black area represents the background. We count the total area that is covered by the white area, i.e. the green strips and thus have an approximate idea of the area covered by each strip. For example, if three strips in total cover an area of 6000 pixels, each strip covers an area that would be approximately equal to 2000 pixels. Now counting the pixel count of this image would help us in determining the traffic density of the lane that is it would help us estimate the number of vehicles covering the green strips, and thus help us in calculating the duration of each signal [7]. Less the pixel count, more are the vehicles covering the green strips and hence more is the density of traffic in that particular lane. So, say if the pixel count is less than 2000, traffic density is high, if it is between 2000 and 4000 pixels, traffic density is medium, and if it is greater than 4000 pixels, traffic density is low. Now, depending upon the traffic density, duration of the traffic signal can be defined as follows:

High Traffic—35 s
Medium Traffic—30 s
Low Traffic—25 s

In the second system designed, the traffic system is given instructions by having a count of the number of vehicles that are present during a given time interval. So, we divide a given area, which is divided into four equal parts, each part in a given direction. Each of the four parts is now divided into two equal parts. We then call the two parts as OUT lane or IN lane, which is decided by the flow of a given vehicle in that part. IN lane is that part of the system which consists of vehicles moving in the direction of the four-lane intersection and OUT lane consists of vehicles moving in the opposite direction. Using blob detection, the vehicles are counted [8]. Now, an algorithm is developed which considers the number of vehicles in the IN lane as the preceding factor that determines the priority of a given lane, whereas the number of seconds for which the signal is green is determined by the number of vehicles in the OUT lane. Now, take an example, we have 30 cars in a given IN lane, and the rest of the IN lanes have 20, 10, 15 number of cars, respectively. Then the 30 car lane is accorded the highest priority. But the signal would be ON, i.e. the signal would be green, only for 20 s if the number of cars in its corresponding OUT lane is between 10 and 15. With increase in the number of cars in an OUT lane, the duration for which the signal is green decreases. Now, consider if an emergency vehicle or an ambulance vehicle for instance is present in a given lane. Highest rung of priority is then provided to the lane with such vehicles. The signal immediately turns green, and changes its colour only after the given emergency vehicle has crossed the system.

## 3 Working Principle

The idea of developing this project led us to develop two models for the same. The first model was a naïve version of the project which equipped us to understand the project better and then later helped us develop an advanced algorithm to enhance the working and reliability of the project.

The first model was designed to calculate the no of pixels and accordingly decide the density of traffic in that particular lane. For that purpose, a set of three green strips were installed on the road. By capturing the photo of the road and performing a set of basic MATLAB commands, we were able to process the image and calculate the number of pixels of the three green bands using green colour subtraction. More the vehicles on those green bands less will be the calculated pixels of the three bands as the vehicles will cover the green bands and that portion of the green bands will not be captured in the image resulting in a low pixel count. More the number of pixels less is the traffic density in that particular lane and vice versa.

In the second model we devised, a far better and advanced algorithm where we considered a four-lane junction and accordingly designed the algorithm. Each lane

**Fig. 1** Traffic junction diagram

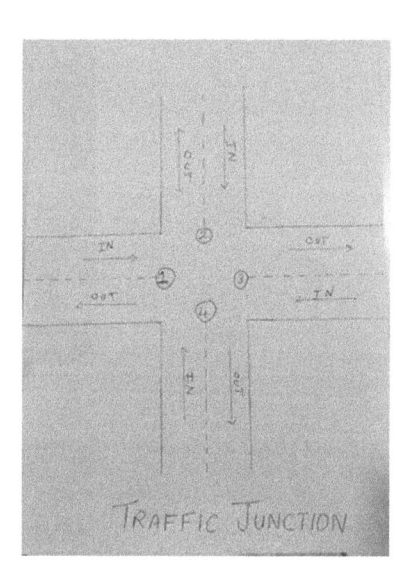

of the four-lane junction is further divided into two lanes namely the IN lane and the OUT lane. The IN lane consists of the vehicles approaching the junction while the OUT lane consists of vehicles moving away from the junction. The priority of lane is decided by the number of vehicles in the IN lane whereas the duration of green signal is decided by the number of vehicles in the OUT lane. More the number of vehicles in the OUT lane less will be the green signal duration (Fig. 1).

The project also detects ambulances and fire on the roads and takes the necessary action. The lane having the ambulance will be given highest priority. When a fire is detected, it will generate a system security alert saying that the fire is detected [9].

## 4 Simulation Environment

### 4.1 Counting the Number of Vehicles in Each Lane

Here, the yellow pieces of papers are used to represent cars. Now, using image processing, a number of cars were detected in each lane. First, the image is captured using a webcam and then by using blog detection algorithm, the vehicles in each of the lanes were detected (Figs. 2 and 3).

**Fig. 2** Experimental setup

**Fig. 3** MATLAB command window

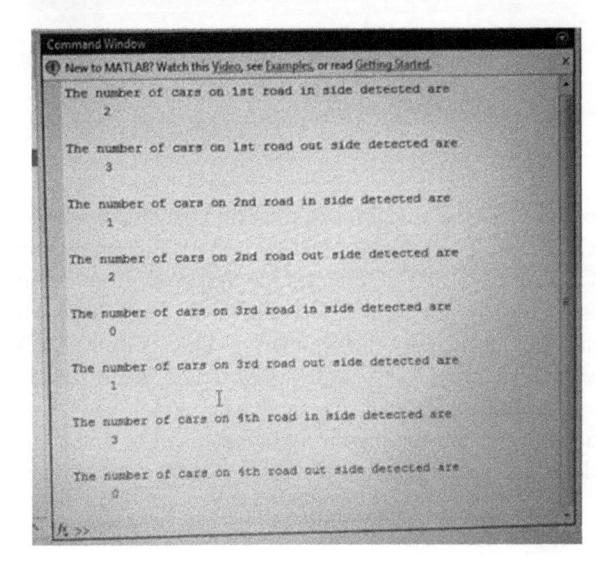

## 4.2 Ambulance Detection

Here, a red piece of paper is used to depict an ambulance. The red light on the top of the ambulance can be used to detect it. The lane in which the ambulance is present is then given highest priority (Figs. 4 and 5).

**Fig. 4** Ambulance in lane

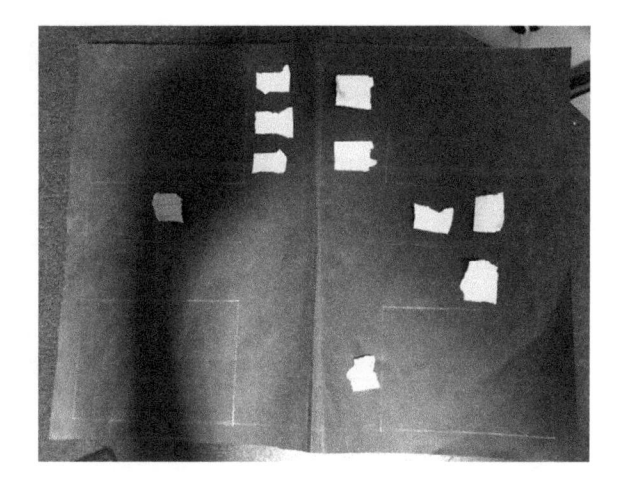

**Fig. 5** MATLAB command window

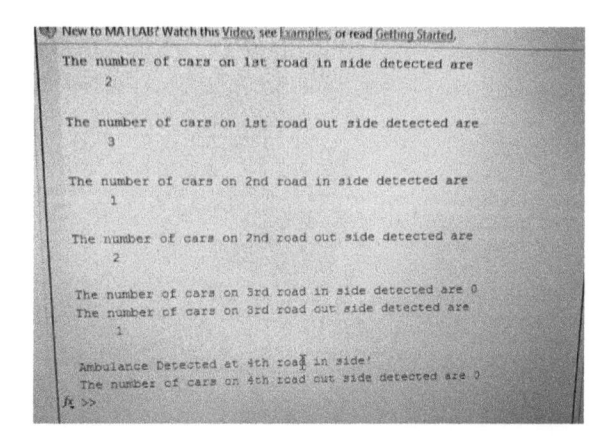

## 4.3 Fire Detection

When fire is detected in any of the lanes, the system will generate a system alert message saying the fire is detected (Figs. 6 and 7).

**Fig. 6** Fire in lane

**Fig. 7** MATLAB command window

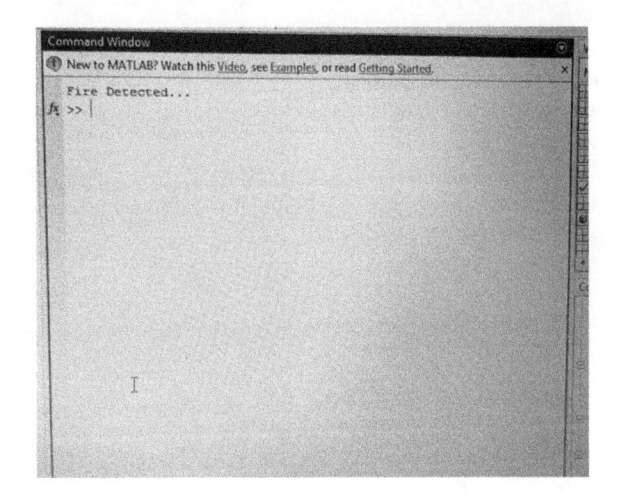

## 5 Future Scope and Benefits

This project can be used for a variety of future applications and the list is endless. However, a few of them are as follows:

This project uses a single camera for all the four lanes. To make the system more efficient, single cameras can be deployed for each of the lanes which will in turn increase the system accuracy. The installation cost of such a system is very meagre and hence is extremely cost-efficient. Moreover, the maintenance of such a system is very less since it only requires live feed which can be obtained through surveillance cameras installed. The other traffic monitoring systems which employ pressure mats suffer wear and tear due to their placement on roads where they are subjected to immense pressure constantly. Whereas this system completely eliminates the wear and tear problem as it does not include much of hardware. The small parts of ambulance detection and fire detection included in this project are add-on bonuses which serve as a great benefit. Also, a GSM module can be used to constantly track the position of the Ambulance.

## 6 Conclusion

This project not only resolves the anomalies of the existing system but also provides a prototype for future developments. With the help of image processing we have successfully designed a system without using much hardware and yet designed a traffic system solution better than the one where pressure mats are used. The development of two models for the same purpose helped us understand the existing system and its anomalies in a much detailed manner. The use of image processing over sensors has provided benefits like low cost, easy setup, and comparatively better accuracy. The project will help improve the existing traffic conditions in densely populated cities where traffic congestion is major problem. Further, to enhance the facilitation of emergency vehicles like ambulances and fire brigades, an add-on feature of ambulance and fire detection has been added which further adds value to the overall use of the project. The possibilities of this project being applicable for future use are vast.

**Acknowledgements** This project helped us bringing many of our theoretical concepts to good practical use. We would hereby like to thank our college Dwarkadas J. Sanghvi, College of Engineering, Mumbai, Maharashtra, India for providing us the infrastructure required for the implementation of this project. Further, we would also like to thank all the faculty members of the Electronics and Telecommunication department of our college who helped us during the course of this project.

## References

1. Abbas, N., Muhammad, T., Tahir Qadri, M.: Real time traffic density count using image processing. Int. J. Comput. Appl. (0975–8887) **83**(9), (2013)
2. Choudekar, P., Banerjee, S., Muju, M.K.: Implementation of image processing in real time traffic light control. In: 2011 3rd International Conference on Electronics Computer Technology, IEEE Explorer, 7th July 2011
3. Jadhav, P., Kelkar, P., Patil, K., Thorat, S.: Smart traffic control system using image processing. Int. Res. J. Eng. Technol. **03**(03), (2016)
4. Andronicus, F., Maheswaran: Intelligent ambulance detection system. Int. J. Sci. Eng. Technol. Res. **4**(5), (2015)
5. Patil, J.: Vehicle identification based on image processing techniques. California State University, Long Beach, ProQuest Dissertations Publishing, 1606096 (2016)
6. Salvi, G.: An automated vehicle counting system based on blob analysis for traffic surveillance. In: Proceedings of the International Conference on Image Processing Computer Vision and Pattern Recognition; vol. 1, pp. 397–402; Image Processing, Computer Vision, and Pattern Recognition; IPCV 2012 by CSREA Press, United States (2012)

7. Hasan, M.M., Saha, G., Hoque, A.: Smart traffic control system with application of image processing techniques. In: International Conference on Informatics, Electronics and Vision (ICIEV), IEEE Explorer, 10th July 2014
8. Tripathi, J., Chaudharya, K., Joshia, A., Jawaleb, J.B.: Automatic vehicle counting and classification. Int. J. Innov. Emerg. Res. Eng. e-ISSN: 2394 – 3343 p-ISSN: 2394 – 5494
9. Poobalan, K., Liew, S.-C.: Fire detection algorithm using image processing techniques. In: International Conference on Artificial Intelligence and Computer Science, At Penang, Malaysia, October 2015

# Energy-Efficient Solar-Powered Weather Station and Soil Analyzer

Aniket Kalkar, Abhiroop Mattiyil, Krupa Modi, Sagar Moharir and Archana Chaudhari

**Abstract** Incorporation of renewable sources of energy has become imperative. Earth receives solar power in abundance. This paper focuses on utilization of the same for powering the Arduino microcontroller board, so as to facilitate garnering of real-time information regarding prevalent weather as well as soil conditions. Area-specific temperature, relative humidity, and soil moisture content can be found out. Furthermore, a beforehand list of crop harvest can be developed depending on the results thus obtained. A weather station and soil analyzing system which is completely independent of nonrenewable energy sources can be developed. The data can be put to use for understanding the pattern of crop growth in the area under consideration.

**Keywords** Solar power · Arduino · Weather · Soil · Moisture

## 1 Introduction

India being an agrarian nation, the economy of our country majorly depends on agriculture. Farmers are the backbone when it comes to boosting the economy. In the present day, it has become imperative to integrate technology with agriculture.

---

A. Kalkar · A. Mattiyil · K. Modi · S. Moharir · A. Chaudhari (✉)
Department of Electronics and Telecommunication, DJSCE, Mumbai, India
e-mail: archana.chaudhary@djsce.ac.in

A. Kalkar
e-mail: aniket.kalkar28@gmail.com

A. Mattiyil
e-mail: abhiroop.mattiyil@gmail.com

K. Modi
e-mail: modikrupa00@gmail.com

S. Moharir
e-mail: sagarm937@gmail.com

Technology as we know helps in facilitating numerous tasks thus effectively reducing the human involvement to a greater extent. Although electrification of villages has been a top priority of the incumbent government, there is a lot of work to be done in the same field. Technology has not been integrated the way it should have been, and this happens to be one of the most important challenges faced in the present day. These were some of the shortcomings incurred thanks to humans. But, in a broader perspective, nature has its own ways of playing tricks. One can never be sure about the prevalent weather conditions, and thus has to effectively utilize the technology for weather forecasting. Weather plays a significant role in determining the yield of the crop. Here, real-time analysis is of utmost importance. Also, predicting the characteristics of soil is imperative as well. With a will to reduce the degree of this problem to a certain extent, an idea can be thought of which encompasses integration of technology in the field of agriculture more effectively. This, in return, shall ameliorate the quality of the yield of the crop, thereby boosting the prospects of efficient farming. The data which is desired to be acquired essentially happens to be about the weather conditions such as temperature and relative humidity as well as the soil characteristics such as moisture content.

## 2 Existing Technology Review

As a part of the literature survey undertaken so as to analyze the existing technology being used, several papers were referred. A comprehensive account of the same is mentioned below:

In "Arduino Based Automatic Plant Watering System", the soil moisture sensor based system was used so as to automatically water the plants using pump. This would facilitate the irrigation. As far as this system was concerned, it was totally dependent on external (nonrenewable) power supply [1].

In the paper "Soil moisture and temperature sensor based intelligent irrigation water pump controlling system using Arduino", the temperature measurement system was also incorporated along with the intelligent irrigation controlling system. Arduino microcontroller board was used so as to make the implementation cost-effective [2].

In "Monitoring moisture of soil using low-cost homemade soil moisture sensor and Arduino UNO", soil characteristic sensing is incorporated so as to examine several parameters and transmit the processed data in the form of wireless communication [3].

A comprehensive study regarding solar panels with respect to various characteristics was discussed [4].

An economic survey regarding the solar panel technology specific to certain regions was performed which was helpful in understanding the monetary aspects as well as methods of cost reduction, in order to make the system cost-efficient [5].

In "Correlation between Soil Resistance and pH using Arduino UNO", a comprehensive study regarding crop-specific information is depicted which helps in understanding the suitable crop growth under the specific moisture as well as soil conditions [6].

## 3 Innovation and Scope

Having performed an extensive research with the help of the aforementioned papers in the purview of this domain, it is understood that all the available systems had used nonrenewable sources of energy. And the systems that used renewable sources were not efficient. In this paper, the proposed system intends to optimize the solar energy to increase the efficiency of the system. This in turn shall make the system completely independent of conventional sources of electricity. This, in a way, should address the shortage of electricity in the rural areas. The proposed system will power the microcontroller, through which the data is to be transmitted in the form, which the farmers are able to effectively decipher. The idea is to integrate several components utilized in the papers reviewed so as to develop a system which is efficient when it comes to making the most out of available solar energy. This shall reduce the extent to which nonrenewable sources are currently being used. Also, real-time updates regarding the prevalent weather conditions and soil characteristics can be acquired displayed on a screen directly, helping farmers understand the aforementioned conditions in an efficient manner.

## 4 Proposed System

DHT11, soil moisture sensor, LCD display, solar panel, and rotating assembly containing servo motors and light detecting resistors are interfaced with microcontroller Arduino Uno as shown in Fig. 1.

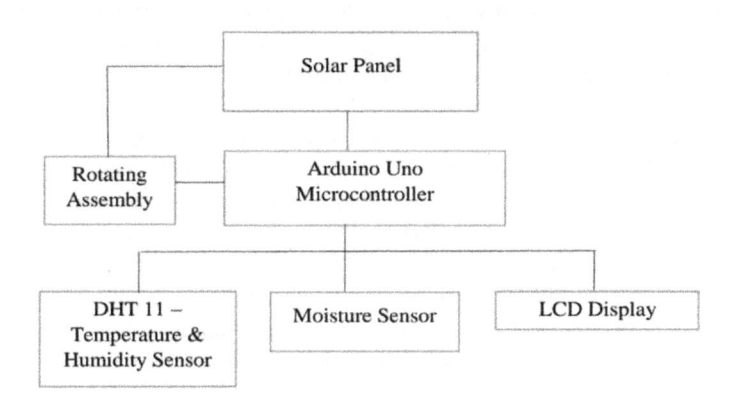

**Fig. 1** Flow diagram

The two parameters such as temperature and relative humidity can be estimated using an analog DHT11 sensor. Also, from the perspective of the farmer, information on the prevalent soil conditions such as moisture content is essential. Therefore, a soil moisture sensor has been used. Depending on the moisture content, the Arduino has been programmed to display outputs based on four conditions namely—high, moderate, and low moisture as well as dry if negligible moisture is present. The aforementioned components have been interfaced with a microcontroller—Arduino Uno 328-p. The data sent by these sensors is thus processed by this microcontroller and the output thus obtained can be displayed on a $16 \times 2$ LCD display.

To drive the Arduino microcontroller, solar power is used. This makes the system energy efficient as well as portable. To utilize the incident solar power to its maximum potential, a rotating assembly consisting of servo motors and light detecting resistors is incorporated. Depending on the light intensity incident on the light detecting resistors placed at four corners, the average intensity of every side can be calculated and the solar panel can be tilted in the direction of maximum intensity. This tilting shall be facilitated with the help of servo motors which in turn are interfaced with the Arduino board. In order to power the board with the help of solar energy, a smart charging circuit containing boost converter and a USB charger has been used. This ensures proper functioning of the complete assembly even in the absence of solar energy.

## 5 System Setup

Rays from the sun consist of two main components—energy beam and diffusion beam. Energy beam contains 90% of the total solar power whereas diffusion beam accounts only for the remaining 10%. In this project, solar power is used to drive the microcontroller Arduino. To make the most out of solar energy, the solar panel must always be in the direction of energy beam. Therefore, as the earth rotates and the sun moves from east to west, the solar panels must also orient accordingly. Also, north–south tilting is provided for better accuracy according to seasons. For three-dimensional rotation, as shown in Fig. 2, two servo motors are used—one for horizontal direction and one for vertical direction. LDRs are used as sensors to determine tilting.

Figure 3 shows the solar circuitry to be used as a power supply for the microcontroller Arduino Uno. In this circuit, diode is used for protection purpose. Battery charger TP 4056, Li-ion 3.7 V battery and voltage booster constitute a smart charging assembly which generates constant 5 V output which is required to power the Arduino microcontroller board.

Figure 4 shows overall practical setup of weather station and soil analyzing system using microcontroller Arduino Uno 328-p. For demonstration purpose, two soil samples of different moisture levels are taken under consideration.

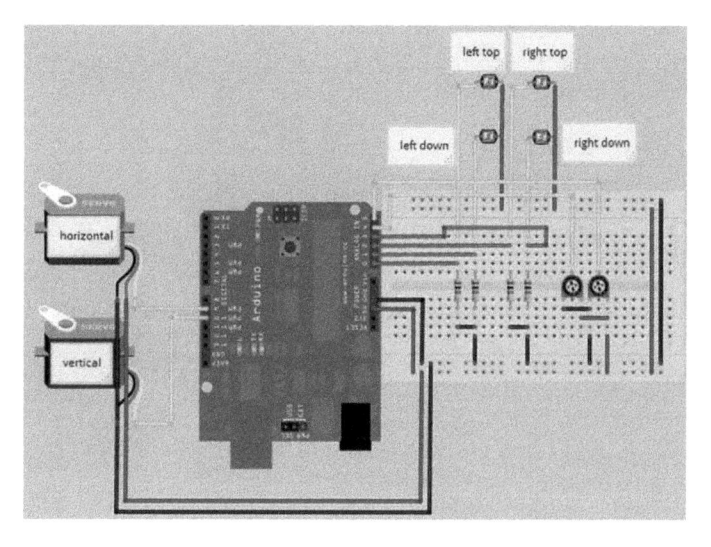

**Fig. 2** Solar panel rotation assembly

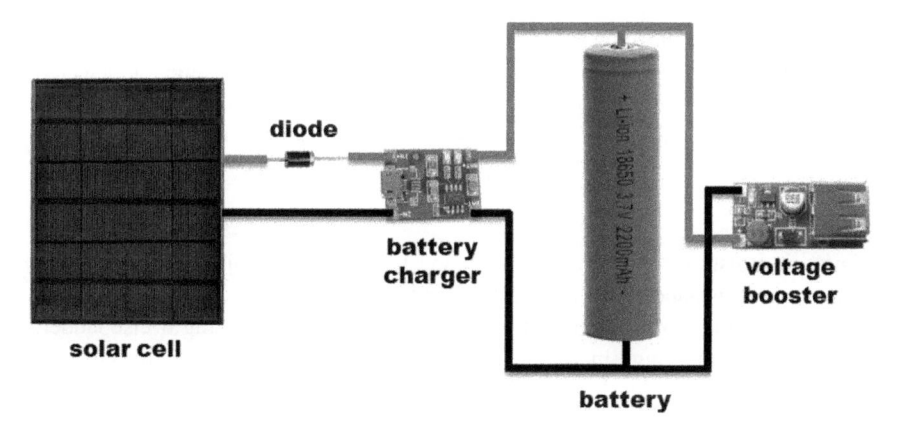

**Fig. 3** Smart solar charging circuit

The DHT11 functions as an analog sensor in order to sense the prevalent temperature and relative humidity conditions. The basic functioning principle of this sensor is that it detects water vapor by determining the electrical resistance. This essentially happens to be the resistance between the two electrodes. The relative humidity can be calibrated with the help of a substrate which holds the moisture when the electrodes are applied to the surface. On the absorption of water, the substrate releases ions which in turn are responsible for increasing the conductivity between the electrodes. This change in resistance helps in determining the relative humidity. As far as measurement of temperature is concerned, the DHT11 incorporates a thermistor mounted on its surface which is built on the same unit.

**Fig. 4** Weather station and soil analyzer

The soil moisture sensor consisting of two probes passes electric current through the soil to measure electrical resistance between the probes and determine the soil moisture content accordingly. Less resistance means more current passing through the soil which implies higher water content. Similarly, high value of detected resistance means less current passing through the soil proving that the soil is dry. This sensor can be interfaced with Arduino using any one of the following pins:

1. Analog
2. Digital

The soil moisture sensor is used in analog mode in this project. Arduino maps the electrical resistance of the sensor to discrete values from 0 to 1023. Higher the value, lesser is the moisture content in the soil. In this project, using if-else ladder, depending upon the range in which the detected value lies, output is displayed in the form of a message on the second line of LCD display.

## 6 Results and Discussion

As seen in Figs. 5 and 6, DHT11, soil moisture sensor and LCD display were interfaced with an Arduino Uno. Two soil samples were taken into consideration with one having high moisture content and the other containing comparatively less moisture. The system is programmed to display real-time temperature and relative humidity on the first line of the LCD screen. The sensor DHT11 is responsible for fetching this information. For moisture content information, probes of soil moisture sensor are inserted in the soil samples. This information regarding soil characteristics is displayed on the second line in the form of a message.

**Fig. 5** Soil sample 1 (high moisture)

**Fig. 6** Soil sample 2 (moderate moisture)

Different soil samples were examined and the subsequent outputs were displayed on an LCD screen. In this demonstration, the Arduino Uno 328-p microcontroller was powered with the help of an external power supply. The supply thus mentioned can be effectively replaced with the solar panel circuitry as discussed in the proposed system. The microcontroller will be powered with the help of solar energy to ensure that the complete system need not rely on the conventional nonrenewable sources of energy. Also, the data was displayed on an LCD screen, and it can be transmitted to the cell phone with the help of a GSM Module. Climatic change and agriculture go hand in hand. Changes in average temperature and rainfall have direct impact on crop productivity. Farmers in different parts of the country need to customize their agricultural patterns depending upon region-specific climatic variations.

## 7 Conclusion

The proposed system would aid in acquiring the following:

1. Real-time weather updates
2. Location-specific soil data

Getting the aforementioned updates real time will not only help the farmers modify their current agricultural practices but also make them free from any risks in deciding the nature as well as the yield of the crop to be harvested. This proposed system focuses mainly on farmers, because in a true sense, the farmer is responsible for the nourishment as well as sustenance of individuals in the society. With a will to pacify the anomalies faced by this most significant unit of existence, facilitation of their livelihood stands as the top priority. Hence, an efficient implementation of the proposal thus mentioned would mean that the rural areas of our country would no longer have to rely on the conventional sources of energy for getting the weather and soil updates, but now can turn toward such renewable sources of energy. Earth receives solar power in abundance. So, the same can be effectively harnessed to generate energy on the required scale thereby helping to save nonrenewable sources of energy. Affordability of the equipment is yet another significant factor as far as socioeconomic conditions of farmers are concerned.

## References

1. Devika, S.V., Khamuruddeen, S., Khamurunnisa, S., Thota, J., Shaikh, K.: Arduino based automatic plant watering system. Dept. of ECE, HITAM, Hyderabad, India, Department of Electronics, HRD, Hyderabad, India
2. Pushpa Latha, D.V., Devabhaktuni, S.: Soil moisture and temperature sensor based intelligent irrigation water pump controlling system using Ardunio; Gokaraju, Zareen, S.G., et al.: IJSRE 4(11), 6077 (2016). Rangaraju Institute of Engineering and Technology, Hyderabad. ISSN: 2319-7277, vol. 1, Issue 3 (2014)
3. Kumar, M.S. et al.: Monitoring moisture of soil using low cost homemade soil moisture sensor and Arduino UNO. J. Adv. Comput. Commun. Syst. (2016)

4. Solar Panel & Its Importance: https://www.scribd.com/document/192588824/Solar-Panel-Research-Paper
5. Timilsina, G.R., Kurdgelashvili, L., Narbel, P.A.: A review of solar energy markets, economics and policies
6. Jose, D., Tripathy, R.P.: Correlation between soil resistance and pH using Arduino UNO. Int. J. Adv. Sci. Technol. Eng. Manage. Sci.

# An Algorithm to Extract Handwriting Feature for Personality Analysis

Anamika Sen, Harsh Shah, Jessie Lemos and Shivani Bhattacharjee

**Abstract** Personality of an individual can be analyzed by their handwriting. Handwriting analysis can quickly reveal such factors as one's character, emotions, intellect, creativity, social adjustment, your desires, fears, weaknesses, strengths, and sexual appetite among others. Features like the size of one's handwriting, the slant, and others help in identifying the particular trait associated with the subject. A handwriting analysis report can help you gain insights into one's own strengths and weaknesses. In this work, we suggest an algorithm to extract one of the features used in graphology, i.e., Tittle over letter i. Image Processing was used for feature extraction using MATLAB.

**Keywords** Graphology · Handwriting · Personality traits · Image processing
Baseline · Slant · Tittle · Word spacing · MATLAB · Guide

## 1 Introduction

Handwriting of an individual provides direct insights into his physical, emotional, and mental condition [1]. Formation of one's handwriting begins from childhood itself and everyone has a distinctive handwriting style. Each stroke and pattern is somehow associated with a personality trait. This is so because a particular neuromuscular movement occurs while writing. And thus handwriting is often termed as "brain-writing" also [2, 3]. Since, while writing, fingers, and arms are in control of the brain, the way in which one writes must bear a direct relationship with the brain. People with erratic baselines have an unstable nature. Graphology is also

---

A. Sen · H. Shah · J. Lemos
Dwarkadas J. Sanghvi College of Engineering, Mumbai 400056, India

S. Bhattacharjee (✉)
Electronics and Telecommunication Department,
SVKM's, Dwarkadas J. Sanghvi College of Engineering,
Vile Parle (W), Mumbai 400056, India
e-mail: shivani.bhattacharjee@djsce.ac.in

© Springer Nature Singapore Pte Ltd. 2018
H. Vasudevan et al. (eds.), *Proceedings of International Conference on Wireless Communication*, Lecture Notes on Data Engineering and Communications Technologies 19, https://doi.org/10.1007/978-981-10-8339-6_35

used in psychoanalysis. Emotional state of a subject under consideration can be correctly identified using graphology [2]. Very small cramped handwriting is one of the symptoms of Parkinson's disease.

The prime features used in graphology are left and right margin, word expansion, letter size, line and word spacing, line skew, slant [4]. Graphology is a very effective tool for helping to make hiring decisions. Humans are prone to error in analysis and thus this paper suggests development of machine algorithms. In addition, it saves time and works with minimum human intervention.

In this paper, one of the handwriting features, i.e., Tittle over I has been discussed and algorithm has been suggested to extract it using Image Processing. Around 75 handwriting samples were obtained in .jpg format and since handwriting is an image, all the image processing operations were performed on numerical vectors and features were extracted. MATLAB has been used for Image Processing and the final GUI application has also been made in MATLAB Guide.

## 2 Related Work

Joshi et al. [1] have suggested machine learning technique to implement the automated handwriting analysis tool.

Kedar et al. [2] discussed the various methodologies to identify personality traits through handwriting.

Hashemi et al. [4] suggest a number of methods for automated extraction of handwriting features from Farsi handwriting. Experimental results on 30 training and 150 test samples are presented and discussed.

Kamath et al. [5] used features like size, slant and pressure, baseline, number of breaks, margins, speed of writing, and spacing between the words. The proposed system was calibrated with manual analysis.

Champa et al. [6] have proposed to predict the personality of a person from the baseline, the pen pressure and the letter **t** as found in an individual's handwriting. These parameters are the inputs to the Artificial Neural Network which outputs the personality trait of the writer. The performance is measured by examining multiple samples. Kedar et al. [7] have explained how handwriting can be used to determine the emotion levels of a person. I would eventually help identifying those people who are emotionally disturbed or depressed and need psychological help to overcome such negative emotions.

## 3 Block Diagram of System

The block diagram of the implemented system can be seen in Fig. 1. The blocks are handwriting samples, image preprocessing, feature extraction, and consequent mapping.

An Algorithm to Extract Handwriting ...

**Fig. 1** Block diagram of the proposed system

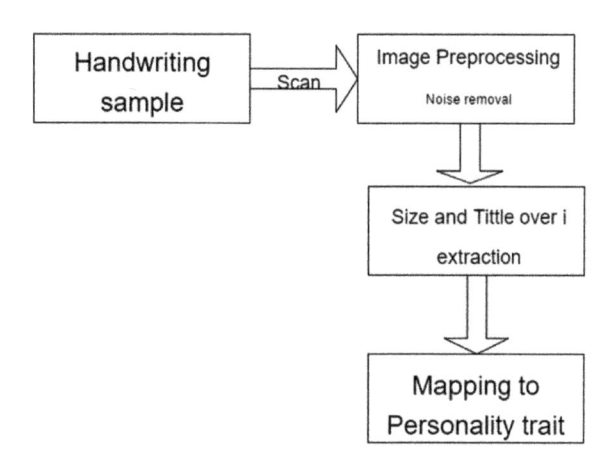

```
Handwriting sample --Scan--> Image Preprocessing / Noise removal
                              ↓
                              Size and Tittle over i extraction
                              ↓
                              Mapping to Personality trait
```

## 4 Image Preprocessing

The handwriting samples were scanned with a scanner with a resolution of 300 dpi. Noise removal was carried out and images were converted into binary images in the preprocessing stage. The handwriting sample is shown in Fig. 2.

## 5 Features

The two important features discussed here are size of the handwriting and Tittle over i. The algorithms for their extraction have been discussed in the succeeding sections. An important processing stage to be done for Tittle over i is to do horizontal segmentation of the handwriting sample.

**Fig. 2** Handwriting sample

## 6 Tittle Over i

The tittle over i which is the dot over i plays an important role in graphology. A slash, a circle-like tittle, and a dot are the different types of tittles. In this paper, algorithm for extracting a dot and circle is only carried out. The different types of

**Table 1** Title over eye and associated characteristics

| Type of tittle | | Corresponding traits |
|---|---|---|
| Tittle is a dot right over i | i | Detail-oriented, organized and emphatic |
| Tittle is a circle | ○i | Visionary and child-like |
| Tittle is a slash | ⁄i | Overly self-critical |
| Tittle is a dot high over i | ˙i | Has a great imagination |
| Tittle is a dot to the left of i | ˙i | Tends to be a procrastinator |

# An Algorithm to Extract Handwriting ...

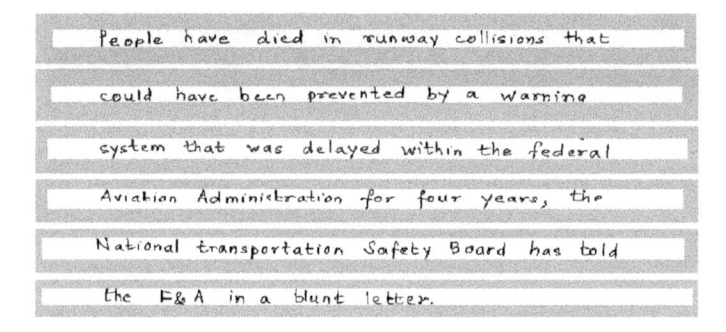

**Fig. 3** Horizontally segmented image

tittles along with the personality trait associated with them, in Graphology, can be described in the Table 1.

The following algorithm is proposed and implemented to estimate the Tittle over letter i:

Step 1: Perform horizontal segmentation on the binary image (handwriting sample) in order to separate out the lines from the paragraph as done in Fig. 3.
Step 2: For each line, resize the image into 128 × 1024 pixels.
Step 3: Invert the image, thus making the background black.
Step 4: Remove objects lesser than 20 pixels.
Step 5: Fill the remaining area.
Step 6: After filling, if number of pixels increase by a certain threshold, the tittle is a circle. Or else it is a dot.

**Fig. 4** Handwriting sample with tittle as circle

> The tittle is a circle and you are visionary and child-like.

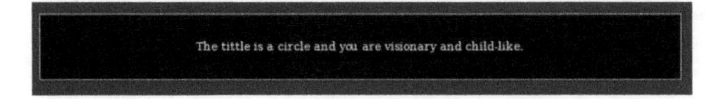

**Fig. 5** Result on GUI

> People have died in runway collisions that could have been prevented by a warning system that was delayed within the federal aviation administration. For four years, the National transportation safety Board has told the F & A in a blunt letter

**Fig. 6** Handwriting sample with tittle as dot

> The tittle is a dot and you are detail-oriented, organised in your approach.

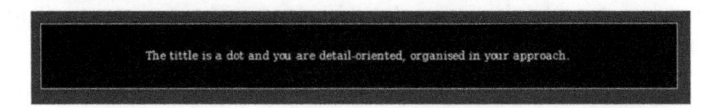

**Fig. 7** Result on GUI

## 7 Implementation and Results

The handwriting analysis system was created in MATLAB Guide. The Image Processing was also done in MATLAB. Figure 4 shows one of the handwriting samples which has characteristics of circle as the tittle over i. The corresponding result can be seen in Fig. 5. Figure 6 shows one of the handwriting samples which has characteristics of dot as the tittle over i. The corresponding result can be seen in Fig. 7. For samples, containing both dot and circle, weights were assigned and the characteristic with the maximum weight was assigned to the handwriting sample.

## 8 Conclusion and Future Scope

Artificial Neural Networks or Machine Learning Approach can be further employed to increase the scope of the system. This system is economical and can be easily deployed in professional hiring. Handwriting analysis is widely used in medicine in diagnosis of diseases like Parkinson's, Alzheimer's, and even cancer through Kanfer Tests [5, 8, 9].

More handwriting features can be integrated into the system for an enhanced analysis.

# References

1. Joshi, P., Agarwal, A., Dhavale, A.: Handwriting analysis for detection of personality traits using machine learning approach. Int. J. Comput. Appl. (0975 – 8887) **130**(15), (2015)
2. Kedar, S., Nair, V., Kulkarni, S.: Personality identification through handwriting analysis: a review. Int. J. Adv. Res. Comput. Sci. Softw. Eng.
3. Rahiman, A., Varghese, D., Kumar, M.: HABIT-handwriting analysis based individualistic traits prediction
4. Hashemi, S., Vaseghi, B., Torgheh, F.: Graphology for Farsi handwriting using image processing techniques. IOSR J. Electron. Commun. Eng. (IOSR-JECE)**10**(3), 01–07 (2015). e-ISSN: 2278-2834, p-ISSN: 2278-8735
5. Kamath, V., Ramaswamy, N., Karanth, P.N., Desai, V., Kulkarni, S.M.: Development of an automated handwriting analysis system. ARPN J. Eng. Appl. Sci.
6. Champa, H.N., Ananda Kumar, K.R.: Artificial neural network for human behavior prediction through handwriting analysis. Int. J. Comput. Appl. (0975 – 8887) **2**(2), (2010)
7. Kedar, S.V., Bormane, D.S., Dhadwal, A., Alone, S., Agarwal, R.: Automatic emotion recognition through handwriting analysis: a review. In: International Conference on Computing Communication Control and Automation (2015)
8. Walton, J.: Handwriting changes due to aging and Parkinsons symdrome. Forensic Sci. Int. **88**, 21–197 (1997)
9. Yan, J.H., Rountree, S., Massman, P., Doody, R.S., Li, H.: Alzheimer's disease and mild cognitive impairment deteriorate fine movement control. J. Psychiatr. Res. **42**(14), 1203–1212 (2008)

# Author Index

**A**
Ambekar, Aarti G., 103, 139, 185
Anjali, Yeole, 23
Awale, R.N., 75

**B**
Bhattacharjee, Shivani, 273, 323

**C**
Chaudhari, Archana, 313
Cheeran, A.N., 111

**D**
Deshmukh, Amit A., 59, 67, 85, 103, 119, 139, 151, 161, 169, 177, 185, 195, 207, 217, 255
Deshmukh, S.B., 217
Doshi, Akshay, 119, 169, 207
Doshi, Megh, 281
Dubey, Akash, 49
Duttagupta, Siddhartha P., 59

**F**
Fafadia, Maitri, 281

**G**
Gala, Divye, 49
Gala, Mohil, 177
Gandhi, Charmi, 281
Garge, Aniruddha, 291
Godse, Manali J., 255
Gosavi, Gauri, 185

**H**
Hemnani, Vishesh, 49

**J**
Jishnu, P., 59

**K**
Kadam, Ameya A., 161
Kadam, Poonam A., 119
Kakatkar, S.S., 111
Kalbande, D.R., 23, 39, 49
Kalkar, Aniket, 313
Kamble, Pritish, 169, 207
Kanawade, Anushka, 39
Karamchandani, Sunil, 281, 291
Khanapuri, Jayashree, 227
Kulkarni, Makarand G., 111

**L**
Labde, Saurabh, 67
Lemos, Jessie, 323

**M**
Mathew, Thomaskutty, 3
Mattiyil, Abhiroop, 313
Mehta, Niharika, 273
Mishra, Anish, 103
Modi, Krupa, 313
Moharir, Sagar, 313

**N**
Narayanan, Priyanka, 185
Nayak, Siddharth, 185
Nema, Shikha, 129
Nichani, Akshita V., 255
Nishad, Archana, 185

## O
Odhekar, Anuja, 207

## P
Pal, Ranjushree, 247
Patel, Aditi, 265
Patel, Stuti, 67
Patil, Pooja, 103
Pattanayak, Arnab, 59
Pattani, Mital, 265
Pawar, Shefali, 85
Pistolwala, Shruti T., 255

## R
Rane, Saurabh S., 301
Rathod, Mansing, 227
Rathod, S.M., 75
Ray, K.P., 75, 95, 111

## S
Sanghavi, Jeet D., 301
Sarang, Devidas, 29
Sarwade, Nisha, 95
Satam, Vandana, 129
Satpute, Abhishek, 265

Sen, Anamika, 323
Shah, Alay M., 301
Shah, Ayush, 49
Shah, Foram, 103
Shah, Harsh, 323
Shah, Hetvi, 103
Shaikh, Bushra Almin, 237
Shekokar, Narendra, 13, 29
Shukla, Megh, 67
Singh, Bharati, 95
Singh, Divya, 151, 195
Suhasaria, Sweta, 291

## T
Thakur, Sanjay Singh, 129
Thomas, Sanu, 3

## V
Varvadekar, Smruti, 39
Venkata, A. P. C., 67, 139, 169
Venkataramanan, V., 265, 301
Verma, Shikhar, 273

## Y
Yadav, Nilesh, 13